FLORA ZAMBESIACA

Flora terrarum Zambesii aquis conjunctarum

VOLUME THREE: PART ONE

ACACIA MACROTHYRSA

FLORA ZAMBESIACA

MOZAMBIQUE

MALAWI, ZAMBIA, RHODESIA

BOTSWANA

VOLUME THREE: PART ONE

Edited by

J. P. M. BRENAN

on behalf of the Editorial Board

J. P. M. BRENAN

Royal Botanic Gardens, Kew

A. W. EXELL

Commonwealth Forestry Institute, Oxford

A. FERNANDES

Junta de Investigações do Ultramar, Lisbon

H. WILD

University College of Rhodesia, Salisbury, Rhodesia

Published by the Managing Committee on behalf of
the contributors to the Flora Zambesiaca:
Portugal, Malawi, Zambia
and Southern Rhodesia
Crown Agents for Oversea Governments and Administrations,
4, Millbank, London, S.W.1
July 30, 1970

Printed at the University Press Glasgow
by Robert MacLehose & Company Limited

CONTENTS

LIST OF FAMILIES INCLUDED IN VOL. III, PART 1

ANGIOSPERMAE

61 Leguminosae (Mimosoideae)

LIST OF NEW NAMES PUBLISHED IN THIS WORK

Acacia robusta Burch. subsp. *clavigera* (E. Mey.) Brenan, comb. nov. *page* 104
Acacia robusta Burch. subsp. *usambarensis* (Taub.) Brenan, comb. nov. *page* 104

61. LEGUMINOSAE

(incl. *MIMOSACEAE*, *CAESALPINIACEAE* and *PAPILIONACEAE*)

By J. P. M. Brenan

Trees, shrubs, woody climbers or perennial or annual herbs; roots usually producing nodules containing bacteria which fix atmospheric nitrogen.* Leaves alternate or rarely (and not with us) opposite, usually pinnate or 2-pinnate, more rarely digitately compound, 1-foliolate or simple or reduced to phyllodes or (not with us) absent; stipules usually present. Flowers actinomorphic or (more frequently) zygomorphic, usually bisexual. Calyx gamosepalous or rarely of wholly free sepals, usually 5-lobed, sometimes bilabiate or spathaceous or even entire and rupturing irregularly. Petals usually 5, rarely reduced to 3, 1 or 0, usually free, sometimes connate (flowers then usually actinomorphic). Stamens sometimes ∞, generally 10, rarely fewer than 10, free or mono- or diadelphous. Ovary superior, almost always 1-carpous, 1-locular, with (1) 2–∞ amphitropous or anatropous or campylotropous ovules on a single placenta along the ventral margin. Fruit usually a pod either dehiscing along both margins or one margin only or indehiscent, sometimes winged, rarely fleshy. Seeds usually with a hard or tough testa and little or no endosperm.

The Leguminosae, with 500–600 genera and 15–20,000 species, are one of the largest and most widely distributed families of flowering plants. Although with a wide range of structure, the family is very natural and is also divided into three very natural subfamilies distinguished below, *Mimosoideae*, *Caesalpinioideae* and *Papilionoideae*, which are frequently considered as the separate families Mimosaceae, Caesalpiniaceae and Papilionaceae respectively. This difference in rank really reflects a rather minor divergence of opinion. There are a number of genera which appear to be genuinely on the border-line. Thus *Erythrophleum* and *Burkea* link *Mimosoideae* and *Caesalpinioideae*, and *Swartzia* and *Cordyla* link *Caesalpinioideae* and *Papilionoideae*. If emphasis is laid on these, then subfamily is the reasonable rank; if on the other hand they are discounted in view of the numerically much larger mass of genera about whose position there is no doubt, then the subfamilies are reasonably considered as families. The groups, however, remain objective and unaltered in general content even if their boundaries may be hard to draw and a few genera hard to place.

Key to the subfamilies

Flowers actinomorphic (radially symmetric) in all organs other than the ovary; leaves mostly 2-pinnate, sometimes simply pinnate or phyllodic:

Corolla valvate in bud; calyx usually valvate (imbricate only in *Parkia*); leaves 2-pinnate or (in exotic genera only) simply pinnate or phyllodic; seeds generally marked with areoles† (except in *Elephantorrhiza, Newtonia* and the giant species of *Entada*) - - - - - - - Subfamily I *Mimosoideae* (p. 8)

Corolla imbricate in bud; calyx usually imbricate, rarely valvate or open; leaves mostly

* Very little information has been hitherto available about the occurrence of these nodules in our area. Recently, however, H. D. L. Corby, Department of Botany, University College of Rhodesia, formerly Officer in Charge of the Grasslands Research Station, Marandellas, Rhodesia, has studied intensively the nodulation of indigenous Rhodesian Leguminosae and has generously made his results available. When nodules are known to occur in a genus the fact is mentioned and is to be taken as based on Corby's results. A few genera, including *Brachystegia*, do not appear to develop nodules.

† The areole is an area on each face of the seed, generally circular to elliptic or oblong, bounded (except usually for a gap opposite the micropyle) by a fine line frequently appearing as a fissure in the testa. It corresponds to the term " pleurogram " as used by Corner in Phytomorphology **1**: 117–150 (1951). For a fuller discussion see Brenan, F.T.E.A. Legum.–Mimos.: 1 (1959).

simply pinnate, sometimes 2-pinnate, rarely 1-foliolate or simple; seeds without areoles (except in *Burkea* and a few species of *Cassia*)
Subfamily II *Caesalpinioideae* (see Part 2 of this vol.)

Flowers zygomorphic (bilaterally symmetric, with at least the uppermost petal differentiated from the others); leaves usually simply pinnate, digitate, 3-foliolate or simple, occasionally 2-pinnate; seeds without areoles (except in *Tamarindus*, *Paramacrolobium* and a few species of *Cassia*):

The uppermost petal (vexillum) inside, i.e. with its margins overlapped by that of the lateral petal adjacent on each side; stamen-filaments free or ± connate below; radicle in the seeds usually straight; flowers ± zygomorphic; mostly trees and shrubs, sometimes herbs - - Subfamily II *Caesalpinioideae* (see Part 2 of this vol.)

The uppermost petal (vexillum) outside, i.e. with its margins overlapping that of the adjacent petal on each side; stamen-filaments usually mono- or diadelphous, usually connate into a tube, or often with the upper (vexillary) stamen ± free, rarely the filaments free or almost so; radicle in the seeds usually curved; flowers usually strongly zygomorphic (" papilionaceous "); trees, shrubs and herbs
Subfamily III *Papilionoideae* (see Part 3 of this vol.)

Subfamily I **MIMOSOIDEAE**

Trees, shrubs or rarely herbs, often prickly or spiny. Leaves 2-pinnate or (in exotic species only) simply pinnate or modified to phyllodes or absent. Inflorescences usually spikes, racemes or heads of sessile or shortly pedicellate, usually small or very small, regular, (3)5(6)-merous flowers. Sepals with valvate or rarely imbricate aestivation, often open from an early stage of bud, usually united to form a toothed or lobed calyx, rarely free. Petals valvate in bud, free or more often connate below into a tube. Stamens 4–10 (as many as or twice as many as the petals) or ∞, free or adnate below to the corolla, or the filaments connate below into a tube, usually ± exserted; anthers small, versatile, sometimes with an apical gland; pollen-grains sometimes simple but frequently compound or united. Pods and seeds various, the latter generally marked with areoles; radicle of embryo in seed usually straight.

Key to genera based on vegetative and fruit characters

Plant armed with prickles, thorns or spines:
- Pod at maturity splitting transversely into segments each containing a seed:
  - Valves of pod falling away at maturity, leaving the persistent margins; stems, and often leaves, with scattered prickles:
    - Pod without bristles or prickles on surface or margins, 2–2·6 cm. wide; inflorescence spicate - - - - - - - **2. Entada** (sp. 12)
    - Pod ± bristly or prickly on surface or on margins only, rarely unarmed but then pods less than 1 cm. wide; inflorescence capitate - - - **10. Mimosa**
  - Valves of pod not falling away from the margins; stems spinous only when young or when mature armed only with paired stipular spines:
    - Lateral veins of leaflets invisible; mature stems usually armed **12. Acacia** (*A. kirkii* and *A. xanthophloea*)
    - Lateral veins of leaflets fine and clearly visible on both surfaces; mature stems and leaves always unarmed; juvenile and sucker shoots spinous **15. Cathormion**
- Pod not splitting tranversely:
  - Spines terminating short branchlets; plant otherwise unarmed; pods densely clustered, indehiscent, often contorted - - - - **8. Dichrostachys**
  - Spines replacing stipules and paired at nodes, or plant armed with prickles either scattered or 1–3 together at stem-nodes; pods various - - - **12. Acacia**

Plant unarmed (except for single minute very inconspicuous prickles below the nodes sometimes present in *Albizia harveyi*, and for some spinescent branches in *A. anthelmintica*):
- Aquatic herb with creeping usually floating and swollen stems; pod 1·3–2·7 (3·8) cm. long - - - - - - - - - **9. Neptunia**
- Trees or shrubs, rarely herbaceous and then not aquatic; pod usually longer than 3·8 cm.:
  - Leaflets alternate to subopposite, petiolulate; pod bluntly tetragonal or subcylindric in section - - - - - - - - **7. Amblygonocarpus**
  - Leaflets opposite, very rarely alternate and then sessile and very narrow; pod usually flattened, rarely ± turgid:
    - Pod splitting transversely into segments each containing a seed:
      - Leaf-rhachis glandular between at least the uppermost pairs of pinnae; inflorescence capitate; pod 1·3–2 cm. wide - - **15. Cathormion**

Leaf-rhachis eglandular; inflorescence spicate; pod usually more than 2 cm. wide - - - - - - - - - **2. Entada**
Pod not splitting transversely:
Pod dehiscent:
Valves of pod separating along one margin only:
Pod 3–8 mm. wide; seeds small, black to brown, unwinged; inflorescence of paniculate or racemose heads - **12. Acacia** (spp. 58, 59)
Pod 1·3–2·5 cm. wide; seeds large, brown, conspicuously winged; flowers in spikes or spiciform racemes - - - - **4. Newtonia**
Valves of pod separating along both margins:
Valves of pod woody, recurving:
Pinnae always one pair per leaf; inflorescence capitate - **6. Xylia**
Pinnae 2–6 pairs per leaf; inflorescence racemose **5. Pseudoprosopis**
Valves of pod membranous to rigidly coriaceous but not woody or recurving:
Leaves reduced to simple entire phyllodes **12. Acacia** (exotic species)
Leaves 2-pinnate:
Inflorescence spicate (elongate axis visible in fruit) **12. Acacia** (spp. 6, 8)
Inflorescence capitate:
Seeds 4–5 mm. wide, with endosperm; leaflets acute at the apex - - - - - - - - - - **11. Leucaena**
Seeds 6–13 mm. wide, without endosperm; leaflets variable **13. Albizia**
Pod indehiscent:
Valves separating from the margins, usually splitting into 2 layers; young branchlets usually glabrous, rarely pubescent; inflorescence spicate **3. Elephantorrhiza**
Valves not separating from the margins and not splitting into layers; young branches glabrous or pubescent; inflorescence capitate:
Peduncles 9–25 cm. long, pendulous; pods 30–60 cm. long (including a 4–10 cm. long stipe) - - - - **1. Parkia**
Peduncles c. 1–8 cm. long, not pendulous; pods 9–34 cm. long, more shortly stipitate:
Pod flattened, usually at most 0·5 cm. thick, usually more than 2 cm. wide (range 1·8–7 cm.); valves without obviously developed mesocarp - - - - - - - - - **13. Albizia**
Pod thick and turgid, usually 1 cm. thick or more but ± constricted between the seeds, 1·6–2 cm. wide; mesocarp ± thick and well-developed - - - - - - - - - **14. Samanea**

Key to genera based on vegetative and floral characters

Plant armed with prickles, thorns or spines:
Inflorescence bicoloured, with the upper part yellow and the lower white or mauve; spines terminating lateral branchlets; plant otherwise unarmed **8. Dichrostachys**
Inflorescence concolorous; plant with scattered or grouped prickles or spinescent stipules:
Flowers in globose or subglobose heads:
Stamens as many as or twice as many as the (3)4–5(6) corolla lobes; flowers mauve or pink - - - - - - - - - **10. Mimosa**
Stamens numerous and indefinite; flowers mostly cream, yellow or white:
Mature stems and leaves normally prickly or spinous; stamen-filaments free or nearly so - - - - - - - - **12. Acacia**
Mature stems and leaves always unarmed; juvenile and sucker shoots spinous; stamen-filaments united below into a tube - - - **15. Cathormion**
Flowers in spikes or spiciform racemes:
Stamens 10; anthers ± elongate, 0·5–1 mm. long; prickles scattered; leaflets 9–23 mm. wide; flowers dark red - - - - **2. Entada** (sp. 12)
Stamens ∞; anthers about as wide as or wider than long, minute, 0·1–0·3 (0·5) mm. across; prickles usually grouped 1–3 together, rarely scattered and then leaflets 0·5–9 mm. wide; flowers very rarely dark red and then prickles paired just below the nodes, not scattered - - - - - **12. Acacia**
Plant unarmed:
Inflorescence capitate, globose or subglobose:
Aquatic herb, with creeping usually floating and swollen stems; flowers of 2 sorts, hermaphrodite in upper part of head, neuter with elongate staminodes in lower part - - - - - - - - - **9. Neptunia**
Trees or shrubs, rarely herbaceous and then not at all aquatic:

Calyx 10–13 mm. long, with imbricate lobes; flower-heads large, claviform, on pendulous peduncles 9–52 cm. long - - - - - - **1. Parkia**

Calyx mostly 1–5, sometimes to 7 mm. long, with valvate lobes; flower-heads smaller, not claviform, on normally non-pendulous peduncles 1–8 cm. long:

Leaves reduced to simple entire phyllodes - - **12. Acacia** (exotic species)

Leaves 2-pinnate:

Pinnae in one pair per leaf:

Stamens 10, free; anthers glandular at the apex; pinnae always in one pair - - - - - - - - **6. Xylia**

Stamens many (19–50), their filaments connate below into a tube; anthers eglandular at the apex; pinnae varying from one to few pairs **13. Albizia**

Pinnae in more than one pair per leaf:

Anthers conspicuously pilose; stamens 10, free; petals free **11. Leucaena**

Anthers glabrous; stamens numerous and indefinite; petals normally connate below into a tube:

Central flower of inflorescence not different from the rest; stamens free - - - - - - - - **12. Acacia**

Central flower of inflorescence usually differing from and often larger than the rest; stamen-filaments united below into an included or exserted tube - - - - - - * { **13. Albizia** / **14. Samanea** / **15. Cathormion** }

Inflorescence elongate, spicate or racemose (if inflorescences are dense spiciform racemes borne on quite leafless woody stems bearing yellow flowers with 10 stamens, then compare 3. *Elephantorrhiza*, or if flowers purple, then compare 2. *Entada*):

Leaflets alternate to subopposite, 0·7–1·9 cm. wide, with petiolules 1·5–3 mm. long **7. Amblygonocarpus**

Leaflets opposite, very rarely alternate and then sessile and less than 2 mm. wide:

Rhachis of leaf with a gland at the insertion of each pair of pinnae or the upper 1–17 pairs of pinnae:

Stamens 10; petals free; petiolar gland 0 - - - **4. Newtonia**

Stamens ∞; petals connate below into a tube; petiolar gland usually present **12. Acacia** (spp. 6, 8)

Rhachis of leaf eglandular:

Petals glabrous outside; ovary glabrous - - † { **2. Entada** / **3. Elephantorrhiza** }

Petals ± puberulous to pubescent outside; ovary hairy **5. Pseudoprosopis**

Conspectus of pod differences

In all genera except *Acacia* the pod is usually rather constant in form and structure.

1. Pod dehiscing into two separate valves:
   Valves woody:
   Acacia
   Pseudoprosopis
   Xylia
   Valves papery to rigidly coriaceous:
   Acacia

* These three genera are separated principally by their pods: see generic descriptions and key to fruiting specimens. It is difficult to separate the three genera concisely by vegetative and floral characters, although the species of *Samanea* and *Cathormion* are easily separable from each individual species of *Albizia*. Both in *Samanea* and *Cathormion* the basal pair of leaflets at the pinna is characteristically asymmetric, the leaflet on the proximal side being well developed while that on the distal side is represented only by a minute stipel. In *Albizia* this seems to occur only rarely and sporadically, although in *A. versicolor* the basal pair of leaflets is often represented only by a pair of stipels. In our area *Cathormion* and *Samanea* are rare trees of evergreen fringing forest, apparently confined to Zambia.

† These two genera are difficult to distinguish without fruit. If the plant in question has glabrous branchlets and produces yellow flowers when leafless, if it is a glabrous suffrutex with many small leaflets, or if it is a quite glabrous shrub or tree with narrow leaflets up to 3·5 mm. wide, then it is likely to be an *Elephantorrhiza*, not an *Entada*. The only *Elephantorrhiza* in our area with puberulous or pubescent branchlets has numerous pinnae and leaflets, the latter very small and with midrib marginal throughout, thus clearly differing from the species of *Entada* in the Flora area.

Albizia
Leucaena
Neptunia

2. Pod dehiscing into two valves which remain attached to one another along one margin:
   Acacia (*A. mearnsii*)
   Newtonia
3. Pod splitting tranversely into segments each containing a seed:
   Acacia (*A. kirkii*, *A. xanthophloea*)
   Cathormion
   Mimosa
   Entada
4. Pod indehiscent, not splitting transversely:
   Valves separating from margins, usually splitting into two layers:
   Elephantorrhiza
   Valves spiral or contorted, but neither separating from margins nor splitting:
   Acacia
   Dichrostachys
   Valves flattened, but neither spiral nor contorted nor separating from margins nor splitting:
   Acacia
   Albizia
   Valves not flattened but convex or angled, neither spiral nor contorted nor separating from margins nor splitting:
   Acacia
   Amblygonocarpus
   Parkia
   Samanea

In addition to the genera fully dealt with in the Flora there are several others occurring only in cultivation.

Adenanthera L., of which *A. pavonina* L. is grown in Mozambique. An unarmed tree with racemes of small yellowish flowers and dehiscent pods satiny-yellow inside and containing hard scarlet seeds.

Prosopis L., of which *P. limensis* Benth. is grown in Rhodesia. An armed tree with spiciform racemes of small yellow flowers and rather thick yellow-brown indehiscent pods. Native of S. America.

Pithecellobium dulce (Roxb.) Benth., a native of S. America, is cultivated in Mozambique. A shrub or small tree with spinescent stipules, a single pair of pinnae, each pinna with one pair of leaflets, small round heads of creamy or yellow flowers, and spirally twisted moniliform dehiscent pods.

1. PARKIA R. Br.

(By J. P. M. Brenan)

Parkia R. Br. in Denham & Clapperton, Trav., app.: 234 (1826).

Trees without spines or prickles. Leaves 2-pinnate; leaflets ± numerous; petiole usually glandular on its upper side. Inflorescence capitate, shortly claviform (with a globose apical part abruptly narrowed into a ± short cylindric neck) or (but not in the African species) globose or constricted in the middle; heads stalked, solitary or paniculate. Flowers in upper part of heads ☿, in lower part ♂ or neuter. Calyx infundibuliform or long-tubular, gamosepalous, with 4–5 imbricate segments, 2 larger and 2–3 smaller, the mouth of the calyx being thus irregular. Corolla with 5 petals, which are free or ± united, not much exceeding the calyx. Stamens 10, all fertile, their filaments connate below into a tube, to which the petals may be adnate; anthers eglandular. Ovary usually stipitate. Pods oblong to linear, straight or curved, dehiscent or not, usually ± thick and often woody, or somewhat fleshy when living. Seeds ellipsoid to ellipsoid-oblong, ± compressed or flattened.

A genus of c. 40 species, widely distributed through the tropics; c. 7 species in Africa and Madagascar, the others in Asia and America.

The Asiatic *P. biglandulosa* Wight & Arn., with two ± collateral and flattened glands a short way above the petiole-base on the upper side and numerous narrow leaflets 0·75–1·5 mm. wide, is cultivated in Mozambique (*Gomes e Sousa* 3476 (COI; K; LM); *Pedro & Pedrógão* 1866 (PRE)).

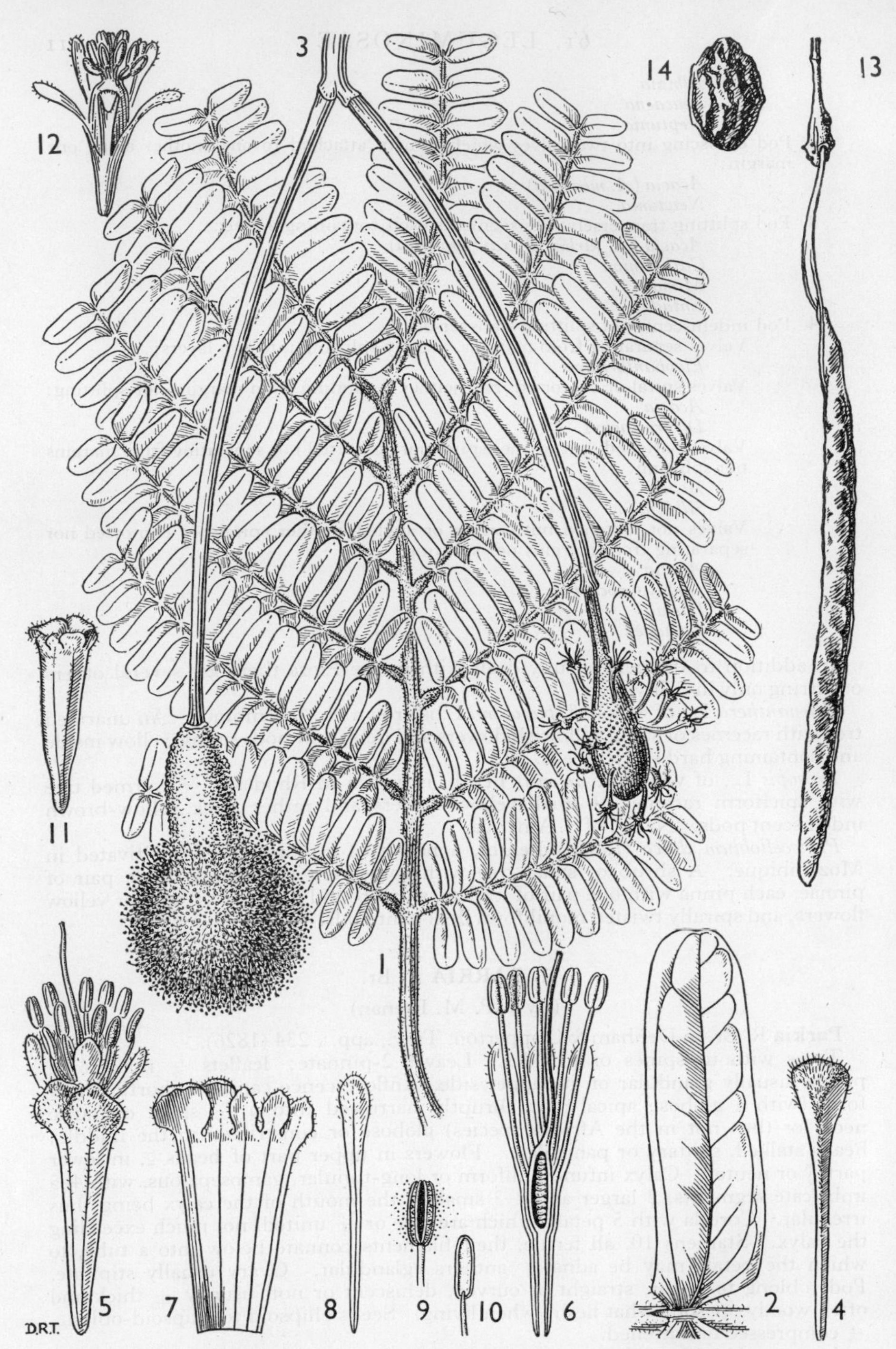

Tab. 1. PARKIA FILICOIDEA. 1, leaf (×½); 2, leaflet, upper side (×2); 3, inflorescences (×½), all from *Stolz* 1676; 4, bract from base of flower (×3); 5, hermaphrodite flower from upper part of capitulum (×3); 6, longitudinal section of hermaphrodite flower, calyx removed (×3); 7, calyx opened out (×3); 8, petal (×3); 9, anther, front view, filament cut off (×4); 10, anther, back view, filament cut off (×4); 11, neuter flower from lower part of capitulum (×3); 12, neuter flower, calyx removed (×3), all from *Faulkner, Pretoria No.* 11; 13, pod (×⅕) *Semsei* in F.H.2906; 14, seed (×1) *Purves* 209. From F.T.E.A.

The W. African *P. biglobosa* (Jacq.) Benth., similar to the last but with a single petiolar gland close to the petiole-base on the upper side, is also grown in Mozambique (*Andrada* 1459 (BM; COI; K; LISC; LMJ)).

Root-nodules not yet recorded.

Parkia filicoidea Welw. ex Oliv., F.T.A. **2**: 324 (1871).—Sim, For. Fl. Port. E. Afr.: 52, t. 44 (1909).—Crété in Trav. Lab. Médic. École Sup. Pharmacie Par. **7**, 4: 42 (1910).—R.E.Fr., Wiss. Ergebn. Schwed. Rhod.-Kongo-Exped. **1**: 65 (1914).—Bak. f., Legum. Trop. Afr. **3**: 781 (1930).—Brenan, T.T.C.L.: 346 (1949).—Gilbert & Boutique, F.C.B. **3**: 141 (1952).—Torre in Mendonça, Contr. Conhec. Fl. Moçamb. 2: 86 (1954).—Williamson, Useful Pl. Nyasal.: 91 (1955).—Torre, C.F.A. **2**: 256 (1956).—Burtt Davy & Hoyle rev. Topham, N.C.L., ed. **2**: 66 (1958).—Brenan, F.T.E.A. Legum.-Mimos.: 7, fig. 1 (1959).—F. White, F.F.N.R.: 93 (1962).—Gomes e Sousa, Dendrol. Moçamb. Estudo Geral, **1**: 224, t. 29 (1966). TAB. **1**. Type from Angola (Cuanza Norte, Pungo Andongo).

Tree 8–30(35) m. high; crown spreading, flat or rounded; bark scaly or smooth, grey to yellow-brown; young branchlets glabrous to puberulous. Leaves: petiole on upper side usually with 2 narrow ± collateral glands; rhachis puberulous or shortly pubescent; pinnae (3)4–11 (14 *fide* Eggeling & Dale, Indigenous Trees of Uganda); leaflets 10–18 pairs (to 28 pairs in juvenile or coppice leaves), oblong, often with a slight sigmoid curve, rounded at the apex, asymmetrically rounded or subtruncate at the base, mostly 1·2–3·2(3·4) cm. long, 5–12(14) mm. wide, glabrous except for puberulence on the margins near the base, 2 longitudinal nerves more distinct than the others. Peduncles c. 9–52 cm. long. Heads pendent, claviform, up to c. 8·8 × 7·5 cm., brick-red to reddish-pink, with a strong pungent smell. Bracteoles linear, enlarged at the apex, up to 15 mm. long. Hermaphrodite flowers: pedicel 3–4 mm. long; calyx 10–14 mm. long, glabrous or nearly so except for the lobes, which are densely tomentellous outside, the larger lobes 1·7–2·5 mm. long, rounded; petals adnate to the staminal tube for c. 2–4 mm. at the base, above this free for c. 10 mm. and linear-spathulate, puberulous at the apex which is c. 0·6–0·75 mm. wide. Pods 30–60 cm. long (including 4–10 cm. long stipe), 1·5–2·8 (3·5) cm. wide, glabrous or nearly so; margins straight or ± constricted between the seeds.

Zambia. N: Lake Bangweulu, Chiluwi I., fl. 13.x.1947, *Brenan & Greenway* 8096 (BM; FHO; K). W: Nchanga, Kafue R., fl. xi.1933, *Duff* 192 (BM). E: between Changwe and Luangwa, fl. & fr. 16.xii.1958, *Robson* 968 (BM; K; LISC; SRGH). **Malawi.** N: Karonga, fl. xi.1887, *Scott* (K). C: Dedza Distr., Mua-Livulezi Forest Reserve, fl. 30.ix.1953, *Adlard* 39 (K; PRE; SRGH). S: Blantyre Distr., Chileka Kuntazi, fl. x.1956, *Jackson* 2117 (K; SRGH). **Mozambique.** N: Ribáuè, fl. 16.x.1948, *Andrada* 1402 (BM; COI; K; LISC; LM; LMJ). Z: Mocuba, Namagoa, fl. & fr. x.1946, *Faulkner* Pretoria No. 11 (BM; K; PRE).

Also in the Congo, Uganda, Kenya, Tanzania and Angola. Occurrence in West Africa very dubious. Generally in fringing forest by rivers or lakes, but said sometimes to occur on termite mounds; 130–900 m.

P. filicoidea is a handsome tree with very distinctive red inflorescences in shape like electric-light bulbs hanging on long pieces of flex. It is likely that the flowers are pollinated by bats.

Purves 209 (Malawi, Zomba) shows two heads of young pods. In one the pods are glabrous, in the other they are covered with a short brown velvety indumentum. Whether these variations are to be looked on as no more than normal fluctuations within the species needs investigation.

2. ENTADA Adans.

(By J. P. M. Brenan)

Entada Adans., Fam. Pl. **2**: 318 (1763).—Brenan in Kew Bull. **20**: 361–378 (1966).

Pusaetha L. ex Kuntze, Rev. Gen. Pl. **1**: 204 (1891).
Entadopsis Britton in N. Amer. Fl. **23**: 191 (1928).

Trees, shrubs, suffrutices or lianes; prickles absent or sometimes present. Leaves 2-pinnate; pinnae each with one to many pairs of leaflets. Inflorescences of spiciform racemes or spikes, which are axillary or supra-axillary, solitary or clustered and often ± aggregated. Flowers ☿ or ♂. Calyx gamosepalous, with 5 teeth. Petals 5, free or nearly so (or ± connate in species not occurring in our area), separated from the ovary-base by a very short perigynous zone composed of stamens adnate to an apparent corolla-tube. Stamens 10, fertile; anthers with a usually very caducous apical gland. Pods straight or curved, flat or rarely spirally twisted, sometimes very large; at maturity the valves (but not the margins) splitting transversely into 1-seeded segments from which the outer layer (exocarp) of the pod-wall often peels off, the inner layer (endocarp) persisting as a closed envelope round the seed; the segments falling away from the margins, which persist as a continuous but empty frame. Seeds (in the African species at least) ± compressed, mostly elliptic or subcircular in outline, deep brown, smooth.

A genus of c. 30 species, widespread and mainly tropical; c. 20 in Africa and Madagascar; only c. 4 in America.

Root-nodules recorded from species Nos. 3 and 4.

Plant unarmed; petiole eglandular; ovary glabrous, sessile or nearly so; petals free (except for 7. *E. dolichorrhachis*; flowers of various colours (Subgenus *Entada*):
- Flowers cream-coloured to yellowish or greenish (uncertain but probable in 4. *E. chrysostachys*); exocarp normally separating from the endocarp at full maturity (? except in *E. dolichorrhachis*):
  - Leaf-rhachis ending in a forked tendril; large lianes with not more than 2 pairs of pinnae per leaf and 3–5 pairs of leaflets per pinna; pods gigantic, woody or very stiff, 0·4–2 m. long; seeds 3·5–5 cm. wide (Sect. *Entada*):
    - Flowers on distinct slender pedicels 1–1·5 (2) mm. long; racemes supra-axillary; pods spirally twisted, less woody than in 2. *E. pursaetha* - - - 1. *gigas*
    - Flowers sessile or nearly so (pedicels to 0·5 mm. long); spikes axillary; pods straight or sometimes curved, but not spirally twisted, woody - 2. *pursaetha*
  - Leaf-rhachis without a tendril; erect trees or shrubs or suffrutices (rarely a climbing shrub, 4. *E. chrysostachys*, but then pinnae 3–5 pairs and leaflets 10–17 pairs); pods at most 0·4 m. long and usually much smaller, papery to coriaceous; seeds 0·7–1·5 cm. wide (Sect. *Neoentada* Harms, spp. 3–6; Sect. *Dolichorrhachis* Brenan, sp. 7):
    - Leaflets in (15)22–55 pairs, 1–3·5 mm. wide; small tree c. 2·5–15 m. high 3. *abyssinica*
    - Leaflets in 6–17 (24) pairs, 2·5–20 mm. wide; shrubs, climbers or suffrutices to about 3 m. high:
      - Pinnae 2–5 pairs; petals free, inserted on a very short perigynous tube not longer than the calyx; habit various:
        - Plants in habit shrubs or climbers 1·2–3 m. high:
          - Young branchlets glabrous or grey-pubescent:
            - Pinnae 3–5 pairs; stipe of pod 0·5–1 cm. long - - 4. *chrysostachys*
            - Pinnae on some leaves 6–10 pairs; stipe of pod c. 1·5 cm. long 5. *bacillaris* var. *plurijuga*
          - Young branchlets with golden to yellowish spreading hairs; stipe of pod c. 1·5–3·5 cm. long - - - - 5. *bacillaris* var. *bacillaris*
        - Plant in habit a suffrutex with annual stems 5–100 cm. high - 6. *nana*
      - Pinnae 10–20 pairs or more; petals connate into a corolla-tube longer than the calyx; suffruticose - - - - - - - 7. *dolichorrhachis*
- Flowers dark red or purple; exocarp apparently not separating from the endocarp; usually slender lianes rarely (11. *E. mossambicensis*) suberect (Sect. *Porphyrostachys* Brenan):
  - Flowers produced when plant is leafless; pedicels 0–0·25 mm. long; leaflets in 18–25 pairs, 1–1·75 mm. wide - - - - - - - - 10. *nudiflora*
  - Flowers produced with the leaves; pedicels 1–2·5 mm. long:
    - Leaflets very numerous, 40–138 pairs, 3–5·5 × 0·75 mm.; racemes (including peduncle) 12–20 cm. long - - - - - - 11. *mossambicensis*
    - Leaflets 4–18 pairs, 8–27 × 1·5–14 mm.; racemes (including peduncle) 3–6 cm. long:
      - Lateral nerves and veins of leaflets distinctly raised and easily visible at least beneath; leaflets 4–5(8) pairs per pinna; stamen-filaments c. 3 mm. long 8. *stuhlmannii*
      - Lateral nerves and veins of leaflets not or scarcely visible on lower surface; leaflets 9–18 pairs per pinna; stamen-filaments c. 4–6·5 mm. long 9. *wahlbergii*

Plant ± armed with scattered deflexed prickles on stems and leaf-rhachides; petiole with a large gland on the upper side above the base; ovary pubescent, on a distinct stipe; petals connate below into a tube; flower dark red (Subgenus *Acanthentada* Brenan) 12. *schlechteri*

1. **Entada gigas** (L.) Fawc. & Rendle, Fl. Jama. **4**, 2: 124 (1920).—Bak. f., Legum. Trop. Afr. **3**: 785 (1930) pro parte.—Brenan in Kew Bull. **10**: 164 (1955); F.T.E.A. Legum.-Mimos.: 11 (1959).—Torre, C.F.A. **2**: 257 (1956).—F. White, F.F.N.R.: 92 (1962). Type from Jamaica.
   Mimosa gigas L., Fl. Jamaic.: 22 (1759). Type as above.
   Entada scandens subsp. *planoseminata* De Wild., Pl. Bequaert. **3**: 85 (1925). Syntypes from the Congo.
   Entada scandens subsp. *umbonata* De Wild., tom. cit.: 86 (1925). Type from the Congo.
   Entada planoseminata (De Wild.) Gilbert & Boutique, F.C.B. **3**: 221 (1952). Syntypes as for *E. scandens* subsp. *planoseminata*.
   Entada umbonata (De Wild.) Gilbert & Boutique, tom. cit.: 222 (1952). Type as for *E. scandens* subsp. *umbonata*.

Large liane up to 25 m. high, unarmed; young branchlets subglabrous to puberulous or sometimes pubescent. Rhachis of leaves with (1)2 pairs of pinnae, and ending in a forked tendril; leaflets (3)4(5) pairs, 1·8–8 × 0·8–4 cm., elliptic to obovate-elliptic, often asymmetric, emarginate at the obtuse or rounded apex, glabrous above except for the puberulous midrib, glabrous also beneath except near the base of the leaflet and (sometimes) for some pubescence along the midrib. Spike-like racemes arising from the stem c. 3–5 mm. above the leaf-axils, solitary, 8–25 cm. long, ± pubescent, on a peduncle 1·5–6 cm. long; pedicels 1–1·5(2) mm. long, slender. Flowers creamy to greenish or yellowish. Calyx 1–1·25 mm. long, somewhat puberulous or glabrous, sometimes pubescent. Petals 2·5–3 mm. long. Stamen-filaments 3·5–6 mm. long. Pods gigantic, 40–120 cm. × 7·5–12 cm., less woody than in *E. pursaetha*, twisted into a single or double lax spiral, with the sides also often twisted; outer layer of pod falling away to expose the thick chartaceous somewhat flexible inner layer. Seeds c. 4–5·5 cm. in diam., hard.

Zambia. N: Lufubu R., fl. 12.x.1958, *Fanshawe* 4934 (K). W: road to Mwinilunga 18 km. W. of Kakoma, fl. 28.ix.1952, *Angus* 564 (K).

Central and West Africa, also in Central America, the West Indies and Colombia. In fringing forest by rivers.

A liane climbing by tendrils and with immense spirally twisted pods developing from (in comparison) extremely small ovaries.

Angus 564 is more strongly pubescent than usual—even the young stems are covered by a short dense pubescence—but there do not seem to be sufficient grounds for separating it, even varietally.

2. **Entada pursaetha** DC., Prodr. **2**: 425 (1825); Mém. Legum.: 421 (1826).—Brenan in Kew Bull. **10**: 164 (1955); F.T.E.A. Legum.-Mimos.: 12 (1959).—Keay, F.W.T.A. ed. 2, **1**: 490 (1958). TAB. **3** fig. D. Type a plant cultivated in Mauritius.
   Adenanthera gogo Blanco, Fl. Filip.: 353 (1837). No type specimen extant.
   Entada scandens sensu Sim, For. Fl. Port. E. Afr.: 52 (1909).
   Entada gigas sensu Bak. f., Legum. Trop. Afr. **3**: 785 (1930) pro parte.—Gilbert & Boutique, F.C.B. **3**: 220 (1952).
   Entada gogo (Blanco) I. M. Johnston in Sargentia, **8**: 137 (1949). Type as for *Adenanthera gogo*.
   Entada phaseoloides sensu Brenan, T.T.C.L.: 344 (1949).—Torre in Mendonça, Contr. Conhec. Fl. Moçamb. 2: 87 (1954).

Large liane said to reach 50 m. or more in length, unarmed; young branchlets glabrous (but see note below). Rhachis of leaves with (1)2 pairs of pinnae, and ending in a forked tendril; leaflets 3–5 pairs, 2·5–9 × 1·1–4 cm., elliptic to obovate-elliptic, emarginate at the obtuse or rounded apex, glabrous or nearly so except for puberulence on the midrib above and near the base of the leaflet beneath. Spikes axillary on lateral branches, which are sometimes leafless and abbreviated, the spikes thus aggregated; spikes 7–23 cm. long, ± pubescent, on peduncles 1–8·5 cm. long; pedicels to c. 0·5 mm. long, or flowers sessile. Flowers creamy to yellow or greenish-yellow. Calyx c. 1·25 mm. long, glabrous (but see note below). Petals c. 2·5 mm. long. Stamen-filaments c. 6 mm. long. Pods gigantic, 0·5–2 m. × 7–15 cm., woody, straight or sometimes curved, but not twisted; outer woody layer of pod falling away to expose the woody rigid inner layer. Seeds c. 5 × 3·5–5 cm., hard.

Malawi. N: by Kakayanga Stream, a tributary of Mtazi Stream, fr. 12.xii.1964, *Desanker* 3 (FHO). **Mozambique.** N: Vila Cabral, Litunde, fl. 25.x.1948, *Pedro &*

Pedrógão 5646 (LMJ). Z: Mopeia, between Mopeia and Marral, fl. 15.x.1941, *Torre* 3657 (BM; K; LISC; LM). MS: Dondo, Vila Machado, fl. immat. viii.1947, *Pimenta* in GHS 17231 (K; SRGH). LM: Maputo, left bank of R. Maputo near Salamanga, fl. 10.x.1947, *Gomes e Sousa* 3628 (COI; K; LM; PRE).

Widely distributed in tropical Africa and extending to S. Africa; also from India to China, the Philippines, Guam and N. Australia. Lowland rain-forest and fringing forest.

For remarks on the identity of *E. phaseoloides* and *E. gigas*, names wrongly used for *E. pursaetha* in Africa, see I. M. Johnston in Sargentia, **8**: 135–138 (1949).

E. pursaetha normally has stems and calyces glabrous; however, a West African variant, not so far known from our area, has them pubescent. In our area *E. gigas* appears to have some pubescence on the young stems and can thus be distinguished from *E. pursaetha* although the difference may not apply elsewhere. *E. pursaetha* is very similar to *E. gigas*, but has sessile flowers, and its pods (though just as gigantic) lack the spiral twist of *E. gigas*. The Malawi specimen is rather poor and thus doubtful.

3. **Entada abyssinica** Steud. ex A. Rich., Tent. Fl. Abyss. **1**: 234 (1847).—Harms in Engl., Pflanzenw. Afr. **3**, 1: 402 (1915).—Bak. f., Legum. Trop. Afr. **3**: 789 (1930).—Brenan, T.T.C.L. 2: 344 (1949); F.T.E.A. Legum.-Mimos.: 13, fig. 2 (1959).—Torre in Mendonça, Contr. Conhec. Fl. Moçamb. **2**: 89 (1954); C.F.A., **2**: 260 (1956).—Burtt Davy & Hoyle rev. Topham, N.C.L., ed. 2: 66 (1958).—F. White, F.F.N.R.: 91 (1962).—Boughey in Journ. S. Afr. Bot. **30**: 158 (1964). TAB. **2**. Type from Ethiopia.

Pusaetha abyssinica (Steud. ex A. Rich.) Kuntze, Rev. Gen. Pl. **1**: 204 (1891). Type as above.

Elephantorrhiza pubescens Phillips in Bothalia, **1**: 190, t. 5 fig. 3 (1923). Type: Zambia, 10 km. below Kafue bridge, *Rogers* 8659 (K; PRE, holotype; SRGH).

Entadopsis abyssinica (Steud. ex A. Rich.) Gilbert & Boutique, F.C.B. **3**: 208 (1952). Type as for *Entada abyssinica*.

Small tree 2·7–10(15) m. high, unarmed; crown spreading, flat or rounded; bark rough or smooth; young branchlets glabrous or sometimes ± pubescent. Leaves with (1)2–20(22) pairs of pinnae; tendrils absent; leaflets (15)22–55 pairs, (3)4–12(14) × 1–3(3·5) mm., mostly linear-oblong, rounded to obtuse and slightly mucronate at the apex, ± appressed-pubescent on both surfaces, sometimes becoming glabrous above, rarely quite glabrous; midrib starting at the upper corner of the subtruncate to rounded base, running obliquely but nearer the upper margin. Racemes shortly supra-axillary, 1–4 together, 7–16 cm. long (including the 0·4–1·5 cm. peduncle); axis pubescent, sometimes subglabrous. Flowers creamy-white fading yellowish, sweetly scented; pedicels 0·5–1 mm. long. Calyx glabrous. Petals 1·5–2 mm. long. Stamen-filaments 3·5–4 mm. long. Pods c. 15–39 × (3·8)5–7·5(9) cm., straight or nearly so, subcoriaceous; joints ± umbonate and often somewhat roughened in the centre. Seeds 10–13 × 8–10 mm.

Zambia. B: Balovale Distr., 40 km. S. of Balovale, fr. 1.iv.1953, *Holmes* 1068 (K). N: Mporokoso Distr., 26 km. N. of Kafulwe Mission on Lake Mweru, on Chiengi road, fl. 5.xi.1952, *Angus* 722 (K). W: Mwinilunga, fl. 5.vi.1955, *Holmes* 1310 (K). C: Mt. Makulu, fr. 6.v.1956, *Angus* 1278 (BM; COI; K; PRE). S: Muckle Neuk, fr. 3.v.1958, *Robinson* 2856 (K; PRE; SRGH). **Rhodesia.** E: Nyamkwarara Valley, Stapleford Forest Reserve, fr. iii.1961, *Armitage* 2/61 (K). **Malawi.** N: Nkata Bay, fl. 18.iii.1960, *Eccles* 1 (K; PRE; SRGH). S: Zomba, fl. xi.1915, *Purves* 251 (K). **Mozambique.** N: W. of Mte. Tchirássulo, 2.iv.1942, *Hornby* 2601 (COI; K; PRE). Z: Mocuba, Namagoa, fl. & fr. xi.1945, *Faulkner* Pretoria No. 327 (K; PRE). T: Moatize, Zóbuè, fr. 16.iv.1941, *Torre* 2903 (LISC; LM). MS: between Vila Pery and R. Buzi, fr. 2.vi.1941, *Torre* 2776 (BM; K; LISC; LM).

From Sierra Leone and Eritrea to Angola, Rhodesia and Mozambique. Woodland and wooded grassland of various sorts (*Brachystegia*, *Combretum*, *Baikiaea*, *Acacia*, etc.), 60–2040 m.

So far, densely pubescent young branchlets are shown by a number of gatherings from Zambia only. Elsewhere they seem always glabrous or nearly so, as they are sometimes in Zambia. In general *E. abyssinica* is an easily recognized species, with comparatively little variation.

4. **Entada chrysostachys** (Benth.) Drake in Grandid., Hist. Madag. 30, **1**, 1: 51 (1902). Type from Madagascar.

Acacia chrysostachys Sweet, Hort. Brit., ed. 2: 167 (1830), *nom. subnud.*

Adenanthera chrysostachys Benth. in Hook., Journ. Bot. **4**: 343 (1841). Type as above.

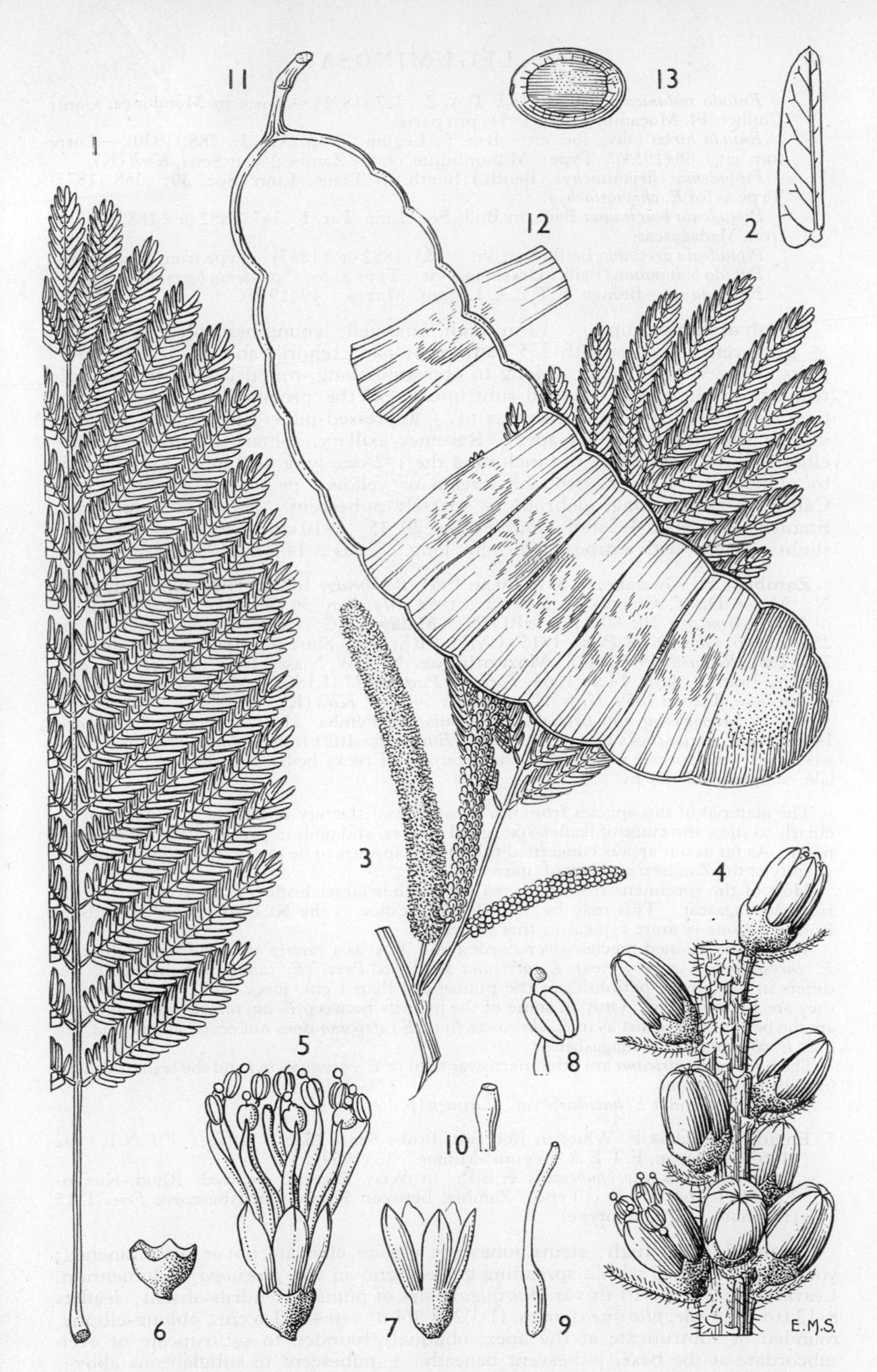

Tab. 2. ENTADA ABYSSINICA. 1, leaf (×⅔); 2, leaflet (×4), both from *Semsei* 865; 3, part of flowering branch (×⅔); 4, part of inflorescence (×8); 5, flower (×8); 6, calyx (×8); 7, petals (×8); 8, anther (×16); 9, ovary (×8); 10, top of style and stigma (×10), all from *Lugard* 600; 11, pod, part fallen away (×⅔); 12, envelope of endocarp containing seed (×⅔); 13, seed (×1½), all from *Semsei* 865. From F.T.E.A.

Entada sudanica sensu Oliv., F.T.A. **2**: 327 (1871).—Torre in Mendonça, Contr. Conhec. Fl. Moçamb. **2**: 89 (1954) pro parte.

Entada kirkii Oliv., loc. cit.—Bak. f., Legum. Trop. Afr. **3**: 788 (1930).—Torre, tom. cit.: 88 (1954). Type: Mozambique, on the Zambezi near Sena, *Kirk* (K).

Piptadenia chrysostachys (Benth.) Benth. in Trans. Linn. Soc. **30**: 368 (1875). Type as for *E. chrysostachys*.

Piptadenia boiviniana Baill. in Bull. Soc. Linn. Par. **1**: 347 (1882 or ? 1883). Type from Madagascar.

Piptadenia greveana Baill., tom. cit.: 353 (1882 or ? 1883). Type from Madagascar.

Entada boiviniana (Baill.) Drake, loc. cit. Type as for *Piptadenia boiviniana*.

? *Entada sp.*—Brenan, F.T.E.A. Legum.-Mimos.: 19 (1959).

Shrub or climber up to c. 12 m. high, unarmed; young branchlets glabrous or ± pubescent. Leaves with 3–5 pairs of pinnae; tendrils absent; leaflets 10–17 pairs, 1·4–2·9 × 0·3–1 cm., oblong to obovate-oblong, rounded to obtuse or subtruncate at the apex, rounded-subtruncate on the proximal side of the base, cuneate on the distal side, glabrous to ± appressed-pubescent on both surfaces; midrib subcentral towards apex. Racemes axillary, solitary or (more usually) clustered, 3–10 cm. long, not including the 1–2 cm. long peduncle; axis glabrous to pubescent. Flowers probably cream or yellow; pedicels 1–1·5 mm. long. Calyx 1–1·5 mm. long, glabrous or sparsely pubescent. Petals c. 3 mm. long. Stamen-filaments c. 4–5 mm. long. Pods c. 20–45 × 5–10 cm., coriaceous, straight or slightly curved, with a stipe c. 0·5–1 cm. long. Seeds c. 14–15 × 12 mm.

Zambia. S: Gwembe, fl. & fr. 10.xii.1964, *Bainbridge* 1023 (FHO; K). **Rhodesia.** N: Mazoe Distr., Chipoli, fl. & fr. 2.xi.1958, *Mowbray* 56 (SRGH). W: Wankie, fr. iii.1960, *Armitage* 105/60 (K; SRGH). **Malawi.** C: Lake Nyasa, Mareli I., fr. 25.viii.1946, *Gouveia & Pedro* 1815 (LMJ; PRE). S: Shire R., Mpatamanga Gorge, fr. 28.ii.1961, *Richards* 14483 (K). **Mozambique.** N: SW. Niassa, fr. iv.1942, *Hornby* 2353 (PRE); Erati, fl. & fr., 14.xii.1963, *Torre & Paiva* 9577 (LISC). MS: Kongone, mouth of Zambese, fl. 26.xi.1859, *Kirk* (K); Sena, fr. iv.1860, *Kirk* (K).

Also in Madagascar and probably Tanzania and Pemba. Habitat uncertain; *Richards* 14483 from " large rocks by river rapids ", *Bainbridge* 1023 from *Colophospermum mopane* woodland; the species is probably characteristic of rocky beds of streams and rivers and lake-margins; 40–460 m.

The material of this species from our area is unsatisfactory and more is needed, particularly to show the range of leaflet-size, good flowers, and pods in various stages of development. As far as our area is concerned the species appears to be restricted to the immediate vicinity of the Zambezi and its tributaries.

Most of the specimens from our area have rather larger leaflets than is usual in those from Madagascar. This may be of little significance as the Kirk specimen cited above from Kongone is more typical in this respect.

The last-mentioned species was recorded in F.T.A. as a variety of *E. sudanica* Schweinf. *E. chrysostachys* is very near *E. africana* Guill. & Perr. (*E. sudanica* Schweinf.), but differs in having the petiolules of the pinnae less than 1 cm. long, while in *E. africana* they are 1–2 cm. long. Only in some of the hybrids between *E. abyssinica* and *E. africana* are the petiolules as short as in *E. chrysostachys*. *E. africana* does not occur in our area, nor does *E. abyssinica* in Madagascar.

The pods of *E. africana* are often narrower than in *E. chrysostachys* and the leaflets usually (not always) glabrous.

See also note under *E. bacillaris* var. *plurijuga* (p. 19).

5. **Entada bacillaris** F. White in Bol. Soc. Brot., Sér. 2, **33**: 5 (1959); F.F.N.R.: 92 (1962).—Brenan, F.T.E.A. Legum.-Mimos.: 13 (1959).

Entada nana var. *pubescens* R.E.Fr. in Wiss. Ergebn. Schwed. Rhod.-Kongo-Exped. **1**: 64 (1914). Type: Zambia, between Katwe and Abercorn, *Fries* 1215 (K, photo; UPS, holotype).

Shrub 1·2–1·8 m. high; stems pubescent, virgate, elongate, not or little branched; young parts clothed with a spreading golden (grey in var. *plurijuga*) indumentum. Leaves with 3–4 (to 10 in var. *plurijuga*) pairs of pinnae; tendrils absent; leaflets 8–13 (to 24 in var. *plurijuga*) pairs, (1·3)2–3·9(4·6) × (0·4)1–1·6 cm., oblong-elliptic, rounded or subtruncate at the apex, obliquely rounded to subtruncate or even subcordate at the base, pubescent beneath, ± pubescent to subglabrous above; midrib subcentral at least in upper part of leaflet. Racemes produced to near the apices of the stems, axillary, 1–3 together, 8–18 cm. long (including the 1·3–4 cm. long peduncle); axis pubescent. Pedicels 1–1·5 mm. long. Calyx 1–2 mm. long, glabrous or slightly pubescent at the apex of the lobes only (rarely on the tube in var. *plurijuga*). Corolla greenish-white to yellow, c. 3·5–4 mm. long; petals 2·5–3·5 mm. long. Stamen-filaments 5–6 mm. long. Pods 26–37 × 8–9 cm., slightly falcate, subcoriaceous; stipe c. 1·5–3·5 cm. long; joints slightly umbonate in the centre. Seeds c. 12–15 × 9 mm.

Var. **bacillaris**

Hairs on young shoots yellowish to golden. Pinnae (3)4 pairs. Leaflets 8–13 pairs per pinna, (2)2·5–4(4·6) × (0·5)1–1·6 cm. Calyx-tube glabrous outside.

Zambia. N: between Abercorn and Mpulungu, fl. 10.x.1936, *Burtt* 6054 (BM; K), fl. 20.x.1947, *Brenan & Greenway* 8167 (K); Kalambo Falls, fl. 14.xi.1952, *Angus* 749 (FHO; K), fl. 15.xi.1960, *Richards* 13564 (K; SRGH).

Also in SW. Tanzania. In escarpment *Brachystegia* woodland on shallow rocky soil, 900–1520 m.

Var. **plurijuga** Brenan in Kew Bull. **20**: 372 (1966). Type: Zambia, Abercorn Distr., Inono Valley, 1 km. from Mpulungu road, *Richards* 2278 (K, holotype).
Entada bacillaris " anomalous gatherings ".—F. White in Bol. Soc. Brot., Sér. 2, **33**: 7–9 (1959).

Hairs on young stems and leaves grey to golden. Pinnae 3–10 pairs. Leaflets (10)11–24 pairs per pinna, (1)1·6–2·7 × 0·4–0·7 cm. Calyx-tube sometimes sparsely hairy outside.

Zambia. N: path to Inono Source, fl. 22.xii.1954, *Richards* 3722 (K; SRGH); Abercorn–Kawimbe road, opposite turning to Ndundu, fl. 28.xi.1958, *Richards* 10236 (K); Abercorn, Chilongowelo Escarpment, fr. 7.vi.1961, *Richards* 15213 (K).

Confined to our area. Ecology apparently similar to that of var. *bacillaris*. Sometimes recorded from sandy soil. Altitude range 1220–1740 m.

The status of this taxon is far from certain. Although occupying much the same area as var. *bacillaris* it is quite distinct in its foliage, and the specimens available are easily identified in spite of slight overlaps in the numbers and sizes given. The characters of var. *plurijuga* are such as might be produced by crossing between var. *bacillaris* and *E. abyssinica*. The problem presented by this plant has been fully discussed by White in Bol. Soc. Brot., Sér. 2, **33**: 7–9 (1959). Although treated here as a variety, its status cannot be considered as finally settled. In fact it is almost as closely related to *E. chrysostachys* as to *E. bacillaris*, and may well prove to be more correctly placed there. It differs in little more than the more numerous pinnae and the longer stipe to the pod.

6. **Entada nana** Harms in Warb., Kunene-Samb.-Exped. Baum: 245 (1903); in Engl., Pflanzenw. Afr. **3**, 1: 403 (1915).—Bak. f., Legum. Trop. Afr. **3**: 787 (1930).—Torre, C.F.A. **2**: 258, t. 51 (1956).—F. White, F.F.N.R.: 92 (1962). Type from Angola (Bié).

Suffrutex with erect annual stems 5–100 cm. high and densely pubescent to subglabrous when young. Leaves with 2–4 pairs of pinnae and a leaf-rhachis 2–14 cm. long; tendrils absent; leaflets 7–13 pairs, (1·2)2–3·5(4) × 0·5–2 cm., narrowly oblong, oblong or obovate-oblong, rounded to emarginate at the apex, very asymmetric at the base with proximal side rounded to cordate and distal side cuneate to cuneate-rounded, ± pubescent at least on the midrib beneath. Racemes axillary, 1–3 together, 6–12 cm. long (including peduncle); axis glabrous to ± pubescent. Pedicels 1–2 mm. long. Calyx 1–1·75 mm. long, glabrous. Corolla pale-cream-coloured; petals c. 3 mm. long; hypogynous zone very short, not exceeding the calyx. Pods variable (see below).

Subsp. **nana**

Stems 0·3–1 m. high. Pods distinctly and strongly falcate, 17–20 × 5–6 cm.; stipe 1·5–2 cm. long. Seeds c. 12·5 × 9 mm.

Zambia. B: Kalabo Distr., Sikongo Forest Reserve, fr. 14.ii.1952, *White* 2075 (K); Machili, fl. 13.x.1960, *Fanshawe* 5841 (K; LISC). S: Livingstone Distr., Lusaka road, fl. 8.xi.1955, *Gilges* 498 (PRE; SRGH). **Rhodesia.** W: Wankie Distr., Ngwashla road, game reserve, fr. 18.ii.1956, *Wild* 4769 (COI; K; LISC; PRE).

Also in Angola and SW. Africa. Woodland (*Baikiaea-Burkea*) on Kalahari Sand, c. 900 m.

It is possible that *E. arenaria* Schinz (in Mém. Herb. Boiss. **1**: 118 (1900)), based on a type from SW. Africa, is conspecific with *E. nana*, and if this were proven it would be the earliest name for the species. The type of *E. arenaria* is, however, so inadequate, consisting only of a few ancient fragmentary pods, that I consider that this name should be rejected as being of uncertain application.

Subsp. **microcarpa** Brenan in Kew Bull. **20**: 373 (1966). Type: Zambia, Mwinilunga Distr., Dobeka Bridge, fr. 8.ii.1938, *Milne-Redhead* 4496 (K, holotype).
Entada sp. 2.—F. White, F.F.N.R.: 92 (1962).

Stems 5–25 cm. high. Pods nearly straight to slightly falcate, 7·5–12 × 1·5–2·8 cm.; stipe 0·5–2 cm. long. Fully mature seeds not seen but probably smaller than in subsp. *nana*.

Zambia. B: 16 km. S. of Chavuma, fr. 1–10.ii.1953, *Holmes* 1046 (K). W: Mwinilunga Distr., Cha Mwana Plain, fl. 14.x.1937, *Milne-Redhead* 2762 (K).
Also in the Congo. In grassland and woodland on Kalahari Sand, c. 1200 m.

It may be that subsp. *microcarpa* is specifically distinct from *E. nana*, as it was considered in F.F.N.R. Recent material has, however, blurred some of the distinctions given there, and it seems more prudent to consider it as a northern subspecies of *E. nana*, especially as the material of subsp. *microcarpa* is still limited, and apparently by no means homogeneous in its pods.

7. **Entada dolichorrhachis** Brenan in Kew Bull. **20**: 374, fig. 1 (1966). Tab. **3** fig. A. Type: Zambia, Abercorn Distr., Lufubu R., Iyendwe Valley, *Richards* 11952 (K, holotype; SRGH).
Entada sp. 1.—F. White, F.F.N.R.: 92 (1962).

Suffrutex with erect annual stems 1–10 cm. high and shortly tomentose or densely pubescent when young. Leaves with 10–20 or more pairs of pinnae and a leaf-rhachis 15–90 cm. long; tendrils absent; leaflets 6–16 pairs, 0·5–2 × 0·25–0·7 cm., ± ovate-oblong, very asymmetric, rounded and mucronate at the apex, at the base with the midrib on the distal margin and the proximal side rounded, subglabrous to ± pubescent especially on the midrib beneath. Racemes axillary, 1–2 together, 4–10 cm. long (including peduncle); axis densely pubescent. Calyx 1·5–2 mm. long, pubescent. Corolla pale dull yellow; petals with their free part c. 1·25–1·75 mm. long, connate below into a corolla-tube 3·5–4 mm. long, longer than the calyx. Pods 3–6·5 × 1·5–1·8 cm., straight; stipe very short, less than 0·3 cm.

Zambia. N: Kawambwa, fl. 15.xi.1957, *Fanshawe* 4043 (K.)
Known only from Zambia. On sandy soil in *Brachystegia* woodland and on open river-banks, 780–1620 m.

This species is very distinct and in some ways extraordinary. The stems, though abbreviated, bear immensely elongate leaves which trail along the ground. The long corolla-tube is also remarkable. The pods are described from *Mutimushi* 6, from Kawambwa, which is somewhat different in facies from the other specimens, and there is therefore a slight doubt of its identity.
A note on the type-specimen states that the flowers are attacked by small caterpillars and ants as soon as they come out, so that it is doubtful if many pods are formed.

8. **Entada stuhlmannii** (Taub.) Harms in Engl., Pflanzenw. Afr. **3**, 1: 401 (1915).—Bak. f., Legum. Trop. Afr. **3**: 788 (1930).—Brenan in Kew Bull. **10**: 170 (1955); F.T.E.A. Legum.-Mimos.: 17 (1959). Types from Tanzania.
Pusaetha stuhlmannii Taub. in Engl., Pflanzenw. Ost-Afr. **C**: 196 (1895). Types as above.
Entada wahlbergii sensu Bak. f., loc. cit., pro parte quoad specim. Allen.—Oliv., F.T.A. **2**: 327 (1871) pro parte quoad specim. Meller.
Entadopsis stuhlmannii (Taub.) Pedro in Bol. Soc. Est. Moçamb. **92**: 10 (1955). Types as above.

Slender woody climber, unarmed, said to have a tuberous root; young branchlets glabrous, often flexuous. Leaves with 2(3) pairs of pinnae; one or more of the pinnae, usually terminal, modified to a tendril, or spirally twisted at the base and bearing leaflets above; leaves of scrambling shoots sometimes with the terminal pair of pinnae without leaflets and tendril-like, and the lower pair much reduced; leaflets 4–5(8) pairs, 1–3 × 0·5–1·5 cm., obovate- or oblanceolate-oblong or sometimes narrowly oblong, rounded to subtruncate and mucronate or not at the apex,

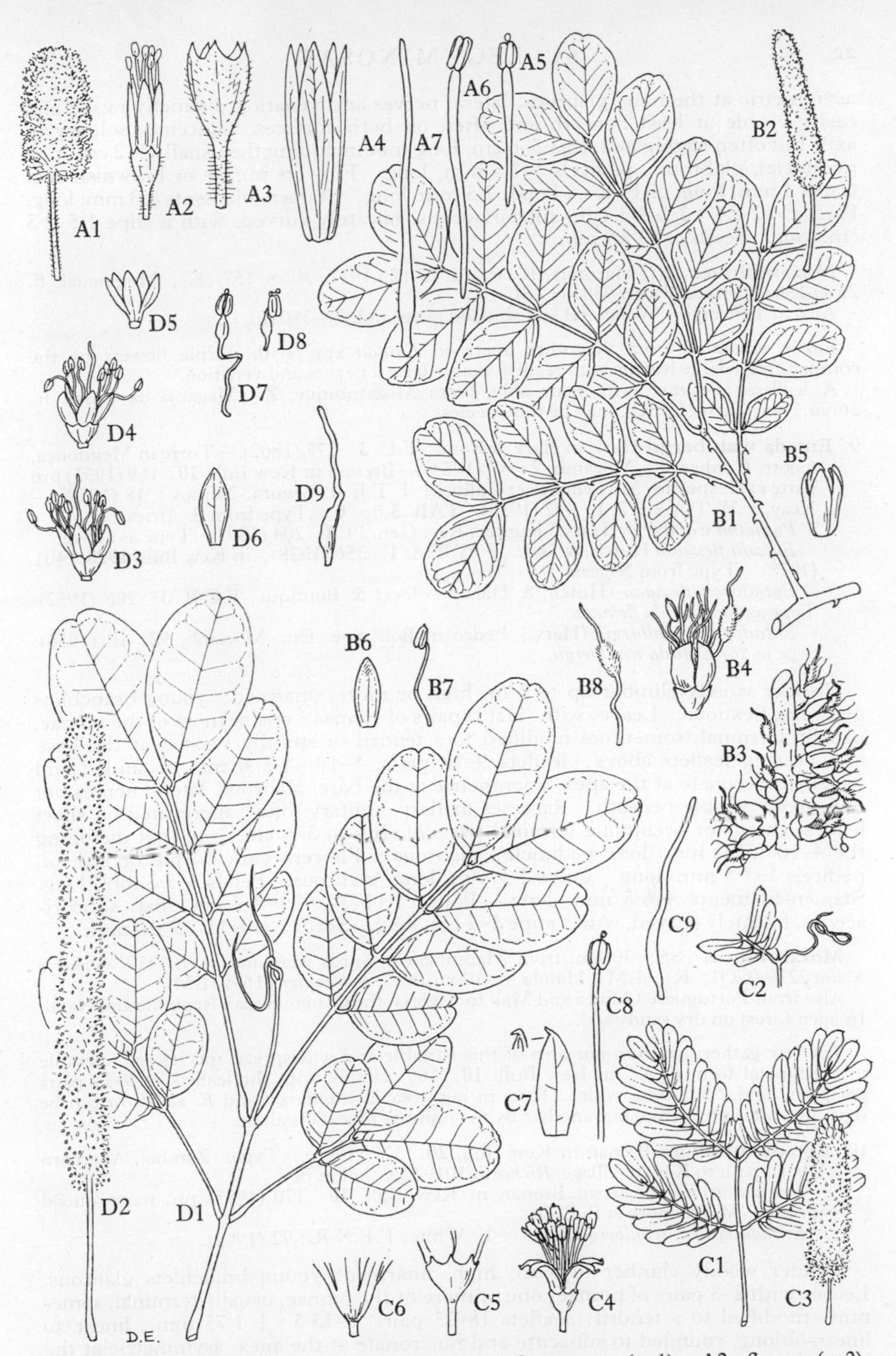

Tab. 3. A.—ENTADA DOLICHORRHACHIS. A1, inflorescence (×½); A2, flower (×2); A3, calyx (×4); A4, corolla (×4); A5, anther with gland (×4); A6, anther without gland (×4); A7, ovary (×4), all from *Fanshawe* 4043. B.—ENTADA SCHLECHTERI. B1, leaf (×½); B2, inflorescence (×½); B3, part of inflorescence (×2); B4, flower (×4); B5, corolla (×4); B6, petal (×4); B7, anther (×4); B8, ovary (×4), all from *Torre* 3903. C.—ENTADA WAHLBERGII. C1, leaf (×½); C2, part of leaf with leaflets and tendril (×½); C3, inflorescence (×2); C4, flower (×3); C5, calyx (×4); C6, base of corolla (×4); C7, petal (×4); C8, anther (×4); C9, ovary (×4), all from *Acocks* 13012. D.—ENTADA PURSAETHA. D1, leaf (×½); D2, inflorescence (×½); D3, D4, flowers (×2); D5, corolla (×2); D6, petal (×4); D7, D8, anthers (×4); D9, ovary (×4) all from *Gomes & Sousa* 3628.

asymmetric at the base, glabrous, lateral nerves and venation distinctly raised and easily visible at least beneath and often on both surfaces. Racemes solitary in axils but often aggregated, (2)3·5–6 cm. long (not including the usually 1–2 cm. long peduncle), glabrous; pedicels 1–1·5 mm. long. Flowers purple or brownish-red. Calyx 1 mm. long, glabrous. Petals 2·5 mm. long. Stamen-filaments c. 3 mm. long. Pods c. 12–24 × 2·7–4·3 cm., subcoriaceous, falcately curved, with a stipe 1·5–2·5 cm. long. Seeds c. 10 × 9 mm.

Mozambique. N: Macomia, R. Msalu, fl. 12.i.1912, *Allen* 157 (K); Mogincual, fl. 25.xi.1963, *Torre & Paiva* 9301 (LISC).
Also in Tanzania. Woodland and wooded grassland, 20–380 m.

Easily separated from the other unarmed *Entada* spp. with purple flowers by the comparatively large leaflets with readily visible lateral nerves and venation.
A leafless specimen with pods only, from Mozambique, Z: Maganja da Costa, fr. 26.vii.1944, *Torre* (LISC), may be this species.

9. **Entada wahlbergii** Harv. in Harv. & Sond., F.C. **2**: 277 (1862).—Torre in Mendonça, Contr. Conhec. Fl. Moçamb. **2**: 87 (1954).—Brenan in Kew Bull. **10**: 169 (1955) pro parte excl. specim. Michelmore et Bullock; F.T.E.A. Legum.-Mimos.: 18 (1959).—Keay, F.W.T.A. ed. 2, **1**: 492 (1958). TAB. **3** fig. C. Type from S. Africa.
Pusaetha wahlbergii (Harv.) Kuntze, Rev. Gen. Pl. **1**: 204 (1891). Type as above.
Entada flexuosa Hutch. & Dalz., F.W.T.A. **1**: 356 (1928); in Kew Bull. **1928**: 401 (1928). Type from Nigeria.
Entadopsis flexuosa (Hutch. & Dalz.) Gilbert & Boutique, F.C.B. **3**: 206 (1952). Type as for *Entada flexuosa*.
Entadopsis wahlbergii (Harv.) Pedro in Bol. Soc. Est. Moçamb. **92**: 10 (1955). Type as for *Entada wahlbergii*.

Slender woody climber up to 3 m. high or more, unarmed; young branchlets glabrous, flexuous. Leaves with (1)2(3) pairs of pinnae; one or more of the pinnae, usually terminal, sometimes modified to a tendril or spirally twisted at the base and bearing leaflets above; leaflets 9–18 pairs, 8–19 × 1·5–6 mm., rounded and usually mucronate at the apex, asymmetric at the base, glabrous, lateral nerves not or scarcely visible beneath. Racemes axillary, solitary, often aggregated on short leafless shoots or occupying terminal parts of shoots, 3–6 cm. long (not including the 4–10 (35?) mm. long peduncle), glabrous. Flowers dark purple or red, on pedicels 1–1·5 mm. long. Calyx 1·5 mm. long, glabrous. Petals 3–3·5 mm. long. Stamen-filaments 4–6·5 mm. long. Pods c. 11–30 × 2·9–4·4 cm., flat, subcoriaceous, falcately curved, with a stipe 1–2 cm. long. Seeds c. 10–11 × 7–10 mm.

Mozambique. SS: 30 km. from Mapinhane towards Mavume, fr. ii.1939, *Gomes e Sousa* 2219 (COI; K). LM: Matola, fr. 10.xii.1897, *Schlechter* 11699 (BM).
Also from Portuguese Guinea and Mali to Nigeria, the Congo, Sudan Republic and Natal. In open forest on dry sandy soil.

The few gatherings from our area of this variable and widespread species are referable to the typical form (" A " in Kew Bull. **10**: 169 (1955)), with the leaflets in 9–12 pairs per pinna and 2·5–5 mm. wide. Thus in our area *E. wahlbergii* and *E. nudiflora* can be separated absolutely from one another by the number of leaflets alone.

10. **Entada nudiflora** Brenan in Kew Bull. **20**: 377 (1966). Type: Zambia, Abercorn Distr., path to Kapata village, *Richards* 10192 (K, holotype).
Entada wahlbergii sensu Brenan in Kew Bull. **10**: 170 (1955) pro parte quoad specim. Michelmore et Bullock.
Entada sp. nr. *wahlbergii* Harv.—F. White, F.F.N.R.: 92 (1962).

Slender woody climber to 3 m. high, unarmed; young branchlets glabrous. Leaves with 2–3 pairs of pinnae; one or more of the pinnae, usually terminal, sometimes modified to a tendril; leaflets 18–25 pairs, 6–13·5 × 1–1·75 mm., linear to linear-oblong, rounded to subacute and mucronate at the apex, asymmetric at the base, glabrous or almost so, lateral nerves not visible beneath. Racemes axillary, solitary, always produced when the plant is leafless, often aggregated on short shoots or occupying terminal parts of shoots, 2·5 cm. long (not including the 10–30 mm. long peduncle), glabrous, dense-flowered. Flowers dark-purple, sessile or almost so (pedicel to 0·5 mm. long). Calyx c. 2·5 mm. long. Petals 3·5 mm. long. Stamen-filaments 6–8 mm. long. Pods c. 25–28 × 3–3·4 cm., flat, subcoriaceous, falcately curved, with a stipe c. 1·5–2 cm. long. Seeds c. 10 × 6·5 mm.

Zambia. N: Chisyera-Chikuka watershed, fl. 24.vi.1933, *Michelmore* 447 (K); Mkupa, fr. 7.x.1949, *Bullock* 1168 (K; SRGH); Abercorn Distr., 16.xi.1952, *White* 3690 (K).
Also in Tanzania (Ufipa Distr.). Deciduous thicket and scrub and *Brachystegia allenii* woodland, often on rocky hillsides, but sometimes also on sandy soil.

Dark-red or purple flowers occur in other species of *Entada* in our area, *E. schlechteri*, *E. stuhlmannii* and *E. wahlbergii*, but they all produce leaves and flowers together, while *E. nudiflora* is leafless while flowering.

11. **Entada mossambicensis** Torre in Mendonça, Contr. Conhec. Fl. Moçamb. **2**: 88 (1954). Type: Mozambique, Nampula, *Torre* 4750 A (BM; K; LISC, holotype).

Suberect shrub, branched from the base, 1–2 m. high; roots fasciculate, thick, fusiform, 20–30 × 4 cm. (*fide* Torre); young branchlets glabrous. Leaves with 3–7 pairs of pinnae; leaflets 40 (usually many more)–138 pairs, 3–5·5 × 0·75 mm., linear-oblong, subacute and mucronate at the apex, asymmetric at the base, glabrous; lateral nerves invisible beneath. Racemes axillary, solitary, elongate, 12–20 cm. long; peduncle 1–2 cm. long; racemes produced together with the leaves. Flowers purple, on pedicels 2–2·5 mm. long. Calyx c. 1 mm. long, glabrous. Petals 4·5–5 mm. long. Stamen-filaments c. 5–6 mm. long. Pods c. 10–12 × 2–2·5 cm., flat, subcoriaceous, falcately curved, with a stipe 1–1·5 cm. long. Seeds c. 12 × 10 mm.

Mozambique. N: Nampula, fl. ix.1936, *Torre* 1140 (COI; LISC; LM); arredores de Nampula, fl. & fr. 3.xi.1942, *Torre* 4750 A (BM; K; LISC).

This is extremely distinct among the unarmed purple-flowered species of *Entada* on account of its very numerous small leaflets and its long racemes of comparatively large flowers.
It would be of value if investigators in the field could ascertain whether the production of tuberous roots is a constant feature of *all* the purple-flowered species of *Entada*.

12. **Entada schlechteri** (Harms) Harms in Engl., Pflanzenw. Afr. **3**, 1: 402 (1915).—Torre in Mendonça, Contr. Conhec. Fl. Moçamb. **2**: 87 (1954). TAB. **3** fig. B. Type: Mozambique, Lourenço Marques, *Schlechter* 11706 (B, holotype †; BM; K).
Piptadenia schlechteri Harms in Engl., Bot. Jahrb. **26**: 260 (1899). Type as above.

Liane, armed with scattered recurved prickles on the stem, and often on the petioles and leaf- and pinna-rhachides as well; branchlets subglabrous to sparsely puberulous. Leaves with 2–3 pairs of pinnae; tendrils absent; leaflets 4–5(6) pairs, 0·8–3·5 × 0·9–2·3 cm., obliquely elliptic to obovate-elliptic, rounded and often slightly emarginate at the apex, cuneate to rounded at the base, glabrous except for some pubescence near the margins and near the base on the lower side. Spikes axillary, 1–3 together, 2–6 cm. long, on peduncles 0·5–2·5 cm. long; axis pubescent. Flowers sessile, " dark red ". Calyx 0·75–1 mm. long, glabrous. Corolla 3–3·5 mm. long; lobes c. 1·5 mm. long. Stamen-filaments c. 5 mm. long. Pods (6)7–16 × 2–2·6 cm., falcately curved, rather thin, with a stipe 0·3–0·5 cm. long; joints slightly umbonate in the centre. Seeds c. 6 × 5 mm.

Mozambique. SS: Gaza, Vila João Belo, fl. 22.i.1942, *Torre* 3903 (BM; K; LISC; LM). LM: Lourenço Marques, Polana, fr. 31.viii.1940, *Hornby* 1001 (K; LISC; LM).
At present known only from southern Mozambique. Thickets.

A very distinct species, related to the S. African *E. spicata* (E. Mey.) Druce but apparently with red not yellowish to white flowers. It is the only prickly *Entada* occurring in our area.
The record of *E. natalensis* Benth. (i.e. *E. spicata* (E. Mey.) Druce) by Sim, For. Fl. Port. E. Afr.: 53 (1909), probably refers to *E. schlechteri* although the description indicates *E. spicata*. The latter species may occur in our area, though it can hardly be accepted at present without confirmation. It is distinguishable from *E. schlechteri* by having c. 6–13 pairs of leaflets per pinna, the individual leaflets being narrow and appressed-puberulous on the lower surface.

3. ELEPHANTORRHIZA Benth.

(By J. P. M. Brenan & R. K. Brummitt)

Elephantorrhiza Benth. in Hook., Journ. Bot. **4**: 344 (1841); Phillips in Bothalia, **1**: 187–193 (1923).

Small trees, shrubs or suffrutices, unarmed. Leaves 2-pinnate, pinnae mostly with many pairs of leaflets. Inflorescences of spiciform racemes which are axillary, solitary or clustered, often ± aggregated; pedicels c. 1–2 mm. long. Flowers normally ☿. Calyx gamosepalous, small, c. 1–2·5 mm. long, with 5 teeth. Petals 5, free. Stamens 10, fertile, free among themselves, slightly adnate to the corolla; filaments c. 4–7·5 mm. long; anthers with a usually very caducous apical gland. Pods straight or somewhat curved, not spirally twisted; at maturity the valves separating from the persistent margins, but not splitting into segments; the outer layer (exocarp) of the pod-wall often peeling off the inner layer (endocarp), the layers remaining intact or breaking irregularly. Seeds ± compressed.

A genus of c. 6–8 species restricted to Africa south of the equator.

Root-nodules recorded from species Nos. 1, 2, 3 and 4.

Plants in habit shrubs or small trees with woody branched aerial stems, up to 7 m. high:
 Leaflets with the midrib marginal throughout; flowers normally (? always) produced with the leaves; minute glands round the pedicel-bases reddish - 2. *suffruticosa*
 Leaflets with the midrib central or nearly so, at least towards the apex; flowers frequently precocious; glands round the pedicel-bases reddish or whitish:
  Leaflets strongly asymmetric at the base, with the proximal side broadly rounded-truncate and the distal side cuneate; leaves usually with (3)14–41 pairs of pinnae (if all with less than 14 pairs then leaflets usually 4 mm. wide or more); glands round the pedicel-bases whitish; flowers usually precocious; pods 1·3–3 cm. wide and up to 44 cm. long - - - - - - - 1. *goetzei*
  Leaflets only slightly asymmetric at the base, with the proximal side cuneate to slightly rounded; leaves with (1)4–8 pairs of pinnae; leaflets 1·5–3·5(4·5) mm. wide; glands round the pedicel-bases reddish; flowers apparently not precocious; pods 2·5–4 cm. wide and up to 19 cm. long - - - - - 3. *burkei*
Plants in habit suffrutices with herbaceous unbranched annual aerial stems up to 0·6 m. high - - - - - - - - - - - 4. *elephantina*

1. **Elephantorrhiza goetzei** (Harms) Harms in Engl., Pflanzenw. Afr. **3**, 1: 400 (1915).—Bak. f., Legum. Trop. Afr. **3**: 802 (1930).—Brenan, T.T.C.L.: 344 (1949); F.T.E.A. Legum.-Mimos.: 19, fig. 4 (1959).—Wild, Guide Fl. Vict. Falls: 149 (1953).—Williamson, Useful Pl. Nyasal.: 52 (1955).—Burtt Davy & Hoyle, rev. Topham, N.C.L. ed. 2: 66 (1958).—F. White, F.F.N.R.: 91 (1962).—Boughey in Journ. S. Afr. Bot. **30**: 158 (1964). TAB. **4**. Type from Tanzania.
 Piptadenia goetzei Harms in Engl., Bot. Jahrb. **28**: 397 (1900). Type as above.

Shrub or small tree 1–7 m. high, deciduous; bark grey-brown to dark dull brown or red; young branchlets glabrous becoming blackish. Leaves up to 53 cm. long (rhachis + petiole), glabrous or nearly so; pinnae 3–41 pairs; leaflets 9–48 pairs, 3·5–22 × 0·7–8 mm., linear-oblong to narrowly oblong, glabrous; midrib starting in the distal corner of the leaflet-base, gradually becoming almost central in the leaflet; proximal side of the base rounded and almost auriculate; apex acute to rounded and mucronate and nearly symmetric; lateral nerves and veins not or scarcely visible. Racemes (5)8–20(23) cm. long (including the peduncle), glabrous; minute whitish mealy glands present round the pedicel-bases. Flowers variously described as yellow, or with brownish-violet petals and yellow stamens.* Petals 2·5–3 mm. long. Pods (15)25–44 × 1·3–3 cm., linear, the seeds showing as bumps at intervals. Seeds 11–20 × 9–18 × 7–12 mm., ellipsoid to lenticular.

Subsp. **goetzei**.—Brenan & Brummitt in Bol. Soc. Brot., Sér. 2, **39**: 189 (1965).
 Elephantorrhiza rubescens Gibbs in Journ. Linn. Soc., Bot. **37**: 441 (1906).—Eyles in Trans. Roy. Soc. S. Afr. **5**: 364 (1916). Type: Rhodesia, Matopo Hills, *Gibbs* 184 (BM, holotype).
 Elephantorrhiza cf. *petersiana*.—Gomes e Sousa, Pl. Menyharth.: 70 (1936).
 Acacia rehmanniana sensu M. A. Exell in Bol. Soc. Brot., Sér. 2, **12**: 16 (1937).

Leaves with (3)14–41 pairs of pinnae, the pinnae themselves 3·5–9·5 cm. long; leaflets in (11)20–48 pairs, 3·5–12 × 0·7–3 mm.

Botswana. N: NE. edge of Makarikari, fr. 12.xii.1929, *Pole Evans* 2591 (K; PRE). **Zambia.** N: Mwunyamadzi R., Luangwa Valley, fl. 5.x.1933, *Michelmore* 639 (K). E: Jumbe, fl. 14.x.1958, *Robson* 107 (BM; K; LISC; SRGH). **Rhodesia.** N: Concession, fl. 18.x.1938, *McGregor* 52/1938 (K). W: Matobo Distr., Besna Kobila, fr. xii.1953,

* Collectors' notes on flower-colour in this species leave it uncertain whether there is variation in the colour.

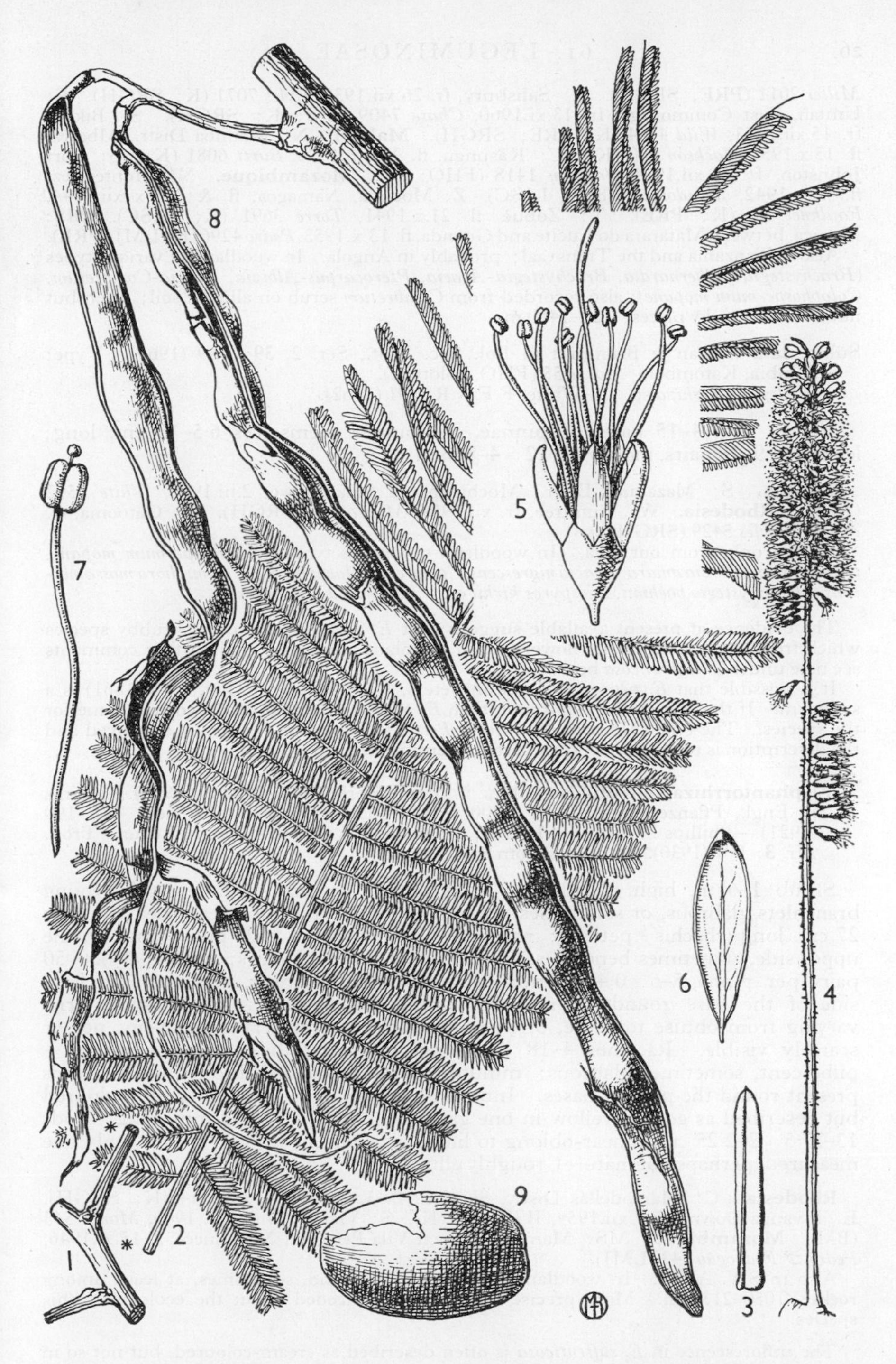

Tab. 4. ELEPHANTORRHIZA GOETZEI. 1, part of branch with petiole-bases ($\times\frac{3}{4}$); 2, leaf, detached from petiole base of 1 ($\times\frac{3}{4}$) 3, leaflet ($\times 6$), all from *Milne-Redhead & Taylor* 9549; 4, flowering raceme ($\times 1\frac{1}{2}$); 5, flower ($\times 9$); 6, petal ($\times 15$); 7, anther ($\times 25$), all from *Andrada* 1452; 8, pods ($\times 1$); 9, seed ($\times 3$), both from *Jackson* 1418. From F.T.E.A.

Miller 2011 (PRE; SRGH). C: Salisbury, fr. 26.xii.1931, *Eyles* 7071 (K; SRGH). E: Umtali, East Commonage, fr. 13.xi.1960, *Chase* 7409 (BM; K; SRGH). S: Bikita, fr. 15.xii.1953, *Wild* 4394 (K; PRE; SRGH). **Malawi.** N: Mzimba Distr., Mbawa, fl. 15.x.1952, *Jackson* 980 (K). C: Kasungu, fl. 21.viii.1936, *Burtt* 6081 (K). S: Fort Johnston, fr. 21.xii.1954, *Jackson* 1418 (FHO; K). **Mozambique.** N: Montepuez, fl. 17.x.1942, *Mendonça* 918 (K; LISC). Z: Mocuba, Namagoa, fl. & fr. ix–xii.1944, *Faulkner* 26 (K; PRE). T: Zóbuè, fl. 21.x.1941, *Torre* 3691 (K; LISC). MS: Manica, between Matarara do Lucite and Goonda, fl. 13.x.1953, *Pedro* 4296 (K; LMJ; PRE).

Also in Tanzania and the Transvaal; probably in Angola. In woodland of various types (*Brachystegia-Julbernardia, Brachystegia-Acacia, Pterocarpus-Albizia, Acacia-Combretum, Colophospermum mopane*); also recorded from *Combretum* scrub on alluvial soil; often but not always in rocky places; 120–1460 m.

Subsp. **lata** Brenan & Brummitt in Bol. Soc. Brot., Sér. 2, **39**: 189 (1965). Type: Zambia, Katombora, *Morze* 55 (FHO, holotype).
Elephantorrhiza sp. 1.—White, F.F.N.R.: 91 (1962).

Leaves with 4–15 pairs of pinnae, the pinnae themselves 6·5–15 cm. long; leaflets in 9–28 pairs, mostly 12–22 × 4–8 mm.

Zambia. S: Mazabuka Distr., Mochipapa to Sinazongwe, 2.iii.1960, *White* 7562 (FHO). **Rhodesia.** W: Plumtree, fr. xii.1954, *Meara* 102 (SRGH). C: Gatooma, st. iv.1926, *Eyles* 5429 (SRGH).

Known only from our area. In woodland of various types (*Colophospermum mopane, Sclerocarya, Albizia amara, Acacia nigrescens, Combretum elaeagnoides, Commiphora mossambicensis, Brachystegia boehmii, Diospyros kirkii* etc., *Julbernardia globiflora*).

The evidence at present available suggests that *E. goetzei* is the only shrubby species which frequently produces its flowers when the plant is leafless. For further comments see note under *E. suffruticosa* below.

It is possible that *E. petersiana* Bolle in Peters, Reise Mossamb. Bot. **1**: 9 (1861) is a synonym. If this were to be confirmed, then *E. petersiana* would be the correct name for the species. The type (Mozambique, Sena, *Peters* (B, holotype †)) is now destroyed and the description is too imperfect for a certain identification.

2. **Elephantorrhiza suffruticosa** Schinz in Mém. Herb. Boiss. **1**: 117 (1900).—Harms in Engl., Pflanzenw. Afr. **3**, 1: 400 (1915).—Dinter in Fedde, Repert. **17**: 190 (1921).—Phillips in Bothalia, **1**: 193, t. 5 fig. 7, t. 6 (1923).—Bak. f., Legum. Trop. Afr. **3**: 801 (1930). Syntypes from SW. Africa.

Shrub 1–6 m. high, deciduous; bark similar to that of *E. goetzei*; young branchlets glabrous, or sometimes puberulous to shortly pubescent. Leaves up to 27 cm. long (rhachis + petiole); rhachides usually with a little pubescence on the upper side, sometimes beneath as well; pinnae (2)15–31 pairs; leaflets (19)30–50 pairs per pinna, 5–6 × 0·4–1·2 mm., linear-oblong to linear, glabrous, proximal side of the base rounded; midrib marginal throughout; apex asymmetric, varying from obtuse to acute, often mucronate; lateral nerves and veins not or scarcely visible. Racemes 4–18 cm. long (including the peduncle), often ± pubescent, sometimes glabrous; minute reddish or reddish-brown mealy glands present round the pedicel-bases. Inflorescence usually said to be cream-coloured but described as golden-yellow in one gathering. Petals 3–3·75 mm. long. Pods 13–27·5 × 2–2·25 cm., linear-oblong to linear. Seeds c. 14 × 11 × 5 mm. (only one measured, perhaps not mature), roughly ellipsoid.

Rhodesia. C: Marandellas Distr., Skipton, fl. 5.xii.1946, *Wild* 1614 (K; SRGH). E: Inyanga Downs, fl. 7.xii.1959, *Wild* 4895 (K). S: Victoria Distr., fl. 1909, *Monro* 543 (BM). **Mozambique.** MS: Manica, between Vila Pery and Macequece, fl. 15.xi.1946, *Pedro & Pedrógão* 243 (LMJ).

Also in SW. Africa. In woodland and *Acacia* grassland, sometimes, at least, among rocks; 1050–2130 m. More precise information is needed about the ecology of this species.

The inflorescence in *E. suffruticosa* is often described as cream-coloured, but not so in *E. goetzei*. This may indicate a difference in flower-colour between these two species, but it needs confirmation in the field.

Only one specimen apparently belonging to *E. suffruticosa* has been collected bearing flowers but no leaves at all: *Hornby* 2448 (K), from Rhodesia (C), Hartley Distr., Poole Farm. This is in contrast with *E. goetzei* which frequently flowers precociously, and may be connected with a possibly earlier flowering-season in that species. Flowering specimens of *E. goetzei* from our area have been collected from August to October, of *E. suffruticosa* from October to December.

The distribution of *E. suffruticosa* appears to be discontinuous between eastern and central Rhodesia and SW. Africa. No satisfactory differences have been so far seen between plants from these two areas except for an inconstant tendency for the leaflets to be more acute and mucronate in Rhodesia than in SW. Africa.

3. **Elephantorrhiza burkei** Benth. in Hook., Lond. Journ. Bot. **5**: 81 (1846).—Phillips in Bothalia, **1**: 192 (1923).—Bak. f., Legum. Trop. Afr. **3**: 801 (1930).—Burtt Davy, F.P.F.T. **2**: 332 (1932).—O. B. Mill. in Journ. S. Afr. Bot. **18**: 31 (1952).—Wild, Guide Fl. Vict. Falls: 149 (1953).—Boughey in Journ. S. Afr. Bot. **30**: 158 (1964). Type from S. Africa (Transvaal).

Shrub or small tree usually 1–6 m. high, occasionally as small as 0·3 m., but then with stems distinctly woody and branched and the inflorescences normally on lateral shoots of the current season; bark dark-grey to reddish; young branchlets glabrous. Leaves 7–18 cm. long (rhachis + petiole), glabrous or almost so when mature; pinnae (1)4–8 pairs; leaflets 9–32 pairs per pinna, 7–15 × 1·5–3·5(4·5) mm., narrowly oblanceolate to very narrowly elliptic or linear-oblong, glabrous, base slightly asymmetric (less so than in *E. elephantina*), with the proximal side rounded to cuneate, apex symmetric, obtuse to rounded, generally mucronate, lateral nerves and veins prominent or not. Racemes 5–10(12) cm. long (including the peduncle), glabrous; minute reddish mealy glands present round the pedicel-bases. Flowers variously described as cream or yellow. Petals 3–4·5 mm. long. Pods 10–19 × 2·5–4 cm. Seeds c. 11–13 × 8–12 × 8 mm., the few seen irregular in shape.

Botswana. SE: Ngwaketse Distr., 1 km. E. of Pharing, near Kanye, fl.x.1947, *Miller* B/514 (COI; K; SRGH). **Rhodesia.** W: Matopo Hills, fl. 22.ix.1909, *Rogers* 5335 (K; SRGH). **Mozambique.** LM: Marracuene, Ricatla, fl. 1917–18, *Junod* 435 (LISC).

Also in the Transvaal. Usually in rocky places, in woodland and grassland, c. 970–1370 m.

E. burkei differs from *E. elephantina* primarily in being a shrub or tree with branched perennial aerial stems and not a suffrutex with annual aerial stems; also in the leaflet-base which is less asymmetric than in *E. elephantina*. The very narrow acute leaflets characteristic of some variants of *E. elephantina* are unknown in *E. burkei*. It is possible that *E. burkei* normally has smaller seeds than *E. elephantina*, but more material is required.

In the Transvaal, the range of leaflet pairs per pinna is 9–22, but in our area it is 17–32. Some specimens from Rhodesia, Matobo Distr., *Davies* 228 (K; SRGH), *Miller* 1935 and 2013 (K; SRGH) and *Hodgson* 18/55 (K; PRE; SRGH), are unusual in having narrow leaflets with the venation not or scarcely prominent even when mature.

O. B. Mill. in Journ. S. Afr. Bot. **18**: 31 (1952) has noted an exceptional pod of *E. burkei* 28 cm. long.

4. **Elephantorrhiza elephantina** (Burch.) Skeels in US. Dept. Agric. Bur. Pl. Industry Bull. **176**; 29 (1910).—Bak. f., Legum. Trop. Afr. **3**: 800 (1930).—Burtt Davy, F.P.F.T. **2**: 332 (1932).—O. B. Mill. in Journ. S. Afr. Bot. **18**: 31 (1952). Type from S. Africa (N. Cape Prov.).

Acacia elephantina Burch., Trav. Int. S. Afr. **2**: 236 (1824). Type as above.

Acacia elephanthorhiza DC., Prodr. **2**: 457 (1825) *nom. illegit*. Type as above.

Elephantorrhiza burchellii Benth. in Hook., Journ. Bot. **4**: 344 (1841) *nom. illegit.*—Harv. in Harv. & Sond., F.C. **2**: 277 (1862).—Harms in Engl., Pflanzenw. Afr. **3**, 1: 400, t. 229 (1915).—Dinter in Fedde, Repert. **17**: 190 (1921).—Phillips in Bothalia, **1**: 189, t. 5 fig. 2 (1923). Type as above.

Suffrutex producing at ground level at the woody end of a ± elongate rhizome annual herbaceous stems c. 20–60 cm. high; aerial stems usually unbranched except for inflorescences (very rarely branched after damage to the main apex); racemes on lower part of stem only; young stems glabrous or very rarely (and not in our area) pubescent. Leaves up to 15(24) cm. long (rhachis + petiole); rhachides glabrous or with very sparse hairs; pinnae 2–4 pairs in lower leaves, increasing upwards to 7–15 pairs; leaflets (7)13–47 pairs per pinna (in ourarea the maximum is usually 30 or more), (4)5–10(15) × (0·3)0·5–2(2·5) mm., linear-oblong to linear, rarely narrowly oblanceolate, glabrous or almost so, base nearly always asymmetric, with the proximal side rounded to cuneate, apex symmetric to asymmetric, acute or rarely obtuse, mucronate, lateral nerves and veins prominent or not.

Racemes (2)4–8(11·5) cm. long (including the peduncle), glabrous or very rarely (not in our area) pubescent; minute reddish mealy glands present round the pedicel-bases. Flowers described as cream or yellow. Petals 3·25–3·5 mm. long. Pods (5)9·5–21 × 3–5·7 cm. Seeds c. 20–25 × 13–18 × 11–13 mm., ± ellipsoid.

Botswana. SW: between Karakobis and Kalkfontein, fr. ii.1952, *de Beer* D 26 (K). SE: Ngwaketze Distr., 2 km. NE. of Pharing, near Kanye, fl. x.1947, *Miller* B/517 (COI; K; SRGH). **Rhodesia.** N: Lomagundi, Mtoroshanga, fl. 25.x.1959, *Leach* 9503 (K; SRGH). W: Matobo, fr. 10.i.1948, *West* 2568 (K; SRGH). C: Salisbury. fl. x.1920, *Eyles* 2654 (K; PRE; SRGH). S: Victoria Distr., *Monro* 2221 (K). **Mozambique.** SS: between Chicomo and Jacobécua, fl. 8.xii.1944, *Mendonça* 3334 (K; LISC).

Also in S. Africa and Swaziland. Grassland and open *Acacia-Combretum* scrub, 1060–1360 m. Sometimes gregarious.

E. elephantina is the only suffruticose species of this genus known to occur in our area. It shows considerable variation, probably genetic, in the number, size and shape of the leaflets.

4. NEWTONIA Baill.

(By J. P. M. Brenan & R. K. Brummitt)

Newtonia Baill. in Bull. Soc. Linn. Par. **1**: 721 (1888)

Trees, often tall, unarmed. Leaves 2-pinnate; pinnae each with one to many pairs of leaflets; rhachis of leaf usually (always in our species) with a gland between each pair of opposite pinnae. Flowers sessile or nearly so, in spikes or spiciform racemes. Calyx gamosepalous, 5-toothed. Calyx and petals pubescent or puberulous outside, sometimes on the margins only. Petals 5, free, separated from the gynophore base by a short perigynous zone. Stamens 10, fertile; anthers with or without an apical gland. Ovary densely pilose outside. Pods straight or somewhat curved, flattened, at maturity dehiscing along one of the margins, the valves remaining attached along the other, splitting neither transversely nor into layers. Seeds flattened, oblong, brown, surrounded by a membranous wing; the body of the seed much elongated in the direction of the length of the pod; cotyledons elongate in the same direction as the radicle; funicle slender, attached at or near one end of the seed.

A genus of 14 or more species, 11 of them over much of tropical Africa, the rest in tropical S. America.

Root-nodules not yet recorded.

Leaflets 6–many pairs, 2–11 × 0·5–3 mm.:
- Pinnae (7)12–23 pairs; leaflets (11)38–67 pairs; anthers with an apical gland soon falling off, but readily seen in bud - - - - - - - - 1. *buchananii*
- Pinnae 4–7 pairs; leaflets 6–19(27) pairs; anthers without any apical gland even in bud - - - - - - - - - - - - 2. *hildebrandtii*

Leaflets (2)4–5 pairs, 15–60 × 5–30 mm. - - - - - - - - 3. *aubrevillei*

1. **Newtonia buchananii** (Bak.) Gilbert & Boutique, F.C.B. **3**: 213 (1952).—Torre. C.F.A. **2**: 261 (1956).—Pardy in Rhod. Agric. Journ. **56**: 964 cum photogr. (1956). —Brenan, F.T.E.A. Legum.-Mimos.: 23 (1959).—F. White, F.F.N.R.: 93 (1962). —Boughey in Journ. S. Afr. Bot. **30**: 158 (1964) TAB. **5** fig. A. Type: Malawi, *Buchanan* 192 (BM; FHO; K, holotype).

Piptadenia buchananii Bak. in Kew Bull. **1894**: 354 (1894).—R.E.Fr. in Wiss. Ergebn. Schwed. Rhod.-Kongo-Exped. **1**: 65 (1914).—Bak. f., Legum. Trop. Afr. **3**: 794 (1930).—Steedman, Trees etc. S. Rhod.: 17 (1933).—Gomes e Sousa, Dendrol. Moçamb. **1**: 182 cum tab. [185] (1948); op. cit. **5**: 107 cum photogr. (1960); Dendrol. Moçamb. Estudo Geral, **1**: 227, t. 32 (1966).—Williamson, Uxful Pl. Nyasal.: 98(1955)—Burtt Davy & Hoyle, rev. Topham, N.C.L. ed. 2: 66 (1958). Type as above.

Tree 10–40 m. high, somewhat buttressed at the base; bark smooth; branchlets densely pubescent when young, often glabrescent later. Leaves 6–27 cm. long (rhachis + petiole); rhachis with a stipitate gland between each pinna-pair; no glands between leaflet-pairs; pinnae (7)12–23 pairs; leaflets (11)38–67 pairs, 2–6(9) × 0·5–1·5(2) mm., linear-oblong, often ± falcate; lateral nerves invisible beneath. Flowers whitish or yellowish, in spikes 3·5–19 cm. long. Anthers with an apical gland that soon falls off. Pod 10–32 × 1·3–2·5 cm. Seeds 4·1–7·5(?8) × 0·9–2·1 cm.

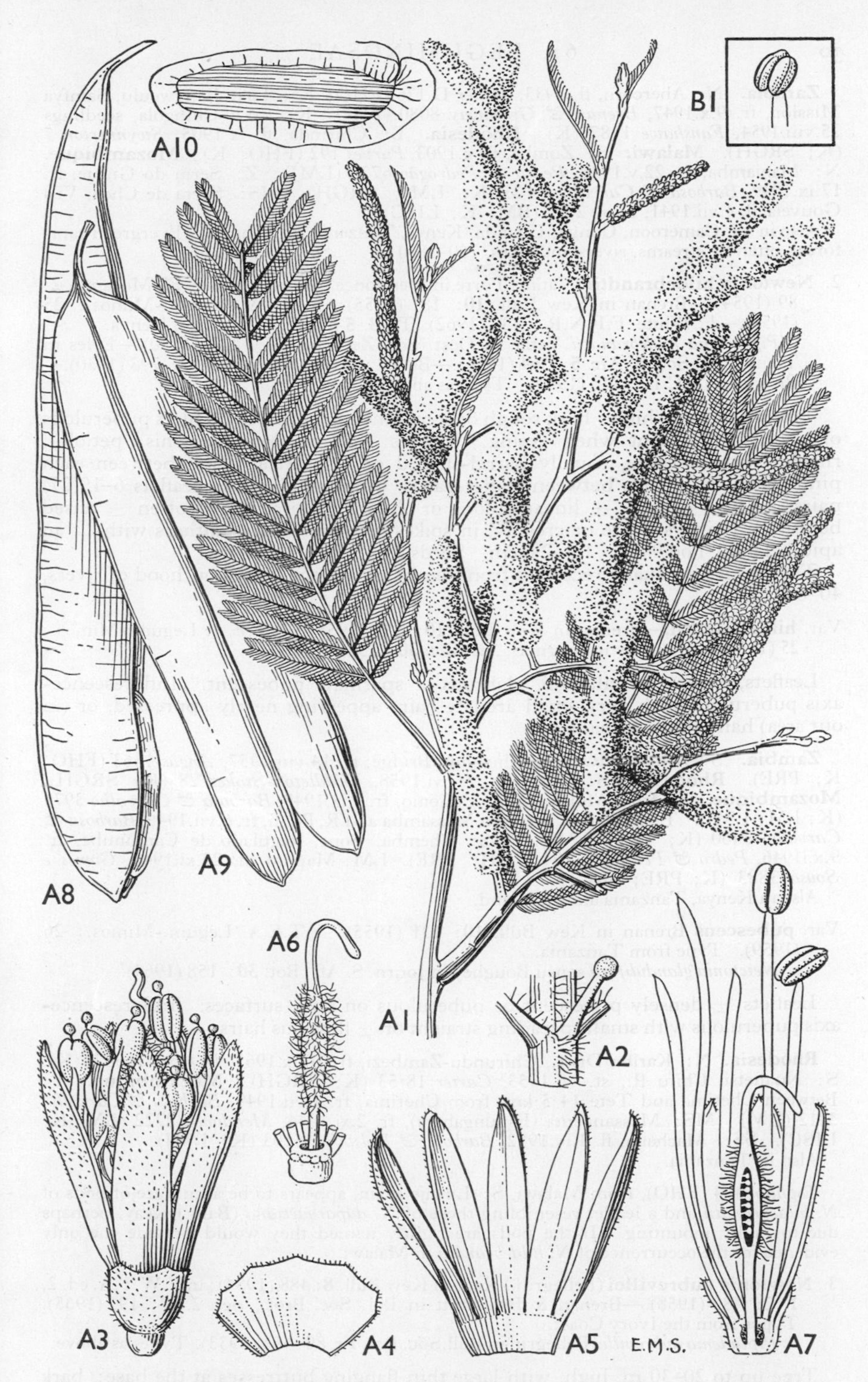

Tab. 5. A.—NEWTONIA BUCHANANII. A1, flowering branch ($\times\frac{2}{3}$); A2, part of leaf-rhachis, showing gland ($\times 6$); A3, flower ($\times 12$); A4, calyx ($\times 12$); A5, corolla ($\times 12$); A6, ovary ($\times 12$); A7, longitudinal section of mature flower ($\times 12$), all from *Eggeling* 3227; A8, pod ($\times\frac{2}{3}$) *Wigg* 3008; A9, pod showing unilateral dehiscence ($\times\frac{2}{3}$) *van Someren* 6771; A10, seed ($\times 1$) *Wigg* 3008. B.—NEWTONIA HILDEBRANDTII. B1, anther, without gland ($\times 12$) *Lewis* 2. From F.T.E.A.

Zambia. N: Abercorn, fl. 1933, *Miller* D 126 (FHO; K); Lake Bangweulu, Samfya Mission, fr. 9.x.1947, *Brenan & Greenway* 8080 (FHO; K). W: Chingola, seedlings 25.viii.1954, *Fanshawe* 1487 (K). **Rhodesia.** E: Chirinda, fl. ix.1905, *Swynnerton* 7 (K; SRGH). **Malawi.** S: Zomba, fl. x.1903, *Purves* 192 (FHO; K). **Mozambique.** N: Maniamba, fr. 22.v.1948, *Pedro & Pedrógão* 6732 (LMJ). Z: Serra do Gúruè, fr. 17.ix.1949, *Barbosa & Carvalho* 4123 (K; LMJ; SRGH). MS: Serra de Choa, Vila Gouveia, fl. 4.vii.1941, *Torre* 2997 (BM; K; LISC).

Also in W. Cameroon, Congo, Uganda, Kenya, Tanzania and Angola. Evergreen rain-forest, often by streams, rivers and lakes, 760–1830 m.

2. **Newtonia hildebrandtii** (Vatke) Torre in Mendonça, Contr. Conhec. Fl. Moçamb. **2**: 89 (1954).—Brenan in Kew Bull. **10**: 181 (1955); F.T.E.A. Legum.-Mimos.: 25 (1959).—F. White, F.F.N.R.: 43/3 (1962). TAB. **5** fig. B. Type from Kenya.
   Piptadenia hildebrandtii Vatke in Oest. Bot. Zeitschr. **30**: 273 (1880).—Eyles in Trans. Roy. Soc. S. Afr. **5**: 364 (1916).—Bak. f., Legum. Trop. Afr. **3**: 793 (1930).—Brenan, T.T.C.L.: 346 (1949). Type as above.

Tree up to 25 m. high; bark rough or sometimes smooth; branchlets puberulous or shortly pubescent when young. Leaves 2–8 cm. long (rhachis + petiole); rhachis with a sessile, usually barrel-shaped or cylindrical gland between each pinna-pair; no glands between leaflet-pairs; pinnae 4–7 pairs; leaflets 6–19(27) pairs, 3–11 × 1–3 mm., ± linear-oblong or oblong; lateral nerves often ± raised beneath. Flowers white or creamy, in spikes 4–9 cm. long. Anthers without an apical gland. Pod 9–30 × 1·4–2·6 cm. Seeds 3·6–5·7 × 1·3–2·1 cm.

The species is usually recorded from sandy soils in the neighbourhood of rivers, 40–600 m.

Var. **hildebrandtii.**—Brenan in Kew Bull. **10**: 181 (1955); F.T.E.A. Legum.-Mimos.: 25 (1959).—Boughey in Journ. S. Afr. Bot. **30**: 158 (1964).

Leaflets, except for ciliation, glabrous or sparingly pubescent. Inflorescence-axis puberulous with very small arcuate hairs appearing nearly appressed, or (in our area) hairs short, spreading.

Zambia. S: Zambezi R., near Chirundu Bridge, fr. 14.viii.1957, *Angus* 1663 (FHO; K; PRE). **Rhodesia.** S: Nuanetsi, fr. vi.1958, *Middleton-Stokes* 28 (K; SRGH). **Mozambique.** Z: from Morine to the Chilomo, fr. 7.ix.1949, *Barbosa & Carvalho* 3979 (K; LM; LMJ). T: Macanga, between Massamba and R. Ponfi, fr. 6.vii.1949, *Barbosa & Carvalho* 3460 (K; LM; LMJ). MS: Chemba, Sone, Regulado de Canhinube, fr. 9.x.i1946, *Pedro & Pedrógão* 104 (LMJ; PRE). LM; Maputo, fl. 22.xi.1948, *Gomes e Sousa* 3883 (K; PRE; SRGH).

Also in Kenya, Tanzania and Zululand.

Var. **pubescens** Brenan in Kew Bull. **10**: 131 (1955); F.T.E.A. Legum.-Mimos.: 26 (1959). Type from Tanzania.
   Newtonia glandulifera sensu Boughey in Journ. S. Afr. Bot. **30**: 158 (1964).

Leaflets ± densely pubescent or puberulous on both surfaces. Inflorescence-axis puberulous with small spreading straight or ± flexuous hairs.

Rhodesia. N: Kariba Distr., Chirundu-Zambezi, fr. 10.ix.1965, *Burrows* 165078 (K). S: Nuanetsi, Chifu R., st. vii.1953, *Carter* 18/53 (K; SRGH). **Mozambique.** T: Between Chetima and Tete 14·5 km. from Chetima, fr. 1.vii.1949, *Barbosa & Carvalho* 3412 (LM). MS: Massangema (Espungabera), fr. 2.xi.1944, *Mendonça* 2722 (BM; K; LISC). SS: Machaíla, fl. 1.iv.1952, *Barbosa & Balsinhas* 5103 (K; LMJ).

Also in Tanzania.

Topham 494 (FHO), from Malawi, S: Lilanje Plain, appears to be a mixture of pods of *N. hildebrandtii* and a leaflet resembling those of *N. duparquetiana* (Baill.) Keay, perhaps due to faulty mounting. If the pods are rightly named they would provide the only evidence for the occurrence of *N. hildebrandtii* in Malawi.

3. **Newtonia aubrevillei** (Pellegrin.) Keay in Kew Bull. **8**: 488 (1954); in F.W.T.A. ed. 2, **1**, 2: 489 (1958).—Brenan & Brummitt in Bol. Soc. Brot., Sér. 2, **39**: 191 (1965). Types from the Ivory Coast.
   Piptadenia aubrevillei Pellegrin in Bull.Soc. Bot. Fr. **80**: 466 (1933). Types as above.

Tree up to 20–30 m. high, with large thin flanging buttresses at the base; bark pinkish-brown, minutely fissured; branchlets shortly and rather densely pubescent when young. Leaves 3–9 cm. long (rhachis + petiole); rhachis with a sessile gland between each pinna-pair; glands present or absent between leaflet-pairs; pinnae

1–4 pairs (1–2 with us); leaflets (2)4–5 pairs, (1·5)2–6 × (0·5)1–3 cm. (to 7 × 3·8 cm. in juvenile shoots), ± irregularly rhombic; lateral nerves slightly raised beneath. Flowers yellowish, in spikes 6–17 cm. long. Anthers with an apical gland that soon falls off. Pods 7–18 × 1·8–2·3 cm. Seeds (3·5)6–8 × 1·3–1·8 cm.

Subsp. **lasiantha** Brenan & Brummitt in Bol. Soc. Brot., Sér. 2, **39**: 191 (1965). Type: Zambia, Mwinilunga, *Holmes* 1343 (K, holotype; SRGH).
Albizia sp. probably *zygia*.—F. White, F.F.N.R.: 90 (1962).

Pinnae 1–2 pairs; pinna-rhachides more sparsely pubescent above than in subsp. *aubrevillei*, sparsely puberulous to subglabrous beneath. Calyx rather densely pubescent outside. Corolla very densely pubescent all over outside.

Zambia. W: Mwinilunga, fl. viii.1959, *Holmes* 1343 (K; SRGH).
This subspecies also occurs in the Congo. In evergreen forest, usually fringing.
The available material of this species is rather limited, particularly from the Congo and Zambia, but it supports the division into two subspecies. Typical subsp. *aubrevillei*, from Sierra Leone, Liberia and the Ivory Coast, differs as follows:—Pinnae (2)3–4 pairs. Pinnae-rhachides pubescent or puberulous all round. Calyx subglabrous outside except for the margins. Corolla subglabrous outside except for the margins and the ends of the petals.
There is a further possible difference. Glands are inconstantly present on the pinna-rhachis between the leaflet-pairs in subsp. *lasiantha* but are apparently always absent in subsp. *aubrevillei*.

5. PSEUDOPROSOPIS Harms

(By J. P. M. Brenan & R. K. Brummitt)

Pseudoprosopis Harms in Engl., Bot. Jahrb. **33**: 152 (1902).

Shrubs or lianes or sometimes small trees, unarmed. Leaves 2-pinnate; petiole and rhachis eglandular; pinnae each with few to many pairs of opposite leaflets. Inflorescences of racemes which are axillary and solitary or up to three together, and often aggregated on ± shortened shoots. Flowers ☿. Calyx gamosepalous with 5 teeth. Petals 5, free or almost so, valvate in bud (or very slightly overlapping in their lower parts in *P. fischeri*). Stamens 10, fertile; anthers each with a caducous apical gland. Ovary hairy. Pods straight or somewhat curved, compressed, woody, oblong, dehiscing from the apex downwards into 2 recurving valves. Seeds lying ± obliquely in the pod, each sunk in a depression in the valve, compressed, brown, glossy, unwinged, elliptic to ± rhombic, without endosperm.
A genus of 4(5) species, confined to tropical Africa.

Leaflets oblong-elliptic, 2–5·5(8) mm. wide, in 7–15 pairs, often persistently puberulous on both surfaces; bracts enlarged and 3-lobed or obtriangular at the apex 1. *fischeri*
Leaflets rhombic-ovate or -obovate, (3)6–14 mm. wide, in 4–9 pairs, glabrous or almost so except on the midrib beneath; bracts linear-lanceolate, acute, not enlarged at the apex - - - - - - - - - - - - - - 2. *euryphylla*

1. **Pseudoprosopis fischeri** (Taub.) Harms in Engl., Bot. Jahrb. **33**: 152 (1902); in Engl., Pflanzenw. Afr. **3**, 1: 406, t. 231 (1915).—Bak. f., Legum. Trop. Afr. **3**: 806 (1930).—Brenan, T.T.C.L.: 347 (1949); F.T.E.A., Legum.-Mimos.: 27, fig. 7, 28 (1959).—Gilbert & Boutique, F.C.B. **3**: 223 (1952).—F. White, F.F.N.R.: 93 (1962). TAB. **6** fig. B. Type from Tanzania.
Prosopis fischeri Taub. in Engl., Pflanzenw. Ost-Afr. **C**: 196 (1895). Type as above.

Shrub or small spreading tree 3–6 m. high, forming coppice, sometimes scrambling; bark silvery-grey, smooth; young branchlets, petioles and rhachides densely puberulous; older branchlets longitudinally striate. Leaves: petiole 0·5–2·5 cm. long; rhachis 1·5–7·5 cm. long; pinnae opposite, 3–5(7) pairs; leaflets 7–15 pairs, 0·5–1·4(1·9) × 0·25–0·55(0·8) cm., oblong-elliptic, rounded or slightly emarginate and mucronate at the apex, normally puberulous on both surfaces, sometimes glabrous except for the midrib beneath. Racemes (2)4–13 cm. long, on peduncles 0·3–1 cm. long; pedicels 1·5–5(7) mm. long; axes and pedicels densely puberulous or shortly pubescent, rarely subglabrous; bracts enlarged and 3-lobed or obtriangular at the apex. Flowers cream-coloured, honey-scented. Calyx 1–1·5 mm. long, puberulous or shortly pubescent. Petals 3·5–5 mm. long, ± puberulous to almost glabrous

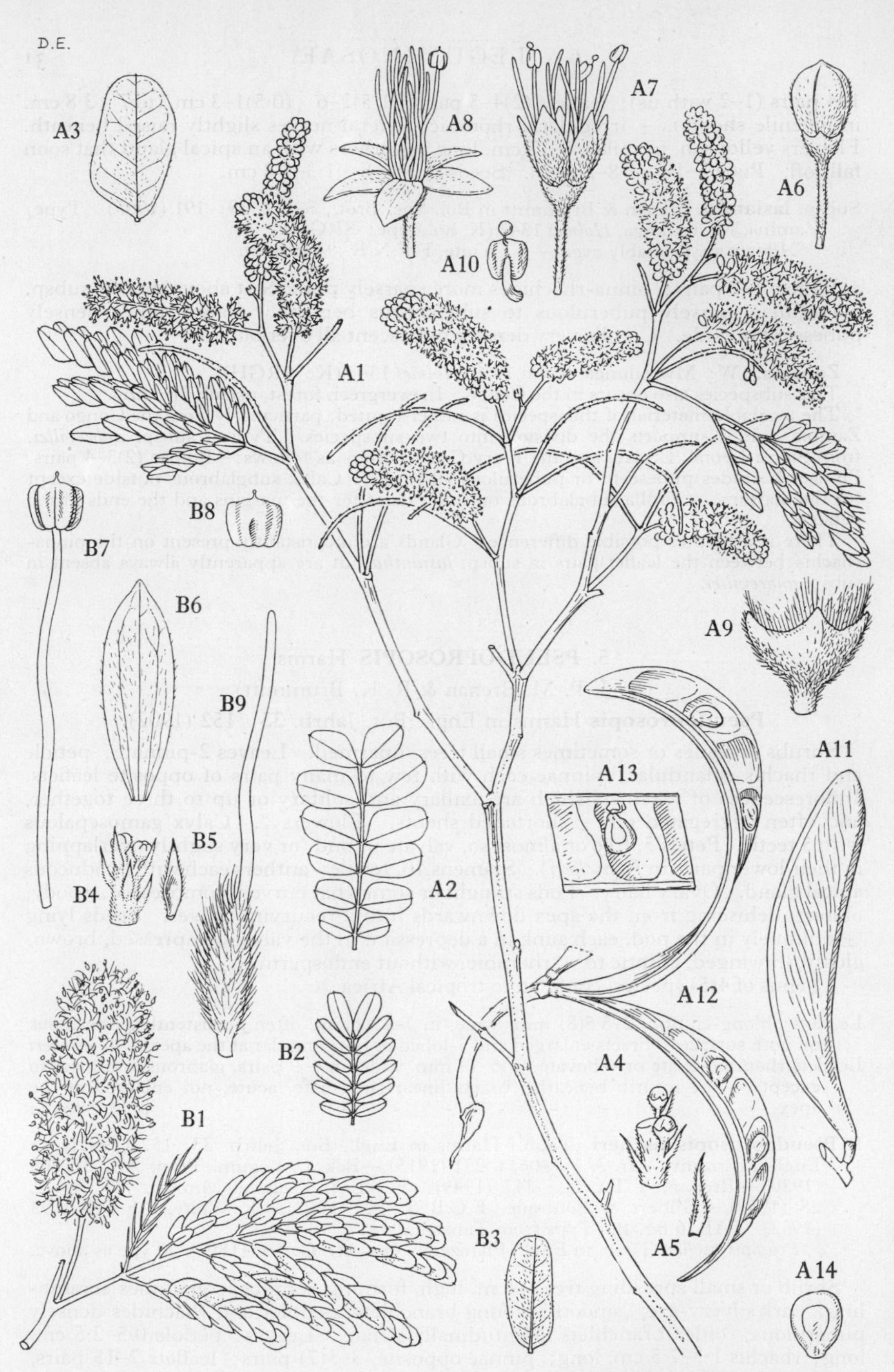

Tab. 6. A.—PSEUDOPROSOPIS EURYPHYLLA. A1, flowering branch ($\times \frac{2}{3}$); A2, pinna of leaf ($\times 2$); A3, leaflet ($\times 4$); A4, part of axis of raceme ($\times 4$); A5, bract ($\times 10$); A6, flower-bud ($\times 4$); A7, flower ($\times 4$); A8, flower, later stage ($\times 4$); A9, calyx ($\times 10$); A10, anther ($\times 10$), all from *Allen* 161; A11, closed pod ($\times \frac{2}{3}$) *Gillman* 1015; A12, open pod ($\times \frac{2}{3}$); A13, section of pod ($\times 1$), all from *Parry* 220; A14, seed ($\times 1$) *Gillman* 1015. B.—PSEUDOPROSOPIS FISCHERI. B1, flowering branch ($\times \frac{2}{3}$); B2, pinna of leaf ($\times \frac{2}{3}$); B3, leaflet ($\times 4$); B4, part of axis of raceme ($\times 4$); B5, bract ($\times 10$); B6, petal (10); B7, stamen ($\times 10$); B8, anther ($\times 10$); B9, ovary ($\times 10$), all from *Richards* 13743.

outside. Stamen-filaments 5–8 mm. long. Ovary densely hairy. Pods 8·5–16 × 1·3–2·2 cm., attenuate below, blackish when dry, ± obliquely longitudinally striate. Seed 10–12 × 7–9 mm.

Zambia. N: Mporokoso Distr., Sumbu, fl. 16.xi.1957, *Savory* 241 (K; PRE; SRGH); Abercorn Distr., Kawa R. gorge, young fr. 14.ii.1959, *Richards* 10901 (K).
Also in the Congo and Tanzania. An important constituent of dense deciduous thicket (mteshi), 760–1000 m.

P. fischeri shows certain divergent tendencies in Zambia when compared with Tanzanian material: racemes often shorter; bracts of flowers with a more elongate central lobe; pedicels longer (4–7 as against 2–4 mm.) often with sparse hairs. These differences do not at present seem sufficiently marked to justify formal taxonomic recognition.

2. **Pseudoprosopis euryphylla** Harms in Engl., Bot. Jahrb. **49**: 419 (1913).—Bak. f., Legum. Trop. Afr. **3**: 806 (1930).—Brenan, T.T.C.L.: 347 (1949); F.T.E.A. Legum.-Mimos.: 28 (1959).—Torre in Mendonça, Contr. Conhec. Fl. Moçamb. **2**: 90 (1954). TAB. **6** fig. A. Types from Tanzania.

Scandent shrub or small tree 3–6 m. high; young branchlets, petioles and rhachides ± puberulous; older branchlets longitudinally striate. Leaves: petiole 0·5–2·9 cm. long; rhachis 1·7–7 cm. long; pinnae opposite, 2–4 pairs; leaflets 4–9 pairs, (0·5)0·7–2·1 × (0·3)0·6–1·5 cm., rhombic-ovate or -obovate, rounded or slightly emarginate and mucronate at the apex. Flowers white. Calyx 0·75–1 mm. long, densely and shortly pubescent. Petals 2·5–3 mm. long, densely puberulous outside. Stamen-filaments 4–5·5 mm. long. Ovary densely hairy. Pods 7–11 × 1·3–1·8 cm., attenuate below, blackish when dry, obliquely ± longitudinally striate. Seeds c. 7–8 × 5–7 mm.

Mozambique. N: mouth of R. Messalo (Msalu), fl. i.1912, *Allen* 161 (K); Nacala, near Fernão Veloso, fl. bud & young fr., 15.x.1948, *Barbosa* 2415 (BM; K; LISC; LMJ).
Also in Tanzania. In evergreen thickets, c. 250 m.

6. XYLIA Benth.

(By J. P. M. Brenan & R. K. Brummitt)

Xylia Benth. in Hook., Journ. Bot. **4**: 417 (1842).

Trees, unarmed. Leaves 2-pinnate; petiole bearing a gland at its apex at the junction of the solitary pair of pinnae; pinnae each with few to many pairs of leaflets. Inflorescences of round heads, pedunculate, axillary, or supra-axillary, solitary or paired or sometimes in threes, sometimes ± racemosely aggregated on short shoots. Flowers ♂ or ☿, sessile or pedicellate. Calyx gamosepalous with 5 lobes. Corolla with 5 lobes ± united below, ± pubescent or puberulous outside. Stamens 10, fertile; anthers each with a caducous apical gland (rarely and only in extra-African species absent). Ovary pubescent. Pods normally obliquely obovate to oblanceolate or dolabriform, compressed, woody, dehiscing from the apex downwards into 2 recurving valves. Seeds lying transversely or obliquely in the pod, each sunk in a depression in the valve, compressed, usually brown, glossy, unwinged, without endosperm.

A genus of c. 13 species in the tropics of the Old World, mostly in Africa and Madagascar.

Root-nodules not yet recorded.

Young branchlets, petioles and leaf-rhachides glabrous; peduncles subglabrous; leaflets glabrous beneath except on the midrib and margins; flowers sessile or almost so - - - - - - - - - - - - - - 1. *mendoncae*
Young branchlets, petioles, leaf-rhachides and peduncles densely pubescent or tomentellous; leaflets velvety all over beneath when young, later densely pubescent or puberulous; flowers on pedicels c. 1 mm. long - - - - - 2. *torreana*

1. **Xylia mendoncae** Torre in Mendonça, Contr. Conhec. Fl. Moçamb. **2**: 94, t. 9 (1954). Type: Mozambique, Vilanculos, *Mendonça* 1913 (BM; K; LISC, holotype; LM).

Small tree 4–7 m. high; branchlets, petioles and leaf-rhachides glabrous. Leaves with petiole 4–6 cm. long; pinna-rhachis 6–10 cm. long; leaflets in (3)4–5 pairs, mostly 3–6 × 1·5–2·5 cm. (smaller ones may occur with the inflorescences)

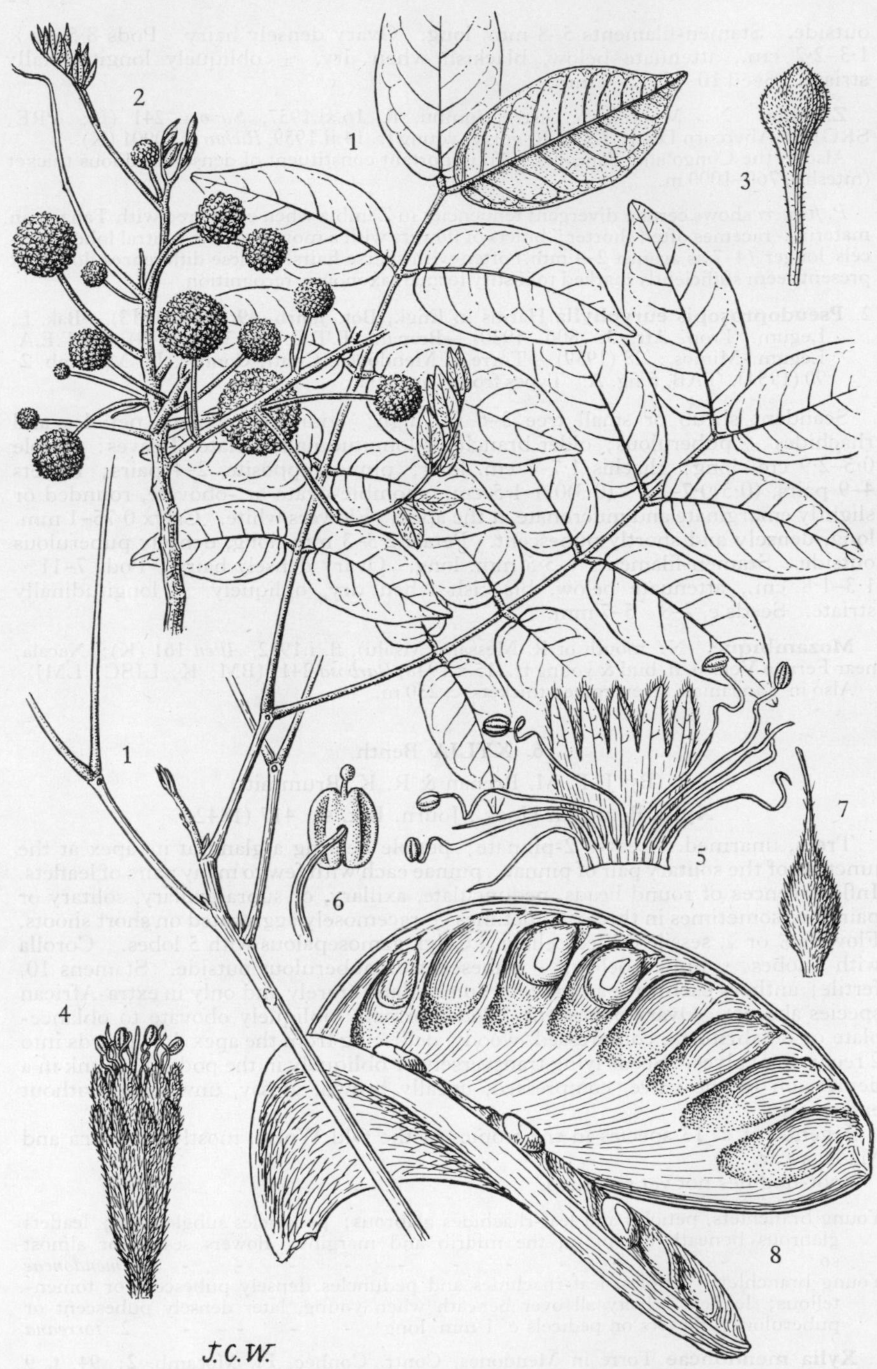

Tab. 7. XYLIA TORREANA. 1, sterile branch ($\times\frac{2}{3}$) *Torre* 2721; flowering branch ($\times\frac{2}{3}$) *Dawe* 463; 3, bract ($\times 8$); 4, flower ($\times 4$); 5, corolla ($\times 4$); 6, anther ($\times 14$); 7, ovary ($\times 4$); 8, fruit ($\times\frac{2}{3}$), all from *Dawe* 463.

elliptic to somewhat ovate- or obovate-elliptic, usually rounded to rounded-cuneate at the base, glabrous beneath except on the midrib and margins. Flowers in heads c. 1·5 cm. in diam. on supra-axillary subglabrous peduncles (2)3·5–8 cm. long; interfloral bracts oblanceolate, 2–3 mm. long. Flowers 6–7 mm. long, sessile or almost so (pedicel to c. 0·25 mm.). Pods 11–12 × 3–3·5 cm., glabrous.

Mozambique. SS: Vilanculos, near Mucoque, fl. 31.viii.1944, *Mendonça* 1913 (BM; K; LISC).
Known only from our area. In *Brachystegia* woodland.

2. **Xylia torreana** Brenan in Kew Bull. **12**: 359 (1958).—Boughey in Journ. S. Afr. Bot. **30**: 158 (1964).—Gomes e Sousa, Dendrol. Moçamb. Estudo Geral, **1**: 228, t. 33 (1966). TAB. **7.** Type: Mozambique, Maringua's village, 10 km. N. of R. Save, 23.vi.1950, *Chase* 2244 (BM; K, holotype; LISC; SRGH).
Xylia africana sensu Torre in Mendonça, Contr. Conhec. Fl. Moçamb. **2**: 93 (1954).

Tree up to 15 m. high, with rough brown to grey bark; branchlets, petioles, leaf-rhachides and peduncles densely brown-pubescent or tomentellous. Leaves with petioles 2–7 cm. long; pinna-rhachides 5–16 cm. long; leaflets in 4–6 pairs, (3·5)4–7·5 × 2–4·2 cm., narrowly ovate or rarely narrowly elliptic, rounded or often slightly cordate at the base, tomentose on both surfaces when young, the upper becoming glabrous, the lower ± densely pubescent at maturity. Flowers in heads c. 1·5 cm. in diam., on axillary peduncles 2–3·5 cm. long; pedicels c. 1·5 mm. long, appressed-pubescent with hairs longer and denser than on the calyx; interfloral bracts 2–3 mm. long, spathulate. Flowers 4·5 (♂)–6 (☿) mm. long, yellow. Pod 9–14 × 3·2–8 cm., brown-tomentellous at least in part.

Rhodesia. N: Mtoko Reserve, Nyangombi R., fr. 29.xii.1950, *Whellan* 490 (K; SRGH). C: Gatooma Distr., Sanyati Reserve, fr. 11.iii.1959, *Cleghorn* 447 (K; SRGH). E: Sabi-Lundi Junction Distr., E. bank of Sabi, fr. 11.vi.1950, *Wild* 3484 (K; SRGH). S: SE. Ndanga Distr., st. 14.vii.1955, *Mowbray* 48 (K; SRGH). **Malawi.** S: Chiradzulu, st. 12.iii.1962, *Adlard* 925 (SRGH). **Mozambique.** MS: Búzi, Madanda Forest, fl. ix–x.1911, *Dawe* 463 (K). SS: Inhambane, between Vilanculos and Funhalouro, fr. 21.v.1941, *Torre* 2721 (BM; K; LISC; LM).
Also in S. Africa (Transvaal). In deciduous woodland, sometimes (or perhaps always) with *Colophospermum mopane*, 80–300 m.

Very little flowering material of this species is available. One specimen at Kew, from Rhodesia, Mtoko, *Lovemore* 439, apparently has some wholly male heads, which are smaller than the hermaphrodite flowers of other specimens including the duplicate of this number at SRGH. Field observations on the occurrence of unisexual flowers are needed.

7. AMBLYGONOCARPUS Harms

(By J. P. M. Brenan & R. K. Brummitt)

Amblygonocarpus Harms in Engl. & Prantl, Nat. Pflanzenfam., Nachtrag 1: 191 (1897).

Unarmed tree. Leaves 2-pinnate, eglandular; pinnae each with several pairs of alternate or sometimes subopposite leaflets. Inflorescences of solitary or paired axillary racemes. Flowers ☿. Calyx gamosepalous, with 5 (rarely and casually 6) teeth. Petals 5 (rarely and casually 6), free. Stamens 10 (rarely and casually 12), fertile; anthers eglandular at the apex, even in bud. Pod straight or nearly so, oblong, woody, indehiscent, in section bluntly tetragonal or even subterete, internally septate between the seeds. Seeds hard, brown, unwinged.
A genus of a single tropical African species.
Root-nodules not yet recorded, believed not to occur.
Immature pods may have four rather prominent ribs, simulating the shape in section of those of the genus *Tetrapleura* Benth. In *Amblygonocarpus* the whole width of the valve is thickened, while in *Tetrapleura* the thickening is restricted to a narrow band running along the face of each valve. The eglandular anthers of *Amblygonocarpus* also distinguish this genus from *Tetrapleura*.

Amblygonocarpus andongensis (Welw. ex Oliv.) Exell & Torre in Bol. Soc. Brot., Sér. 2, **29**: 42 (1955).—Torre, C.F.A. **2**: 264 (1956).—Keay, F.W.T.A. ed. 2, **1**: 492 (1958).—Brenan, F.T.E.A. Legum.-Mimos.: 34, fig. 9 (1959).—Fanshawe, Fifty Common Trees N. Rhodesia: 22 cum tab. (1962).—F. White, F.F.N.R.: 90 (1962).—

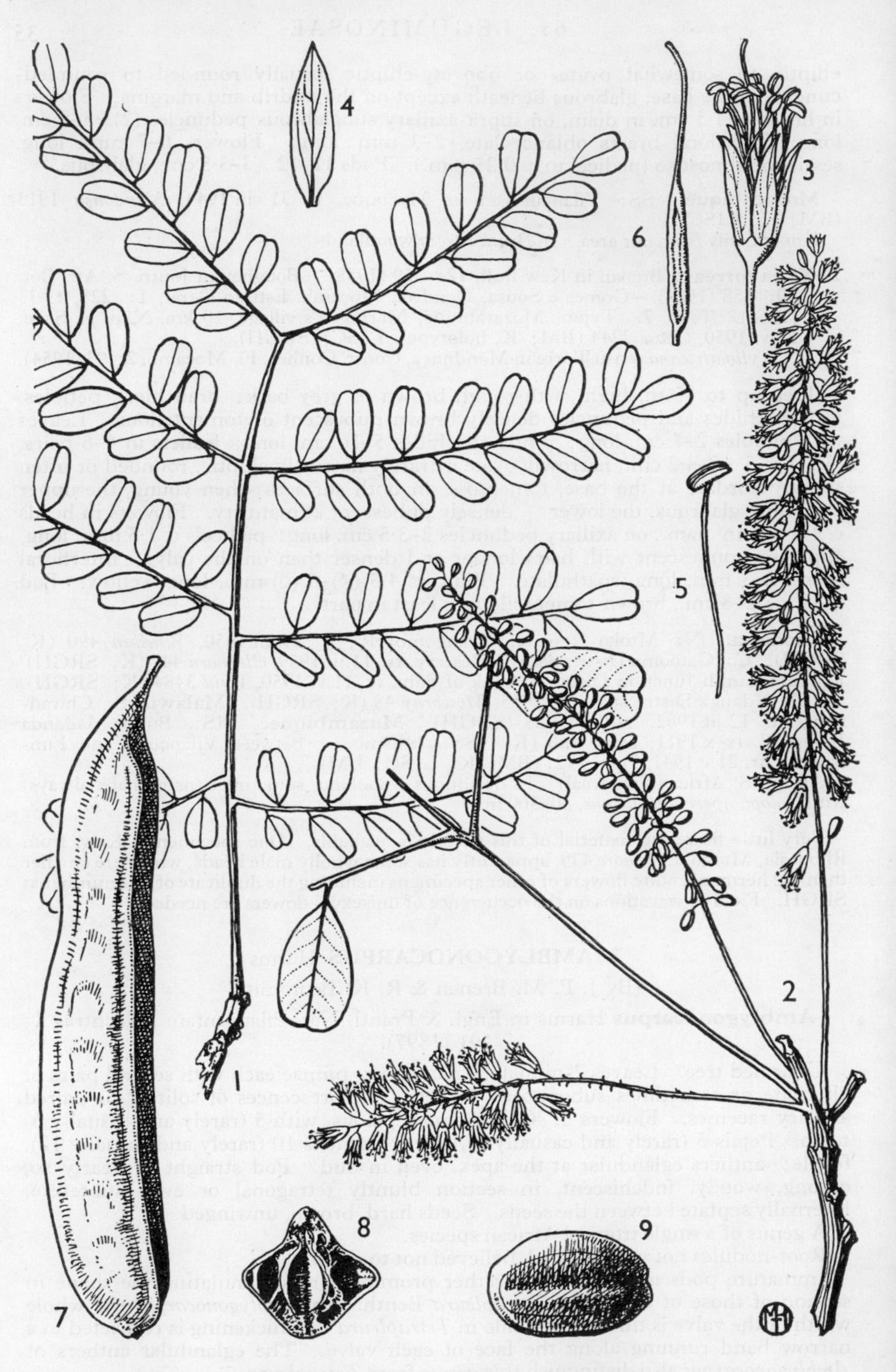

Tab. 8. AMBLYGONOCARPUS ANDONGENSIS. 1, leaf ($\times\frac{1}{2}$) *Eggeling* 6409; 2, part of flowering branch ($\times 1$); 3, flower ($\times 4$); 4, petal ($\times 6$); 5, stamen ($\times 6$); 6, ovary ($\times 6$), all from *Eggeling* 3421; 7, pod ($\times\frac{2}{3}$); 8, cross-section of pod ($\times\frac{2}{3}$); 9, seed ($\times 2$), all from *Dalziel* 26. From F.T.E.A.

Boughey in Journ. S. Afr. Bot. **30**: 158 (1964). TAB. **8**. Type from Angola (Cuanza Norte).

Tetrapleura andongensis Welw. ex Oliv., F.T.A. **2**: 331 (1871).—Bak. f., Legum. Trop. Afr. **3**: 803 (1930). Type as above.

Tetrapleura obtusangula Welw. ex Oliv., loc. cit. Type from Angola (Cuanza Norte).

Amblygonocarpus schweinfurthii Harms in Engl., Bot. Jahrb. **26**: 255 (1899); in Engl., Pflanzenw. Afr. **3**, 1: 396, t. 277 (1915).—Eyles in Trans. Roy. Soc. S. Afr. **5**: 363 (1916).—Bak. f., Legum. Trop. Afr. **3**: 804 (1930). Types from the Sudan and Angola.

Amblygonocarpus obtusangulus (Welw. ex Oliv.) Harms in Engl., Bot. Jahrb. **26**: 256 (1899).—Eyles, loc. cit.—Bak. f., loc. cit.—Gomes e Sousa, Dendrol. Moçamb. **2**: 100 cum tab. (1948); Dendrol. Moçamb. Estudo Geral, **1**: 226, t. 31 (1966).—Brenan, T.T.C.L.: 343 (1949).—Gilbert & Boutique, F.C.B. **3**: 217 (1952).— O. B. Mill. in Journ. S. Afr. Bot. **18**: 28 (1952).—Pardy in Rhod. Agric. Journ. **53**: 509 cum photogr. (1956).—Burtt Davy & Hoyle, rev. Topham, N.C.L. ed. 2: 66 (1958). Types as for *Tetrapleura obtusangula*.

Tetrapleura sp.—Eyles, loc. cit.

Tree 6–15(25) m. high, altogether glabrous; bark reticulate or scaly, grey or brown or blackish. Leaves: petiole 4–9 cm. long; rhachis 2–18 cm. long; pinnae 2–5(6) pairs, opposite or subopposite; leaflets 5–9(10) on each side of a pinna, 1·2–3 × 0·7–1·9 cm., elliptic to obovate-elliptic, usually emarginate at the apex, on petiolules 1·5–3 mm. long. Racemes (3)6–18 cm. long, on peduncles 1–4·5 cm. long. Flowers white except for yellow anthers, fading to yellow, on pedicels 1·5–3·5(5) mm. long. Calyx 0·5–1 mm. long. Petals 3–4·75 × 0·8–1·5 mm. Stamen-filaments 5–6 mm. long. Pods 8–17(20) × 2–3·5 cm., brown, glossy, blunt or ± pointed at the apex. Seeds 10–13 × 7–8 × 4–5 mm.

Caprivi Strip. E. of Cuando R., fl. x.1945, *Curson* 974 (PRE). **Botswana.** N: Kabulabula, Chobe R., fr. vii.1930, *Van Son* 28884 (BM; PRE). **Zambia.** B: Balovale, near Chitokaloki Mission, fl. 11.x.1952, *White* 3466 (BM; COI; K; PRE). N: Lake Bangweulu, N. of Samfya Mission, fl. 7.x.1947, *Brenan & Greenway* 8058 (BM; K). W: Solwezi Distr., Mutanda Bridge, st. 3.vii.1930, *Milne-Redhead* 653 (K). C: Chisamba, fr. 9.vii.1956, *Clarke* 126 (PRE). S: Livingstone, fr. 29.v.1930, *Milne-Redhead* 407 (K). **Rhodesia.** W: Victoria Falls, near station, fr. 7.vii.1930, *Hutchinson & Gillett* 3427 (BM; K; LISC; SRGH). E: Melsetter, fr. vii.1934, *Eyles* 7913 (K). **Malawi.** C: between Lilongwe and Fort Manning, fr. 31.vii.1936, *Burtt* 6241 (BM; K). **Mozambique.** N: Alto Molócuè, between Meluco and Muaguide, fl. 1.x.1948, *Barbosa* in *Mendonça* 2315 (BM; K; LISC; LM; LMJ). Z: Quissanga, between Alto Molócuè and Gilé, fl. 15.x.1949, *Barbosa & Carvalho* in *Barbosa* 4445 (K; LM; LMJ; PRE; SRGH). MS: Chimaio, between Vila Pery and Amatongas Mission, 4.xi.1941, *Torre* 3755 (BM; K; LISC; LM). SS: Govuro, near Mabote, fr. 29.iii.1952, *Barbosa & Balsinhas* 5070 (LM).

Widely distributed in the savanna regions of tropical Africa from Ghana to the Sudan, and southwards to our area. In deciduous woodland of various types, 120–1370 m.

8. DICHROSTACHYS (DC.) Wight & Arn.

(By J. P. M. Brenan & R. K. Brummitt)

Dichrostachys (DC.) Wight & Arn., Prodr. Fl. Ind. Or.: 271 (1834) *nom. conserv.*

Desmanthus Willd. sect. *Dichrostachys* DC., Mém. Legum. **12**: 428 (1825).

Shrubs or small trees; spines present or absent; prickles absent. Leaves 2-pinnate; rhachis glandular at the insertion of some at least of the pinnae; pinnae each with several to many pairs of leaflets. Inflorescences of axillary spikes, solitary or appearing clustered; upper part of spike cylindric, of ☿ flowers, lower part broader, of differently coloured neuter flowers. Calyx shortly 5-toothed. Petals 5, ± united below. Stamens 10, all fertile in ☿ flowers; anthers (in our species) with a stalked apical gland which is caducous. Staminodes of neuter flowers elongate, without anthers. Pods clustered, coriaceous, narrowly oblong or linear, compressed, usually irregularly contorted or spiral, indehiscent or opening irregularly or (not in our species) dehiscent. Seeds (in the African species at least) ± compressed, ovoid to ellipsoid, smooth.

A genus of c. 20 species in the tropics of the Old World from Africa to Australia, most in Madagascar. The generic limits require revision, however.

Root-nodules not yet recorded.

Dichrostachys cinerea (L.) Wight & Arn., Prodr. Fl. Ind. Or.: 271 (1834).—Brenan in Kew Bull. **12**: 357–8 (1958); F.T.E.A. Legum.-Mimos.: 36, fig. 11 (1959).—F.

White, F.F.N.R.: 432 (1962).—Mitchell in Puku, **1**: 130 (1963).—Brenan & Brummitt in Bol. Soc. Brot., Sér. 2, **39**: 61 (1965).—Gomes e Sousa, Dendrol. Moçamb. Estudo Geral, **1**: 225, t. 30 (1966).—Volk in Journ. S.W.A. Wiss. Gesell. Windh. **20**: 47, fig. 9 (1966). Type from Ceylon.

Mimosa cinerea L., Sp. Pl. **1**: 520 (1753); Syst. Nat. ed. 10, **2**: 1312 (1759) non *M. cinerea* L., Sp. Pl. **1**: 517 (1753). Type as above. See Brenan, Kew Bull. **12**: 357–58 (1958) for an explanation of the nomenclature.

Mimosa glomerata Forsk., Fl. Aegypt.-Arab.: 177 (1775). Type from Arabia.

Dichrostachys glomerata (Forsk.) Chiov. in Ann. Bot. Rom. **13**: 409 (1915).—Hutch. & Dalz. ex Greenway in Kew Bull. **1928**: 204 (1928).—Bak. f., Legum. Trop. Afr. **3**: 807 (1930).—Brenan, T.T.C.L.: 344 (1949); in Mem. N.Y. Bot. Gard. **8**: 429 (1954). —Pardy in Rhod. Agric. Journ. **48**: 316 cum tab. (1951).—Suesseng. & Merxm. in Trans. Rhod. Sci. Ass. **43**: 16 (1951).—Gilbert & Boutique, F.C.B. **3**: 202 (1952).—O. B. Mill. in Journ. S. Afr. Bot. **18**: 31 (1952).—Wild, Guide Fl. Vict. Falls: 149 (1953); Common Rhod. Weeds: fig. 54 (1955).—Torre in Mendonça, Contr. Conhec. Fl. Moçamb. **2**: 91 (1954); C.F.A. **2**: 265 (1956).—Mogg in Macnae & Kalk, Nat. Hist. Inhaca I.: 145 (1958).—F. White, F.F.N.R.: 91 (1962). Type as above.

Shrub or small tree 1–8(12) m. high, sometimes suckering and thicket-forming or (*fide* Greenway) even scandent, with rough bark and armed with spines terminating short lateral spreading twigs which often bear leaves and flowers; young branchlets usually ± pubescent, sometimes puberulous or even glabrous. Leaves with (2)5–19(21 *fide* F.C.B.) pairs of pinnae; rhachis (with petiole) 0·5–20 cm. long, with one stalked or occasionally sessile gland between each pair, or at least the distal and basal pairs, of pinnae; leaflets 9–41 pairs, 1–11(14) × 0·3–4(5·5) mm., linear to oblong. Inflorescences yellow in apical hermaphrodite part, mauve or pink or sometimes white in lower neuter part, 2–5 cm. long, pendent on solitary or apparently fascicled peduncles 1–9 cm. long. Calyx 0·6–1·25 mm. long. Corolla 1·5–3 mm. long. Stamen-filaments of ☿ flowers 3·25–3·5 mm. long; staminodes 4–17 mm. long. Pods 2–10 × 0·5–2·6 cm. Seeds 4–6 × 3–4·5 mm., deep-brown, glossy.

A very variable and taxonomically complex species, widespread in Africa and Asia, reaching Australia.

It is likely that the majority of specimens can be correctly placed by the following key, but intermediates occur between most of the taxa, and these will cause difficulty. Further taxa may be found in our area. Typical subsp. *cinerea* is confined to Asia.

Peduncles subglabrous to sparsely or densely puberulous; young branchlets glabrous to sparsely appressed-puberulous; leaves (at least larger ones) with 4–8 pairs of pinnae
subsp. *forbesii*

Peduncles ± densely spreading-pubescent; young branchlets ± densely pubescent; leaves with up to 4–19 pairs of pinnae:

Leaflets all or many 2 mm. or more wide; pinnae mostly 4–8(10) cm. long; leaves often large, (8)10–20 cm. long; peduncles usually fascicled: - - subsp. *nyassana*

Leaflets normally less than 2 mm. wide; pinnae 1–4 cm. long; leaves usually smaller than above; peduncles single or sometimes fascicled:

Larger leaves with 4–7 pairs of pinnae; pods 4–8 mm. wide
subsp. *argillicola* var. *hirtipes*

Larger leaves with 7–19 pairs of pinnae; pods 8–11(14) mm. wide:

Leaflets mostly 1 mm. or more wide (up to 1·7(2) mm.); pinnae 8–13 pairs; leaflets usually not glossy above:

Surfaces of leaflet (as distinct from the ciliate margins) glabrous or only sparsely hairy - - - - - - - - subsp. *africana* var. *africana*

Surfaces of leaflet densely pubescent - - subsp. *africana* var. *pubescens*

Leaflets up to 1 mm. wide; pinnae 7–19 pairs; leaflets glossy or not above:

Glands on leaf-rhachis sessile or very shortly (to 0·3 mm.) stipitate, present between all pairs of pinnae - - - subsp. *africana* var. *setulosa*

Glands on leaf-rhachis stipitate or columnar, 0·5–2 mm. tall, present between all pairs of pinnae or absent from middle ones:

Leaflets very narrow, 0·3–0·7 mm. wide; venation obscure
subsp. *africana* var. *plurijuga*

Leaflets less narrow, mostly 0·7–1 mm. wide; venation usually (not always) ± prominent beneath:

Leaflets ± glossy above (at least when dry), usually shortly and rather sparsely ciliate at least in the upper part of each leaflet
subsp. *africana* var. *tanganyikensis*

Leaflets not glossy above, with strongly ciliate margins
subsp. *africana* var. *lugardiae*

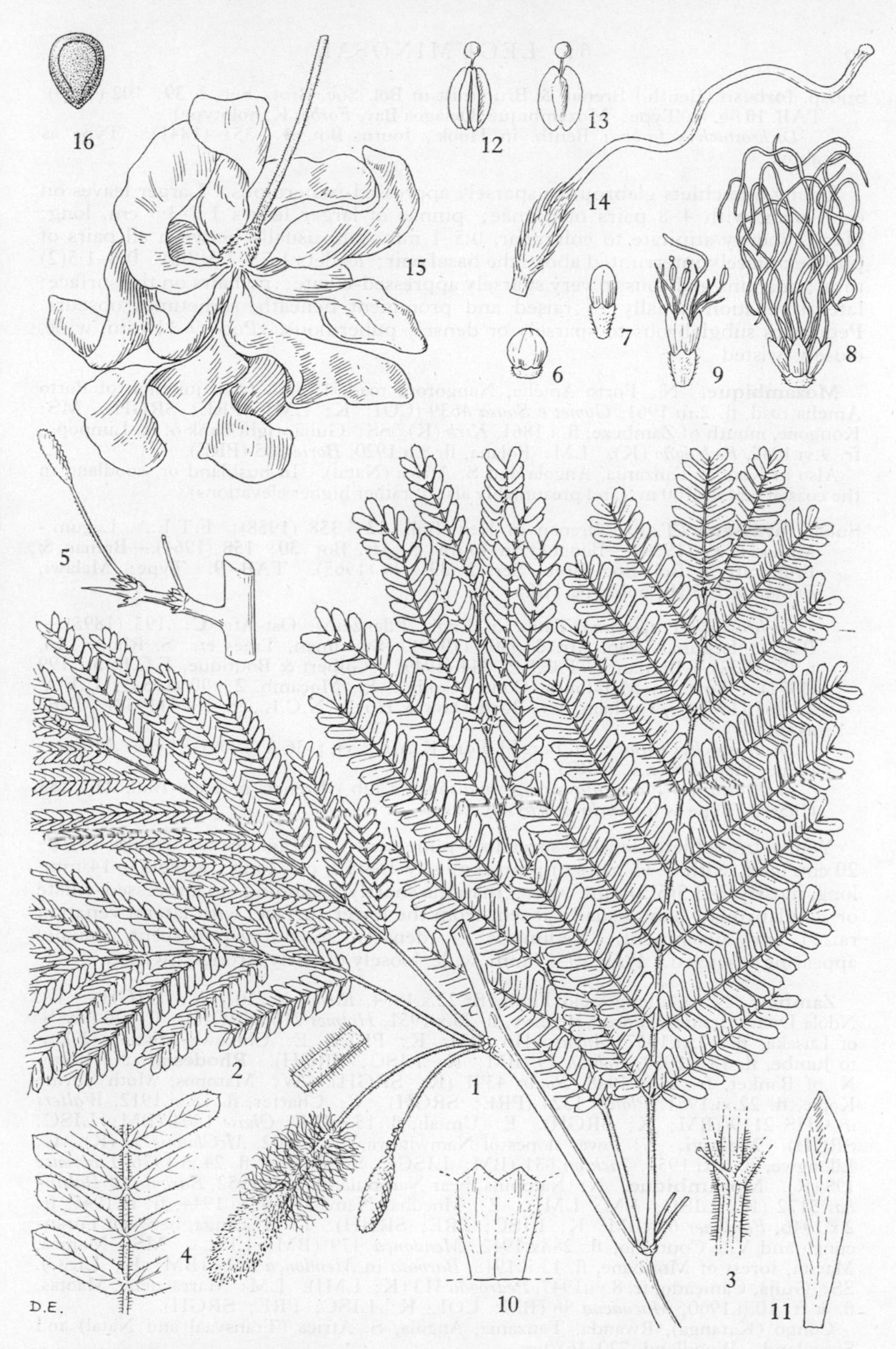

Tab. 9. DICHROSTACHYS CINEREA SUBSP. NYASSANA. 1, flowering branch (× ⅔); 2, portion of petiole, showing indumentum (× 4); 3, part of rhachis of leaf, showing gland (× 4); 4, part of pinna (× 4), all from *Duff* 45; 5, spine (× ⅔) *Lusaka Natural History Club* 162; 6, neuter flower, bud (× 4); 7, fertile flower, bud (× 4); 8, neuter flower (× 4); 9, fertile flower (× 4); 10, part of calyx (× 12); 11, corolla-lobe (× 12); 12, & 13, two views of anther (× 12); 14, ovary (× 12), all from *Duff* 45; 15, cluster of pods (× ⅔) *White* 2486; 16, seed (× 2) *Boaler* 889.

Subsp. **forbesii** (Benth.) Brenan & Brummitt in Bol. Soc. Brot., Sér. 2, **39**: 102 (1965). TAB. **10** fig. A. Type: Mozambique, Delagoa Bay, *Forbes* K, holotype).

Dichrostachys forbesii Benth. in Hook., Journ. Bot. **4**: 353 (1841). Type as above.

Young branchlets glabrous to sparsely appressed-puberulous. Larger leaves on each plant with 4–8 pairs of pinnae; pinnae of larger leaves 1·7–3·5 cm. long; glands shortly stipitate to columnar, 0·5–1 mm. tall, usually between all pairs of pinnae or rarely interrupted above the basal pair; leaflets 1·75–7·5(8·5) × 0·6–1·5(2) mm.; margins glabrous to very sparsely appressed-ciliate; no hairs on the surface; lateral venation usually ± raised and prominent beneath, sometimes obscure. Peduncles subglabrous to sparsely or densely puberulous. Pods 9–11 mm. wide, closely twisted.

Mozambique. N: Porto Amélia, Nangororo road, 15 km. from junction of Porto Amelia road, fl. 2.iii.1961, *Gomes e Sousa* 4639 (COI; K; LMJ; PRE; SRGH). MS: Kongone, mouth of Zambeze, fl. i.1861, *Kirk* (K). SS: Guijá, right bank of R. Limpopo, fr. 9.vi.1947, *Pedrógão* (K). LM: Polana, fl. 8.ii.1920, *Borle* 305 (PRE).

Also in Kenya, Tanzania, Angola and S. Africa (Natal). In bushland or woodland in the coastal belt, 0–100 m. (and presumably also at rather higher elevations).

Subsp. **nyassana** (Taub.) Brenan in Kew Bull. **12**: 358 (1958); F.T.E.A. Legum.-Mimos.: 39 (1959).—Boughey in Journ. S. Afr. Bot. **30**: 158 (1964).—Brenan & Brummitt in Bol. Soc. Brot., Sér. 2, **39**: 96 (1965). TAB. **9.** Type: Malawi, *Buchanan* 195 (B, holotype †;K).

Dichrostachys nyassana Taub. in Engl., Pflanzenw. Ost-Afr. **C**: 195 (1895).—Bak. f., Legum. Trop. Afr. **3**: 807 (1930).—Steedman, Trees etc. S. Rhod.: 16, t. 12 (1933).—Brenan, T.T.C.L.: 344 (1949).—Gilbert & Boutique, F.C.B. **3**: 199 (1952).—Torre in Mendonça, Contr. Conhec. Fl. Moçamb. **2**: 90 (1954); C.F.A. **2**: 265 (1956).—Topham, rev. Burtt Davy & Hoyle, N.C.L., ed. 2: 66 (1958). Type as above.

Dichrostachys major Sim, For. Fl. Port. E. Afr.: 54, t. 36 A (1909). Type: Mozambique, *Sim* 6248 (n.v. ?PRE).

Dichrostachys glomerata subsp. *nyassana* (Taub.) Brenan in Kew Bull. **11**: 188 (1956). Type as for *Dichrostachys nyassana.*

Young branchlets densely spreading-pubescent. Larger leaves usually (8)10–20 cm. long, with 6–11 pairs of pinnae mostly 4–8(10) cm. long; leaflets 6–14 mm. long, mostly 2–3·5(5·5) mm. wide; margins shortly ascending- to appressed-ciliate or subglabrous; hairs usually absent from the lower surface, with lateral venation raised and prominent. Peduncles ± densely spreading-pubescent, usually appearing fascicled. Pods 10–17 mm. wide, loosely to fairly strongly twisted.

Zambia. N: Chilongowelo Farm, fl. 22.x.1954, *Richards* 2101 (K; SRGH). W: Ndola Distr., Ichimpi Forest Reserve, fl. 25.x.1951, *Holmes* 216 (SRGH). C: 20 km. S. of Lusaka, fl. 12.xi.1957, *Angus* 1416 (BM; K; PRE). E: Chikowa Mission, 2 km. to Jumbe, fl. 13.x.1958, *Robson* 79 (BM; K; LISC; SRGH). **Rhodesia.** N: 8 km. N. of Banket, fr. 23.iv.1948, *Rodin* 4391 (K; SRGH). W: Matopos, Moth Shrine Kopje, fl. 22.xi.1951, *Plowes* 1324 (PRE; SRGH). C: Charter, fl. 15.xi.1912, *Walters* in GHS 2119 (BM; K; SRGH). E: Umtali, fl. 15.x.1949, *Chase* 1804 (BM; LISC; SRGH). **Malawi.** N: lower slopes of Namwitawa, fl. ix.1902, *McClounie* 93 (K). C: Lilongwe, fl. 9.xi.1951, *Jackson* 631 (BM; LISC). S: Zomba, fl. 24.xi.1936, *Lawrence* 198 (K). **Mozambique.** N: Nampula, near Nampula, fl. 18.x.1952, *Barbosa & Balsinhas* 5172 (K; LISC; LM; LMJ). Z: Mocuba, Namagoa, fl. xi.1944, fr. iv.1945, fr. 2.x.1946, *Faulkner* 61 (BM; K; LISC; PRE; SRGH). T: Macanga, between Furancungo and Vila Coutinho, fl. 28.ix.1942, *Mendonça* 479 (BM; LISC). MS: Manica, Mavita, forest of Moribane, fl. 13.ii.1948, *Barbosa* in *Mendonça* 1029 (BM; K; LISC). SS: Guijá, Caniçado, fr. 8.vi.1947, *Pedrógão* 313 (K; LMJ). LM: Marracuene, Maotas, fl. & fr. 10.ii.1960, *Macuácua* 86 (BM; COI; K; LISC; PRE; SRGH).

Congo (Katanga), Rwanda, Tanzania, Angola, S. Africa (Transvaal and Natal) and Swaziland. Woodland, 720–1680 m.

When extreme, distinct and easily recognized, but the diagnostic characters are insufficient and too inconstant for it to be maintained as a species. Intermediates with subsp. *africana* occur.

The surface of the leaflets is generally glabrous or nearly so. In two specimens: Malawi, without more exact locality, *Benson* 5 (PRE) and Zambia, Fort Jameson, fl. & fr. 1.xi.1950, *Gilges* 5 (PRE) it is rather densely short-pubescent above and beneath.

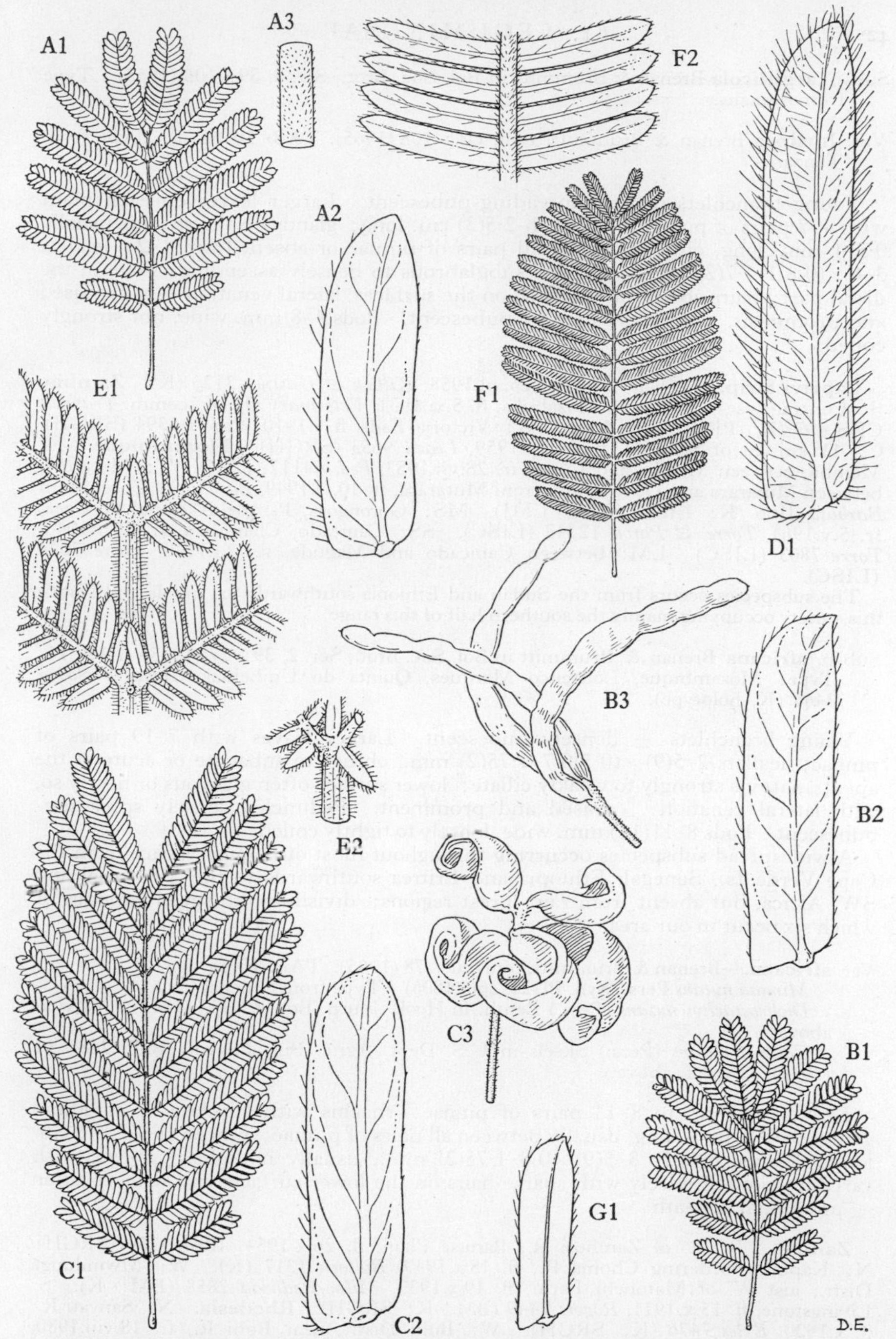

Tab. 10. DICHROSTACHYS CINEREA. A.—SUBSP. FORBESII. A1, leaf (×⅔); A2, leaflet (×10), all from *Pedrógão* 277; A3, indumentum of peduncle (×4) *Gomes & Sousa* 4552. B.—SUBSP. ARGILLICOLA VAR. HIRTIPES. B1, leaf (×⅔); B2, leaflet (×10); B3, pods (×⅔), all from *Barbosa & Carvalho* 3106. C.—SUBSP. AFRICANA VAR. AFRICANA. C1, leaf (×⅔); C2, leaflet (×10), all from *Torre* 4140; C3, pods (×⅔) *Faulkner* 2808. D.—SUBSP. AFRICANA VAR. PUBESCENS. D1, leaflet (×10) *Gilliland* 201. E.—SUBSP. AFRICANA VAR. SETULOSA. E1, E2, parts of leaf-rhachis, showing glands (×3) *Nelson* 509. F.—SUBSP. AFRICANA VAR. PLURIJUGA. F1, leaf (×⅔); F2, leaflets (×10), all from *Barbosa & Carvalho* 3157. G.—SUBSP. AFRICANA VAR. LUGARDIAE. G1, leaflet (×10) *Robertson & Elffers* 86.

Subsp. **argillicola** Brenan & Brummitt in Bol. Soc. Brot., Sér. 2, **39**: 106 (1965). Type from Somalia.

Var. **hirtipes** Brenan & Brummitt, tom. cit.: 108 (1965). TAB. **10** fig. B. Type from Tanzania.

Young branchlets densely spreading-pubescent. Larger leaves on each plant with 4–7 pairs of pinnae; pinnae 1–2·5(3) cm. long; glands on the rhachis usually 0·5–1 mm. long, either between all pairs of pinnae or absent from 1–2; leaflets 3–6 × (0.8)1–1·7(2) mm.; margins subglabrous to densely ascending-ciliate in the distal part; hairs absent or nearly so on the surfaces, lateral venation usually raised and prominent. Peduncles densely pubescent. Pods 4–8 mm. wide, not strongly coiled.

Caprivi Strip. Linyanti area, fl. 26.xii.1958, *Killick & Leistner* 3126 (K). **Zambia.** S: Central Research Station, Mazabuka, fl. 5.xi.1931, *Veterinary Officer*, comm. *Trapnell* CRS 497 (K; PRE). **Rhodesia.** W: Victoria Falls, fl. i.1910, *Rogers* 5394 (SRGH). C: 11 km. S. of Selukwe, fl. 24.xii.1959, *Leach* 9653 (SRGH). **Mozambique.** N: Memba, between Memba and Lúrio, fr. 28.vii.1953, *Pedro* 4112 (LMJ). T: Mutarara, between Mutarara and Dôvo, 8 km. from Mutarara, fr. 10.vi.1949, *Barbosa & Carvalho* in *Barbosa* 3106 (K; LISC; LM; LMJ). MS: Gorongosa, Parque Nacional de Caça, fr. 5.v.1964, *Torre & Paiva* 12242 (LISC). SS: Caniçado, Chamusca, fr. 19.v.1948, *Torre* 7863 (LISC). LM: between Caniçado and Magude, fl. 2.i.1948, *Torre* 7024 (LISC).

The subspecies occurs from the Sudan and Ethiopia southwards to Angola and Natal, this variety occupying mainly the southern half of this range.

Subsp. **africana** Brenan & Brummitt in Bol. Soc. Brot., Sér. 2, **39**: 77 (1965). TAB. **11**. Type: Mozambique, Lourenço Marques, Quinta do Umbeluzi, *Gomes e Sousa* 3466 (K, holotype).

Young branchlets ± densely pubescent. Larger leaves with 7–19 pairs of pinnae; leaflets 2–5(9) × (0·5)0·7–1·75(2) mm., obtuse to subacute or acute at the apex; margins strongly to weakly ciliate; lower surface often glabrous or nearly so, with lateral venation ± raised and prominent. Peduncles densely spreading-pubescent. Pods 8–11(14) mm. wide, loosely to tightly coiled.

A widespread subspecies occurring throughout most of tropical Africa from the Cape Verde Is., Senegal, Ethiopia and Eritrea southwards to the Transvaal and SW. Africa, but absent from rain-forest regions; divisible into seven varieties of which six occur in our area.

Var. **africana**.—Brenan & Brummitt, tom. cit.: 78 (1965). TAB. **10** fig. C.

Mimosa nutans Pers., Syn. Pl. **2**: 266 (1806). Type from Senegal.

Dichrostachys nutans (Pers.) Benth. in Hook. Journ. Bot. **4**: 353 (1841). Type as above.

Cailliea nutans (Pers.) Skeels in U.S. Dept. Agric. Bur. Pl. Ind. Bull. **248**: 61 (1912). Type as above.

Leaves with up to 8–13 pairs of pinnae; rhachis with stipitate or columnar glands 0·75–2 mm. long, usually between all pairs of pinnae; pinnae (1·5)2–4·5 cm. long; leaflets mostly 3–5(9) × 0·8–1·75(2) mm., usually not glossy above, with variable ciliation, rarely with sparse hairs on the lower surface, and with venation ± prominent beneath.

Zambia. B: E. of Zambezi R., Barotse Plain, fl. 26.x.1954, *West* 3259 (SRGH). N: Kaputa, bordering Choma R., fl. 18.x.1949, *Bullock* 1317 (K). W: Mwinilunga Distr., just W. of Matonchi Farm, fl. 19.x.1937, *Milne-Redhead* 2858 (BM; K). S: Livingstone, fl. 15.x.1911, *Rogers* 7449 (BM; K; SRGH). **Rhodesia.** N: Sanyati R., fl. x.1920, *Eyles* 7476 (K; SRGH). W: Bubi Distr., near Bubi R., fr. 18.viii.1930, *Pardy* 4945 (SRGH). C: Salisbury Distr., roadside near Prince Edward Dam wall, fl. 22.xi.1960, *Rutherford-Smith* 361 (K; SRGH). E: Umtali Distr., Battery Spruit, fl. 19.i.1945, *Hopkins* 13227 (K; SRGH). S: Victoria Distr., ± 8 km. NW. of Zimbabwe, fr. 7.v.1963, *Leach* 11668 (K; SRGH). **Malawi.** N: Karonga Distr., fr. iii.1954, *Jackson* 1266 (K). **Mozambique.** N: Macomia, between Ingoane and Quiterajo, fl. ix.1948, *Barbosa* 2082 (LISC; LM; LMJ; PRE). Z: Alto Molócuè, between Gilé and Alto Ligonha, 37·7 km. from Gilé, fl. 11.x.1949, *Barbosa & Carvalho* in *Barbosa* 4375 (K; LM; LMJ). T: Mutarara, between km. 148 and Régulo Fortuna, 17·7 km. from

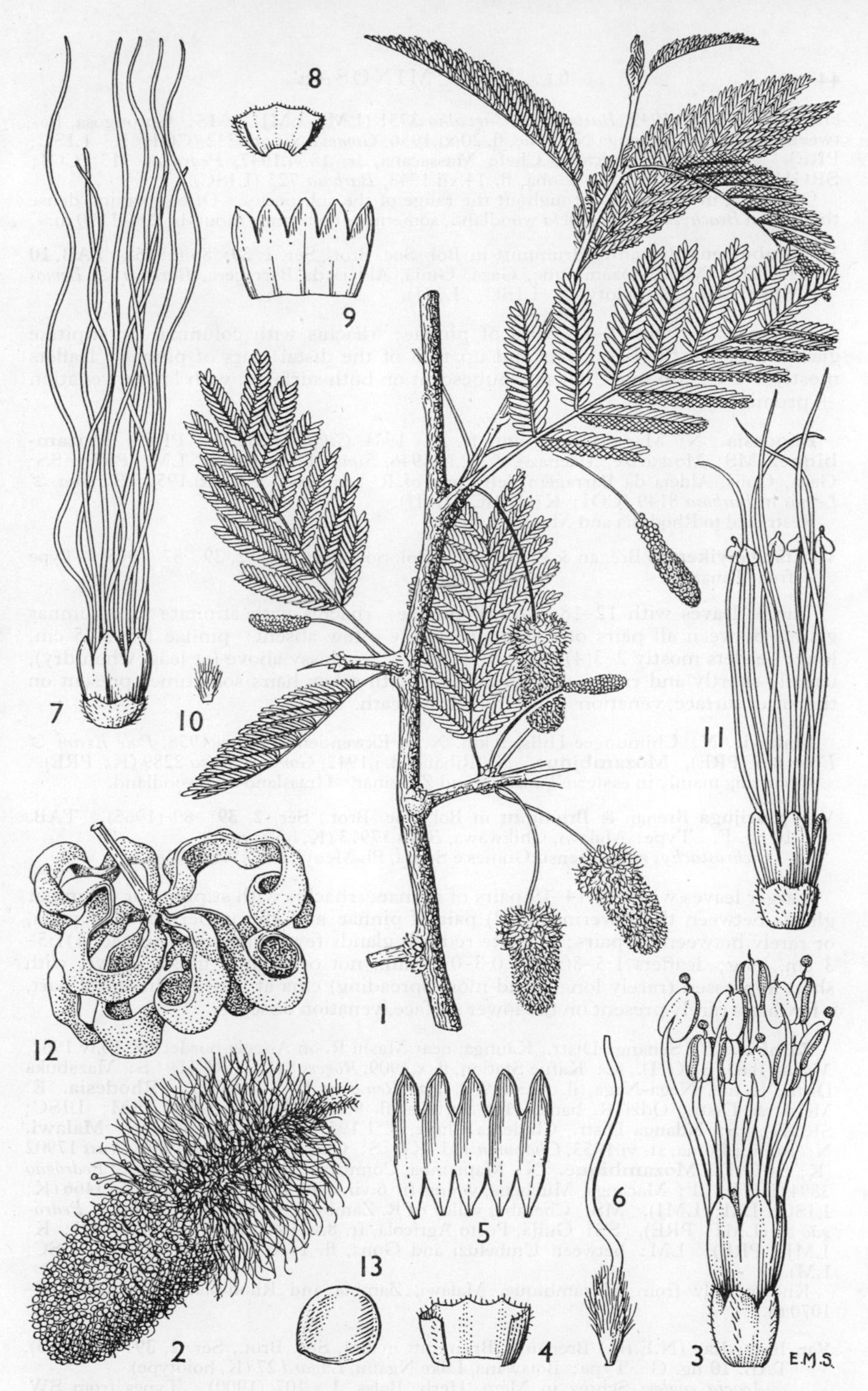

Tab. 11. DICHROSTACHYS CINEREA SUBSP. AFRICANA. 1, flowering branch (×$\frac{2}{3}$); 2, inflorescence (×2); 3, flower (×12); 4, calyx (×12); 5, corolla (×12); 6, ovary (×12); 7, neuter flower (×12); 8, calyx of neuter flower (×12); 9, corolla of neuter flower (×12); 10, rudimentary ovary of neuter flower (×12); 11, neuter flower showing intermediate stage in reduction of stamens (×12), all from *Drummond & Hemsley* 1178; 12, cluster of pods (×$\frac{2}{3}$); 13, seed (×3), all from *Burtt* 1762. From F.T.E.A.

km. 148, fr. 25.vii.1949, *Barbosa & Carvalho* 3751 (LM; LMJ). MS: Gorongosa, between Zangorga and Kanga N'Thole, fl. 20.xi.1956, *Gomes e Sousa* 4332 (COI; K; LISC; PRE). SS: Guijá, Caniçado, Chefe Massacana, fr. 18.vi.1947, *Pedrógão* 313 (COI; SRGH). LM: Namaacha, Goba, fl. 14.xii.1948, *Barbosa* 725 (LISC).

Occurring more or less throughout the range of the subspecies. Often forming dense thickets in *Brachystegia* or *Acacia* woodland, sometimes on termite mounds, 200–1190 m.

Var. **pubescens** Brenan & Brummitt in Bol. Soc. Brot., Sér. 2, **39**: 86 (1965). TAB. **10** fig. D. Type: Mozambique, Gaza, Guijá, Aldeia da Barragem, *Barbosa & Lemos* 8149 (COI; K, holotype; LISC; LMJ).

Larger leaves with 8–11 pairs of pinnae; rhachis with columnar or stipitate glands between the basal pair and up to 5 of the distal pairs of pinnae; leaflets mostly 4–7 × 1–1·6 mm., densely pubescent on both surfaces, with lateral venation ± prominent.

Rhodesia. N: Mazoe Dam, young fr. 24.v.1934, *Gilliland* 204 (K; PRE). **Mozambique.** MS: Mossurize, Machaze, fl. 12.ix.1946, *Simão* 1168 (LISC; LM; PRE). SS: Gaza, Guijá, Aldeia da Barragem, left bank of R. Limpopo, fl. 16.xi.1957, *Barbosa & Lemos* in *Barbosa* 8149 (COI; K; LISC; LMJ).

Restricted to Rhodesia and Mozambique.

Var. **tanganyikensis** Brenan & Brummitt in Bol. Soc. Brot., Sér. 2, **39**: 87 (1965). Type from Tanzania.

Larger leaves with 12–18 pairs of pinnae; rhachis with stipitate or columnar glands between all pairs of pinnae, or rarely a few absent; pinnae 1·5–2·75 cm. long; leaflets mostly 2–3(4) × 0·7–0·9(1) mm., ± glossy above (at least when dry), usually shortly and rather sparsely ciliate, with a few hairs sometimes present on the lower surface, venation ± prominent beneath.

Malawi. N: Chimungee Hills, 5 km. N. of Ekwendeni, fr. 9.vi.1938, *Pole Evans & Erens* 667 (PRE). **Mozambique.** N: Ribáuè, fl. i.1942, *Gomes e Sousa* 2289 (K; PRE).

Occurring mainly in eastern Tanzania and Zanzibar. Grassland and woodland.

Var. **plurijuga** Brenan & Brummitt in Bol. Soc. Brot., Sér. 2, **39**: 89 (1965). TAB. **10** fig. F. Type: Malawi, Chikwawa, *Brass* 17913 (K, holotype).

Dichrostachys nutans sensu Gomes e Sousa, Pl. Menyharth.: 70 (1936).

Larger leaves with (11)14–19 pairs of pinnae; rhachis with stipitate or columnar glands between the lowermost 1(2) pair of pinnae and the uppermost 1–4 pairs, or rarely between all pairs; minute reddish glands few or absent; pinnae (1)1·5–3 cm. long; leaflets 1·5–3(3·5) × 0·3–0·7 mm., not or slightly glossy above, with short appressed (rarely longer and more spreading) cilia at least in the distal part, a few hairs rarely present on the lower surface, venation obscure.

Zambia. B: Senanga Distr., Kaunga, near Mashi R. on Angola border, fr. 10.iv.1962, *Mubita* B 83 (SRGH). C: Kafue Station, fl. x.1909, *Rogers* 8410 (SRGH). S: Mazabuka Distr., Kafue, Nega-Nega, fl. 30.xi.1962, *Van Rensburg* 1014 (SRGH). **Rhodesia.** E: Melsetter Distr., Odzi R. bank, Hot Springs, fl. 9.xi.1950, *Chase* 3092 (BM; LISC; SRGH). S: Ndanga Distr., Chidema Clinic, fl. 1.1959, *Farrell* 25 (SRGH). **Malawi.** N: Rumpi Boma, st. vi.1953, *Chapman* 113 (K). S: Chikwawa, fr. 2.x.1946, *Brass* 17902 (K; SRGH). **Mozambique.** N: Maniamba, Coimbra, fr. 24.v.1948, *Pedro & Pedrógão* 3894 (LMJ). T: Macanga, Muchena, young fr. 6.vii.1949, *Barbosa & Carvalho* 3466 (K; LISC; LM; LMJ). MS: Chemba, valley of R. Zambeze, fl. 8.xi.1946, *Pedro & Pedrógão* 87 (LMJ; PRE). SS: Guijá, Posto Agrícola, fr. 8.vi.1947, *Pedrógão* 260 (COI; K; LMJ; PRE). LM: between Umbeluzi and Goba, fl. 14.xi.1940, *Torre* 1962 (LISC; LM).

Known only from Mozambique, Malawi, Zambia and Rhodesia. Woodland, 60–1070 m.

Var. **lugardiae** (N.E.Br.) Brenan & Brummitt in Bol. Soc. Brot., Sér. 2, **39**: 91 (1965). TAB. **10** fig. G. Type: Botswana, Lake Ngami, *Lugard* 27 (K, holotype).

Acacia engleri Schinz in Mém. Herb. Boiss. **1**: 107 (1900). Types from SW. Africa.

Dichrostachys lugardiae N.E.Br. in Kew Bull. **1909**: 106 (1909) (" lugardae "). Type as for var. *lugardiae*.

Dichrostachys arborea N.E.Br., loc. cit.—O. B. Mill. in Journ. S. Afr. Bot. **18**: 31 (1952). Types: Botswana, Ngamiland, *Lugard* 42, *Mrs. Lugard* 78, (K, syntypes).

Larger leaves with up to 8–12(13) pairs of pinnae; rhachis with stipitate or columnar glands between the lowermost 1(2) pair of pinnae and usually the uppermost 1–4(5) pairs; minute reddish glands inconspicuous or absent; pinnae 1–3 cm. long; leaflets 2·5–3·5 × (0·6)0·7–0·8(0·9) mm., not glossy above, strongly ciliate with cilia mostly appressed or ascending, with hairs often present on the lower surface only, venation obscure to somewhat prominent.

Botswana. N: Chobe Distr., Serondela, near Chobe R., fl. 31.vii.1950, *Robertson & Elffers* 86 (PRE). SE: Mahalapye Distr., Morale Pasture Station, 16.iii.1961, *Yalala* 131 (K; SRGH). **Zambia.** C: Quíen Sabe, Chilanga Distr., fl. 16.x.1929, *Sandwith* 65 (K; SRGH). S: Choma Distr., along the Masuku Mission road, fl. 23.ii.1963, *Astle* 1685 (SRGH). **Rhodesia.** W: Wankie Distr., Zambezi banks, xi.1959, *Armitage* 177/59 (SRGH). C: Marandellas Distr., Sabi R. x–xi.1931, *Myres* 148 (K). E: 16 km. N. of Hot Springs, 6.x.1961, *Lord Methuen* 277 (K). S: Gwanda Distr., Tuli Pasture Research Sub-Station, xii.1958, *Howden* 17/58 (SRGH). **Mozambique.** SS: Limpopo, between Caniçado and Mapai, 13.xii.1940, *Torre* 2390 (K; LISC; LM). LM: between Maputo and Goba, fl. 8.i.1947, *Pedro & Pedrógão* 467 (LMJ).

Also in Angola, SW. Africa, S. Africa (Transvaal, Bechuanaland) and Swaziland. Woodland, 970–1370 m.

Var. **setulosa** (Welw. ex Oliv.) Brenan & Brummitt in Bol. Soc. Brot., Sér. 2, **39**: 93 (1965). TAB. **10** fig. E. Type from Angola.

Dichrostachys nutans var. *setulosa* Welw. ex Oliv., F.T.A.: **2**: 333 (1871). Type as above.

Acacia kalachariensis Schinz in Mém. Herb. Boiss. **1**: 114 (1900). Type from the Kalahari, without exact locality.

Larger leaves with up to 7–16 pairs of pinnae; rhachis with sessile or very shortly stipitate (to 0·3 mm.) glands between all pairs of pinnae; minute dark reddish glandular hairs present particularly around the large single glands; pinnae 1–1·5(2) cm. long; leaflets 2–3·5 × 0·5–0·8 mm., not glossy above, strongly and densely ciliate, with hairs sometimes also present on the lower surface only, venation obscure or absent.

Botswana. SW: Kaotwe, fr. 10.iv.1930, *Van Son* 28862 (K). **Rhodesia.** W: Gwaai, fr. iv.1953, *Davies* 544 (SRGH).

Mainly in SW. Africa and S. Africa (Bechuanaland, Transvaal) but also occurring in Angola and central Tanzania. Scrub, 1070 m.

9. NEPTUNIA Lour.

(By J .P. M. Brenan & R. K. Brummitt)

Neptunia Lour., Fl. Cochinch.: 653 (1790).

Herbs, aquatic or terrestrial, unarmed. Leaves 2-pinnate; pinnae each with several to numerous pairs of leaflets. Inflorescences solitary and axillary, of globose to ellipsoid heads. Flowers in upper part of head ☿, in lower part of head ♂ or neuter with ± elongate staminodes. Calyx 5-toothed. Petals 5, free or ± united. Stamens 5 or 10, free, all fertile in ☿ flowers; anthers glandular or not at the apex. Pods clustered, membranous to subcoriaceous, oblong to subcircular, compressed, not contorted or spiral, dehiscent. Seeds ± compressed, oblong-ellipsoid to obovoid, smooth.

A genus of 14 or more species, widely distributed and mostly tropical but only one in Africa.

Root-nodules not yet recorded.

Neptunia oleracea Lour., Fl. Cochinch.: 654 (1790).—Oliv., F.T.A. **2**: 334 (1871).—Benth. in Trans. Linn. Soc. **30**: 383 (1875).—Torre in Mendonça, Contr. Conhec. Fl. Moçamb. **2**: 93 (1954); C.F.A. **2**: 267 (1956).—Keay, F.W.T.A. ed. 2, **1**: 496 (1958).—Brenan, F.T.E.A. Legum.-Mimos.: 40, fig. 12 (1959).—Mitchell in Puku, **1**: 150 (1963). TAB. **12**. Type from Cochin-China.

Mimosa prostrata Lam., Encycl. **1**: 10 (1783) excl. *β M. natans* L.f., *nom. illegit.* Syntypes from India.

Neptunia prostrata (Lam.) Baill. in Bull. Soc. Linn. Par. **1**: 356 (1883).—Bak. f., Legum. Trop. Afr. **3**: 809 (1930).—Gilbert & Boutique, F.C.B. **3**: 198 (1952). Types as above.

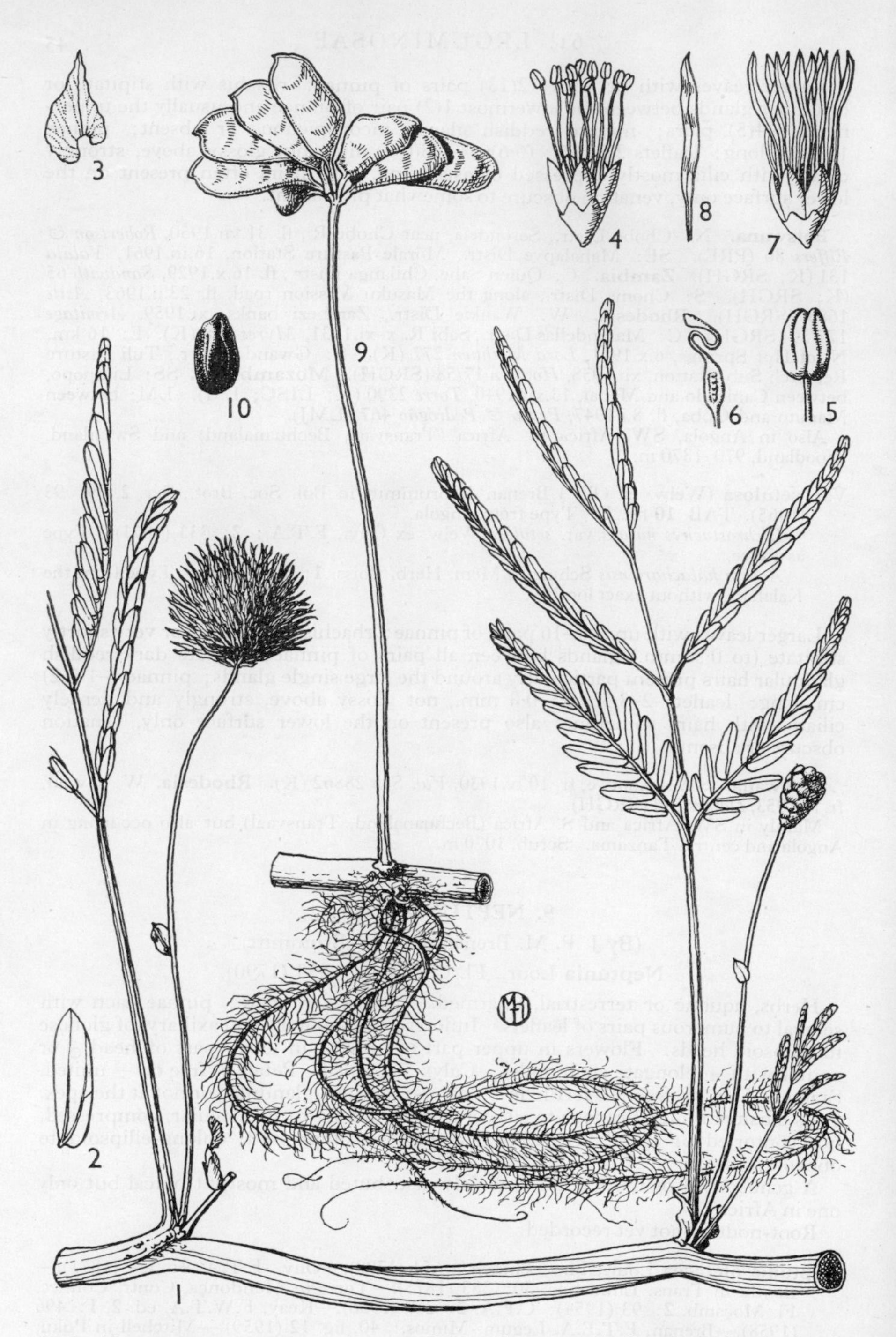

Tab. 12. NEPTUNIA OLERACEA. 1, part of flowering stem (×2); 2, leaflet (×2); 3, stipule (×2); 4, hermaphrodite flower (×2); 5, anther (×10); 6, ovary (×5); 7, neuter flower (×2); 8, staminode from neuter flower (×2), all from *Bally* 6133; 9, part of fruiting stem (×2); 10, seed (×2), all from *Peter* 44973. From F.T.E.A.

Aquatic herb with creeping stems usually floating, swollen, and rooting especially at the nodes, glabrous or rarely puberulous when young. Leaves very sensitive; stipules obliquely ovate, 5–9 × 3–5 mm., thin; petiole 2·5–9 cm. long; rhachis 1·1–4·2(6·5) cm. long; pinnae 2–4 pairs; leaflets 7–22 pairs, 5–20 × 1·5–4 mm., oblong, basal ones smaller, glabrous or with a few hairs on the margins. Flowers yellow, in heads 1·5–2·5 cm. long; peduncles 6·5–23(30) cm. long. Calyx 1–3 mm. long. Corolla c. 3–4 mm. long. Stamens 10; anthers eglandular at the apex, even in bud; staminodes up to 17–21 mm. long. Pods bent at an angle to the short basal stipe, 1·3–2·7(3·8) × 1–1·2 cm., shortly oblong. Seeds 5–5·5 × 3–3·5 mm.

Zambia. B: Sesheke, in the Zambezi, fr. ii.1911, *Macaulay* 489 (K). N: Mpika Distr., M'Fuwe Game Camp, Luangwa R., fl. & fr. 24.iii.1963, *Verboom* 800 (K). E: Lundazi, Luangwa R., fl. & fr. v.1962, *Finney* 646 (K; SRGH). S: Namwala, Musa-Kafue confluence, fl. 2.iv.1962, *Mitchell* 13/77 (K). **Rhodesia.** W: Wankie Game Reserve, Nyamandhlovu Pan, fr. 22.x.1958, *West* 3744 (K; SRGH). Chipinga Distr., Chibuwe Irrigation Scheme, Sabi Valley, fl. & fr. 2.iii.1966, *Plowes* 2759 (K). S: pan W. of Nuanetsi R. 5 km. downstream from Malipate, fl. & fr. 24.iv.1961, *Drummond & Rutherford-Smith* 7528 (K; SRGH). **Malawi.** S: Shire R., Elephant Marsh, st. 1863, *Kirk* (K). **Mozambique.** Z: Sisitso Station, R. Zambeze, fr. 9.vii.1950, *Chase* 2603 (BM; K; SRGH). SS: Caniçado, near Chamusca, fr. 19.v.1948, *Torre* 7884 (LISC). LM: near Chobela, Magude, fl. & fr. 12.ii.1948, *Torre* 7300 (BM; LISC; LM).

Tropics of Old and New Worlds. In and by fresh water of pools, lakes and swamps.

10. MIMOSA L.

(By J. P. M. Brenan & R. K. Brummitt)

Mimosa L., Sp. Pl. **1**: 516 (1753); Gen. Pl. ed. 5: 233 (1754).

Mostly herbs or shrubs, rarely trees, sometimes scrambling or climbing; prickles usually present. Leaves 2-pinnate, or the pinnae seeming almost digitate on account of the very short rhachis, rarely (not in our species) absent or modified to phyllodes; pinnae each with few to many pairs of leaflets. Inflorescences of ovoid or subglobose heads or (not in our species) spikes, which are axillary, solitary or more usually clustered and often ± aggregated. Flowers ☿ or ♂, small, sessile. Calyx very small, irregularly laciniate or denticulate in our species. Corolla gamopetalous, 4- or sometimes 3-, 5- or 6-lobed. Stamens as many as or twice as many as the corolla-lobes, fertile. Anthers without any apical gland. Pods straight to circinate, flat, in our species ± bristly or prickly; at maturity the valves between the margins splitting ± transversely into 1-seeded segments or rarely (not in our species) remaining entire; exocarp (at least in our species) not separating from the endocarp; margins persistent.

A genus of c. 450–500 species, widely distributed through the tropics, but the vast majority of the species in S. America.

Root-nodules recorded in spp. 1 and 6.

The following species from S. America is cultivated in the Flora area:—*M. scabrella* Benth., an unarmed tree with dense minute stellate or branched indumentum, and yellow flowers (**Rhodesia.** E: Vumba, fr. 8.xi.1957, *Chase* 6741 (K; SRGH)).

Pinnae in 1–2 pairs, subdigitately arranged on the very short rhachis which is much exceeded by the petiole; stamens 4; leaves without prickles on the petiole or rhachis (though bristly hairs may be present) - - - - - - - 6. *pudica*

Pinnae in (2)3–21 pairs, pinnately, not subdigitately arranged along the rhachis which is usually as long as or longer than the petiole; stamens (7)8(10); leaves usually prickly but sometimes unarmed:

Leaves without prickles on the petiole or rhachis:

Heads of flowers arranged " racemosely " along a single main axis; petiole 2–8 cm. long - - - - - - - - - - - - 4. *invisa* var. *inermis*

Heads of flowers arranged paniculately, the main inflorescence-axes branched; petiole 0·3–1·7 cm. long - - - - - - - - - - 5. *bimucronata*

Leaves ± prickly on the petiole and/or rhachis:

Rhachis of the leaf with a short or long, straight, erect or forward-pointing, slender prickle at the junction of each of the pairs of pinnae, often with other prickles also; leaflets often setulose on the margins, in 15–42 pairs; pods densely bristly all over, 0·9–1·4 cm. wide - - - - - - - - - - - 1. *pigra*

Rhachis of the leaf without prickles at the junctions of the pinna-pairs, though often prickly elsewhere; leaflets not setulose; if pods with bristles all over, then not more than 0·45 cm. wide:

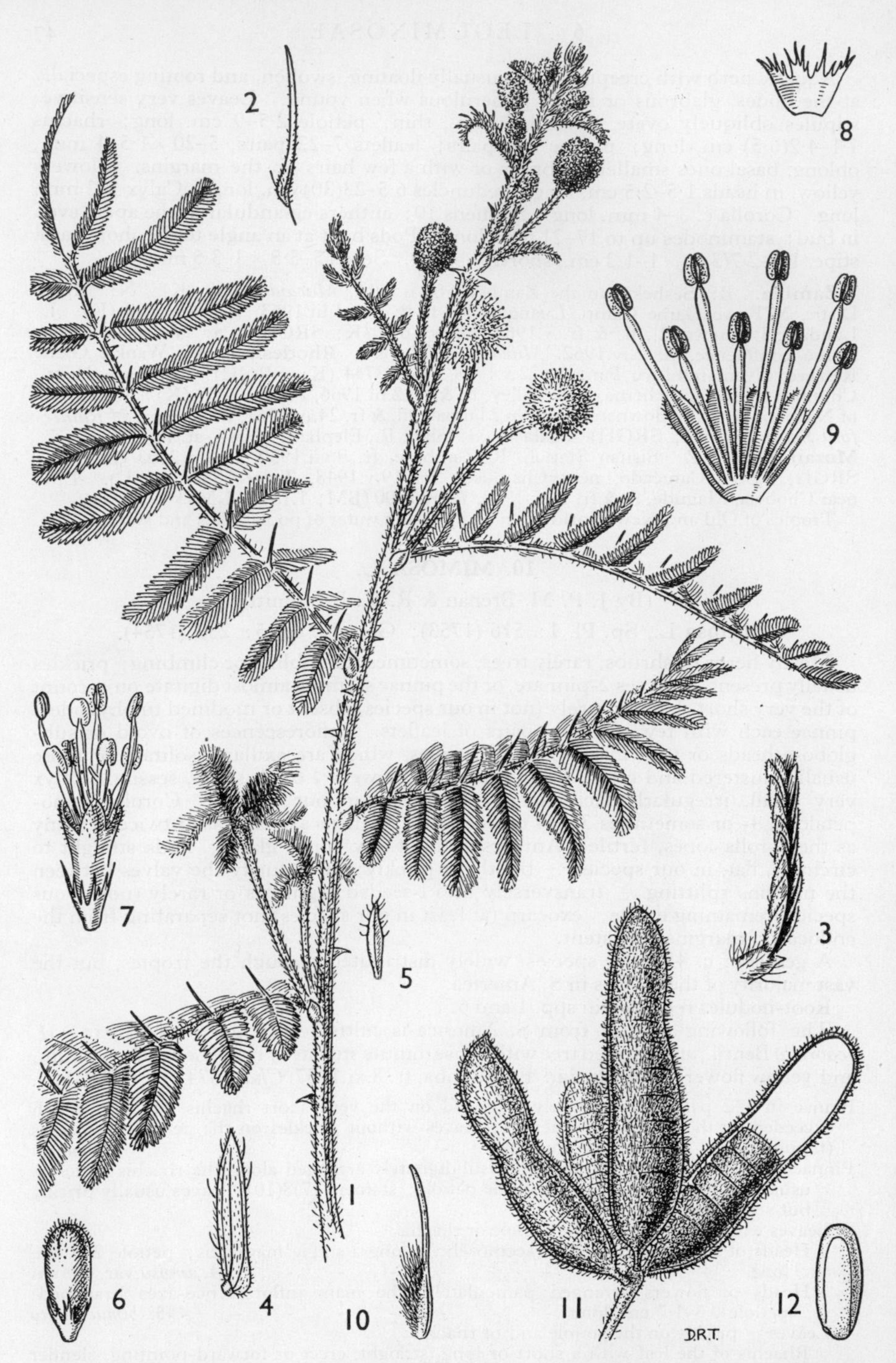

Tab. 13. MIMOSA PIGRA. 1, flowering branch (×1); 2, setiform hair from peduncle (×6); 3, part of pinna showing leaflets closed up in " sleeping " condition (×4); 4, leaflet (×4); 5, bract subtending flower (×6); 6, flower bud (×6); 7, flower (×6); 8, calyx, opened out (×6); 9, corolla and stamens, opened out (×6); 10, ovary (×6); 11, pods (×1); 12, seed (×3), all from *Harris* 45. From F.T.E.A.

Leaflets (2)5–10(12) mm. wide; petiole (0·5)3·5–4·5 cm. long; pinnae (2)3–7 pairs - - - - - - - - - - - - 2. *busseana*
Leaflets 0·7–2 mm. wide; petiole 0·2–8 cm. long; pinnae 3–15 pairs:
Petiole 0·2–0·6 cm. long; rhachis 8–19 cm. long; pinnae 10–15 pairs; pods with prickles on the margin only, 7–8 × 1·2–1·4 cm. 3. *mossambicensis*
Petiole 2–8 cm. long; rhachis 1·3–9 cm. long; pinnae 3–10 pairs; pods prickly on the margins and the surface, 1·5–3·5 × 0·4–0·5 cm. 4. *invisa* var. *invisa*

1. **Mimosa pigra** L., Cent. Pl. **1**: 13 (1755).—Fawc. & Rendle, Fl. Jam. **4**: 135 (1920).—Gomes e Sousa, Pl. Menyharth.: 69 (1936) (" nigra ").—Brenan, T.T.C.L.: 346 (1949); in Mem. N.Y. Bot. Gard. **8**, 5: 429 (1954); F.T.E.A. Legum.-Mimos.: 43 (1959).—Gilbert & Boutique, F.C.B. **3**: 230 (1952).—O. B. Mill. in Journ. S. Afr. Bot. **18**: 34 (1952).—Wild, Guide Fl. Vict. Falls: 149 (1953).—Torre in Mendonça, Contr. Conhec. Fl. Moçamb. **2**: 94 (1956); C.F.A. **2**: 268 (1956).—Keay, F.W.T.A. ed. 2, **1**: 495 (1958).—F. White, F.F.N.R.: 93: (1962).—Mitchell in Puku, **1**: 149 (1963).—TAB. **13**. Type a plant cultivated at Amsterdam.
Mimosa asperata L., Syst. Nat. ed. 10, **2**: 1312 (1759).—R.E.Fr., Wiss. Ergebn. Schwed. Rhod.-Kongo-Exped. **1**: 63 (1914).—Eyles in Trans. Roy. Soc. S. Afr. **5**: 363 (1916).—Bak. f., Legum. Trop. Afr. **3**: 812 (1930).—Burtt Davy, F.P.F.T. **2**: 333 (1932). Type a specimen of unknown origin in the Linnean Herbarium.

Shrub 0·6–3(4·5) m. high, sometimes scandent or rambling; stems armed with broad-based prickles up to 7 mm. long, also usually ± appressed- or sometimes spreading-setose. Leaves sensitive; petiole 0·3–1·5(1·8) cm. long; rhachis 3·5–12(18) cm. long, with a straight, ± erect or forward-pointing, slender prickle (sometimes short) at the junction of each of the 6–14(21) pairs of pinnae, sometimes with other stouter, spreading or deflexed prickles between the pairs; leaflets 15–42 pairs, 3–8(12·5) × 0·5–1·25(2) mm., linear-oblong, ± appressed-pubescent, particularly on the lower surface; venation nearly parallel with the midrib, margins often setulose. Flowers mauve or pink, in subglobose pedunculate heads c. 1 cm. in diam., 1–2(3) together in the upper axils. Calyx minute, 0·75–1 mm. long, laciniate. Corolla c. 2·25–3 mm. long. Stamens 8. Pods clustered, brown, 3–8·5 × 0·9–1·4 cm., bristly all over, breaking up transversely into segments 3–6 mm. long, the margins persisting as an empty frame.

Caprivi Strip. Katima Mulilo Distr., fl. & fr. 23.x.1954, *West* 3246 (SRGH). **Botswana.** N: Chobe Distr., Kasane Boma, fr. vii.1949, *Miller* B/885 (K; PRE). **Zambia.** B: Nangweshi, fl. 22.vii.1952, *Codd* 7143 (BM; COI; K; PRE; SRGH). N: Mkupa Katandula, near Mofwe R., fl. & fr. 10.vi.1950, *Bullock* 2930 (K). C? Kafue R. Gorge, fl. & fr. 6.x.1957, *Angus* 1746 (K; PRE; SRGH). E: Petauke Distr., Luangwa R. above Beit Bridge, fl. & fr. 5.ix.1947, *Brenan & Greenway* 7801 (K). S: Victoria Falls, fl. & fr. 20.xi.1949, *Wild* 3125 (K; LISC; SRGH). **Rhodesia.** N: Urungwe Distr., W. end of Kariba Gorge, fl. & fr. 25.xi.1953, *Wild* 4262 (K; LISC; SRGH). W: Wankie Distr., 16 km. W. of Binga, fl. & fr. 8.xi.1958, *Phipps* 1404 (K; SRGH). **Malawi.** N: Karonga, fl. 10.i.1959, *Robinson* 3137 (K; SRGH). C: Kota Kota Distr., Benga, fl. & fr. 2.ix.1946, *Brass* 17481 (K; SRGH). S: Port Herald Distr., Shire R. at Chiromo, fl. 24.iii.1960, *Phipps* 2681 (K; SRGH). **Mozambique.** N: Mecufi, Chiure, Quedas do Lúrio, fl. & fr. 20.viii.1948, *Barbosa* 1823 (COI; LISC; LM; LMJ). Z: between Quelimane and Mopeia, fl. & fr. 11.x.1941, *Torre* 3619 (BM; K; LISC; LM). T: Tete, Boroma, fl. & fr. 22.ix.1942, *Mendonça* 365 (BM; K; LISC). MS: Chemba, on the road to Tambara, fl. & fr. 23.iv.1960, *Lemos & Macuácua* 146 (BM; COI; K; LISC; LMJ; SRGH). SS: between Chibuto and Caniçado, fr. 2.x.1957, *Barbosa & Lemos* 7999 (COI; K; LISC; LMJ). LM: Marracuene, Vila Luisa, fl. & fr. 2.x.1957, *Barbosa & Lemos* 7909 (COI; K; LISC; LMJ).

Widespread in tropical Africa and America, also in Madagascar and Mauritius; in Asia apparently only a rare introduction; not in Australia. On sand or alluvium by rivers and lakes and in swamps, 30–1200 m.

The setose hairs clothing the stems and the rhachides of the leaves and pinnae are normally appressed, but are ± spreading in some specimens. These are all from the northern and eastern part of our area (Zambia, N; Malawi, N, C; Mozambique, N, Z, T, SS). Similar plants are also known from Uganda (see Brenan, F.T.E.A. Legum.-Mimos.: 43 (1959)).

In Rhodesia *M. pigra* appears to be restricted to the Zambezi valley.

2. **Mimosa busseana** Harms in Engl., Bot. Jahrb. **49**: 419 (1913).—Bak. f., Legum. Trop. Afr. **3**: 813 (1930).—Brenan, T.T.C.L.: 345 (1949); F.T.E.A. Legum.-Mimos.: 45 (1959). Types from Tanzania.

Scandent shrub 1·5–2 m. high or possibly more; stems shortly pubescent and densely armed with downwardly hooked prickles up to 2·5 mm. long. Leaves with petiole (0·5)3·3–4·5 cm. long; rhachis c. (2·5)3·5–5·5(9) cm. long, prickly; pinnae (2)3–7 pairs; leaflets 3–6 pairs, (5)8–18 × (2)5–10(12) mm., ± obovate- to oblong-elliptic, ± appressed-pubescent or puberulous on both surfaces, not setulose, venation pinnate, ± visible beneath. Flowers pink or mauve, in subglobose heads c. 0·7–1·5 cm. in diam., on pubescent unarmed or sparingly prickly peduncles 1–3 cm. long, 1–6 together from axils, usually aggregated into a panicle. Calyx 0·5–0·8 mm. long. Corolla (2)2·5–3 mm. long. Stamens 7–8. Pods ± curved or almost straight, with scattered recurved prickles on margins only (sometimes prickles very sparse), 5–10 × 1·2–2·5 cm.

Mozambique. N: without more exact locality, fl. 12.i.1912, *Allen* 134 (K); Palma, between Nangade and Pundanhar, fr. 17.ix.1948, *Barbosa* in *Mendonça* 2184 (BM; LISC; LM; LMJ).

Also in SE. Tanzania. Habitat uncertain, perhaps showing a preference for sandy soil; c. 10 m.

3. **Mimosa mossambicensis** Brenan in Kew Bull. **10**: 189 (1955). Type: Mozambique, Sena, *Peters* (B, holotype †; K).

Mimosa violacea Bolle in Peters, Reise Mossamb. Bot. **1**: 8 (1861) *nom. illegit.* non *M. violacea* Bonpl. ex Ten. (1843).—Torre in Mendonça, Contr. Conhec. Fl. Moçamb. **2**: 96 (1954). Type as above.

Mimosa cf. *violacea* Gomes e Sousa, Pl. Menyharth.: 69 (1936).

Climber; stems slender, ± appressed-puberulous, slowly glabrescent, sparsely to densely armed with downwardly hooked prickles up to 2·5 mm. long. Leaves with petiole 0·2–0·6 cm. long, prickly; rhachis 8–19 cm. long, prickly; pinnae 10–15 pairs; leaflets 8–13 pairs, 3–6 × 1–2 mm., narrowly oblong, with a very few appressed hairs on both surfaces towards the base, otherwise glabrous, venation pinnate, ± visible beneath. Flowers sweetly scented, in subglobose heads c. 1 cm. in diam., on puberulous very sparsely prickly peduncles c. 2 cm. long, 1–3 together from axils, not conspicuously aggregated into a panicle. Calyx c. 0·5 mm. long. Corolla c. 2–3 mm. long. Stamens c. 8. Pods 7–9 × 1·2–1·6 cm., ± curved, with recurved prickles, sometimes sparse, on margins only.

Mozambique. T: Mutarara, opposite Sena, fl. 1860, *Kirk* (K); Tete, fr. 5.v.1948, *Mendonça* 4097 (BM; K; LISC). ?MS or T: between Lupata and Tete, fl. ii.1859, *Kirk* (K)

Known only from Mozambique. Among trees on river-banks.

4. **Mimosa invisa** Mart. ex Colla, Herb. Pedem. **2**: 255 (1834); in Flora, **20**, Beibl.: 121 (1837).—Gilbert & Boutique, F.C.B. **3**: 231 (1952).—Torre in Mendonça, Contr. Conhec. Fl. Moçamb. **2**: 96 (1954).—Brenan, F.T.E.A. Legum.-Mimos.: 45 (1959). Type from Brazil.

Shrub up to c. 1 m. high, often scandent or prostrate, with long whip-like stems which are densely armed with very downwards-bent prickles 1·5–5 mm. long, and ± pubescent as well, rarely (var. *inermis*) unarmed. Leaves sensitive; petiole 2–8 cm. long, prickly or (var. *inermis*) unarmed; rhachis 1·3–9 cm. long, prickly or not; pinnae 3–10 pairs; leaflets 11–30 pairs, 2–6 × 0·7–1·5 mm., linear-oblong, pubescent at least on the margins but not setulose, venation not or scarcely visible; lowest pair of leaflets on each pinna modified to a pair of unequal filiform structures. Flowers pink, in subglobose or shortly ovoid heads c. 0·5–1 cm. in diam., on prickly (or unarmed in var. *inermis*) and pubescent peduncles (0·3)0·5–1·3(1·6) cm. long, 1–3 together from axils. Calyx minute, c. 0·3–0·4 mm. long. Corolla 1·5–2 mm. long. Stamens 8. Pods clustered, 1·5–3·5 × 0·4–0·5 cm., narrowly oblong, slightly curved, pubescent and with short prickly bristles on the margins and surface of the valves.

Var. **invisa.**

Prickles plentifully present on stems, leaf-rhachides and petioles.

Mozambique. Z: Inhangulue, c. 16 km. N. of Quelimane, fl. & fr. 24.viii.1962, *Wild* 5880 (K; SRGH).

Native of tropical America, introduced here and there in the Old World. A weed of coconut plantations and cultivations in our area.

Var. **inermis** Adelb. in Reinwardtia, **2**: 359 (1953). Type a plant cultivated in Java.

Plant unarmed.

Mozambique. Z: Quelimane, Madal Plantations, fl. 25.viii.1962, *Wild & Pedro* 5892 (K; SRGH).

Appears to be a variant which has arisen under cultivation. A garden weed in the Flora area.

5. **Mimosa bimucronata** (DC.) Kuntze, Rev. Gen. Pl. **1**: 198 (1891).—Burkart in Darwiniana, **8**: 108, t. 15 fig. B (1948); Las Leguminosas Argentinas, ed. 2: 121 (1952). Type from Brazil.

Acacia bimucronata DC., Prodr. **2**: 469 (1825). Type as above.

Mimosa sepiaria Benth. in Hook. Journ. Bot. **4**: 395 (1842).—Bak. f., Legum. Trop. Afr. **3**: 813 (1930). Syntypes from Brazil.

Mimosa stuhlmannii Harms in Engl., Bot. Jahrb. **26**: 254 (1899). Syntypes: Mozambique, Mossuril and Cabaceira, *Rodrigues de Carvalho* (B, syntype †; COI); probably Zambezia, without exact locality, *Stuhlmann* Coll. I, 276 (B, syntype †).

Shrub or small tree up to 10 m. high; stems varying from densely pubescent or puberulous to almost glabrous, and also ± sparsely armed with scattered straight or slightly recurved prickles 2–10 mm. long. Leaves unarmed, petiole 0·3–1·7 cm. long; rhachis 1·3–9·5 cm. long, with 3–9 pairs of pinnae; leaflets 10–30 pairs (the lowest pair very reduced, ± equal, subulate), 4–12 × 0·8–2·6 mm., linear-oblong, venation basal and pinnate, prominent beneath; margins ± ciliate, not setulose. Flowers whitish, in subglobose pedunculate heads 0·7–1·7 cm. in diam., clustered 1–4 together along the leafless branches of a terminal panicle; clusters leafless or with pinnate bracts up to c. 3 mm. long. Calyx 0·3–0·5 mm. long. Corolla c. 1·5–2·5 mm. long. Stamens 8(10). Pods brown, 2–6 × 0·5–0·7 cm., without bristles or prickles, glabrous or almost so, breaking up transversely into segments c. 5–7 mm. long, the margins persisting as an empty frame.

Mozambique. N: Mossuril, Mossuril and Cabaceira, 1884–85, *Carvalho* (B†; COI). Z: without exact locality (" Festland "), i.1889, *Stuhlmann* Coll. I, 276 (B†).

Native of S. America. Introduced.

The reduction of *M. stuhlmannii* to *M. sepiaria* (whose correct name has been shown to be *M. bimucronata*) was made by Harms himself in Engl., Bot. Jahrb. **30**: 77 (1901).

There can be no doubt that *M. bimucronata* is an introduction in our area. It has not been re-collected, its status in our area is wholly doubtful, and it is retained in the Flora only with much hesitation.

6. **Mimosa pudica** L., Sp. Pl. **1**: 518 (1753).—Bak. f., Legum. Trop. Afr. **3**: 812 (1930).—Brenan, T.T.C.L.: 346 (1949); in Kew Bull. **10**: 184 (1955); F.T.E.A. Legum.-Mimos.: 46 (1959).—Gilbert & Boutique, F.C.B. **3**: 229 (1952).—Keay in F.W.T.A. ed. 2, **1**: 495 (1958). Type a specimen of a cultivated plant in Hort. Cliff. (BM).

Annual or perennial herb, sometimes woody below, up to c. 1 m. high, often prostrate or straggling; stems ± sparsely armed with prickles c. 2·5–5 mm. long, in addition varying from densely hispid to subglabrous. Leaves sensitive, unarmed; petiole 1·5–5·5 cm. long; rhachis very short, so that the 2 (rarely only 1) pairs of pinnae are subdigitate; leaflets 10–26 pairs, 6–12·5(15) × 1·2–2·75(3) mm.; venation diverging from and not nearly parallel with the midrib; margins setulose. Flowers lilac or pink, in shortly ovoid, pedunculate heads c. 1–1·3 × 0·6–1 cm., 1–4(5) together from the axils. Calyx minute, c. 0·2 mm. long. Corolla 2–2·25 mm. long. Stamens 4. Pods clustered, 1–1·8 × 0·3–0·5 cm. (excluding the prickles), densely setose-prickly on the margins only.

Var. **tetrandra** (Humb. & Bonpl. ex Willd.) DC., Prodr. **2**: 426 (1825).—Brenan in Kew Bull. **10**: 187 (1955); F.T.E.A. Legum.-Mimos.: 47 (1959). Type from S. America.

Mimosa tetrandra Humb. & Bonpl. ex Willd. in L., Sp. Pl. ed. 4, **4**, 2: 1032 (1806). Type as above.

Stems densely hispid or almost glabrous. Stipules 4–8 (rarely 9–10) mm. long. Bracteoles c. 1–1·5 mm. long, usually shorter than or as long as the corolla in bud or sometimes longer; setiform hairs on the margins few and short (to c. 0·75 mm.) or absent. Outside of the corolla grey-puberulous above.

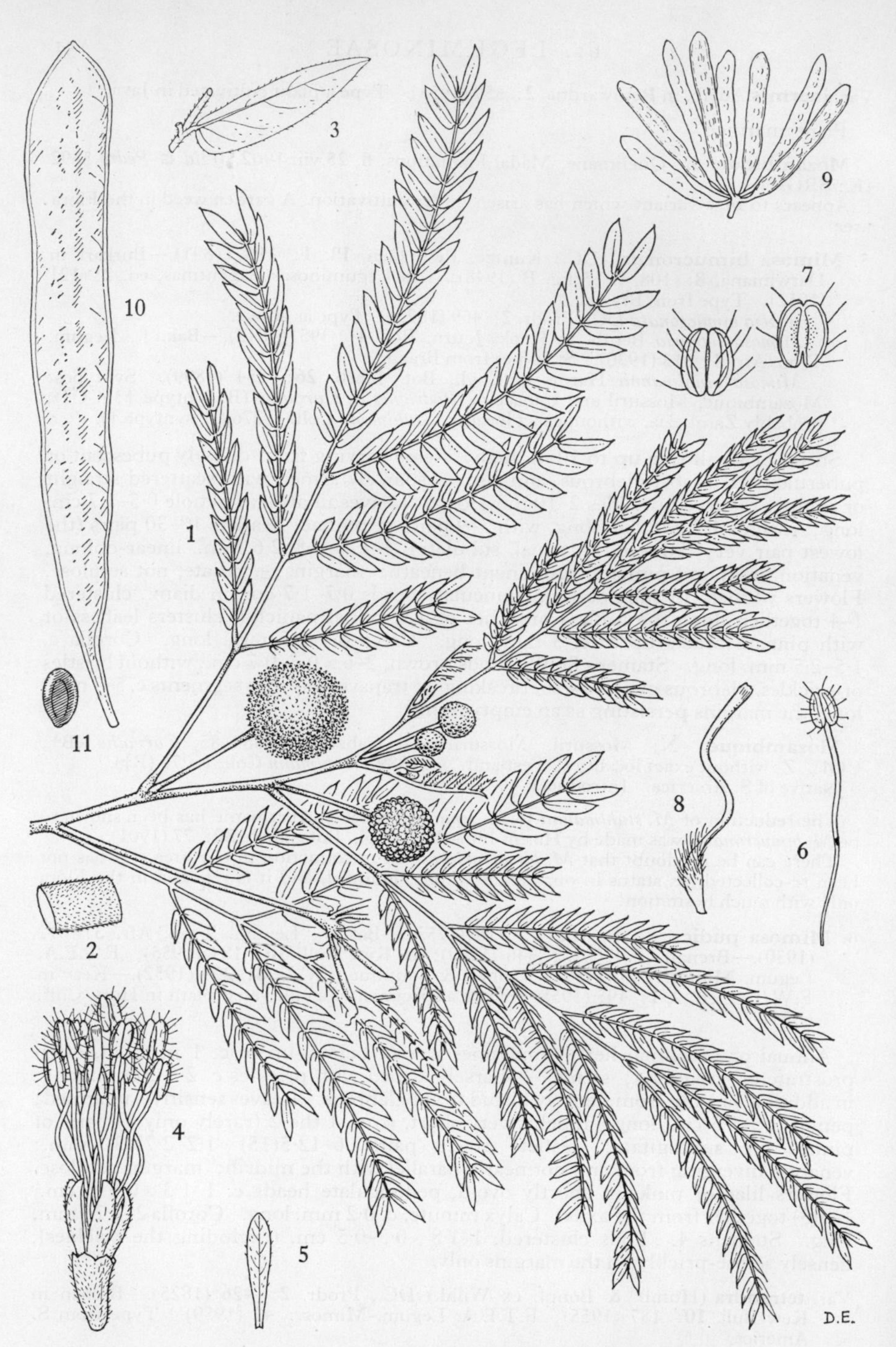

Tab. 14. LEUCAENA LEUCOCEPHALA. 1, flowering branch (× $\frac{2}{3}$) *Lemos & Balsinhas* 22; 2, part of stem to show indumentum (× 4); 3, leaflet (× 3), all from *Faulkner* 576; 4, flower (× 4); 5, petal (× 4); 6, stamen (× 4); 7, two views of anther (× 8); 8, ovary (× 4), all from *Lemos & Balsinhas* 22; 9, cluster of pods (× $\frac{1}{6}$); 10, pod (× $\frac{2}{3}$); 11, seed (× 1), all from *Faulkner* 576.

Malawi. N: Nkata Distr., Tchombe Tea Estate, fl. & fr. 21.ii.1961, *Richards* 14445 (K; SRGH).

Native of S. America, now pantropical. A weed in our area.

Var. *hispida* Brenan, differing in the constantly densely hispid stems, the stipules 8–14 mm. long and particularly in the flower-heads in bud appearing densely bristly owing to projecting setiform hairs, has been grown in our area (Rhodesia, " Salisbury Forest Nursery, reputed to be from Zambesi ", fl. iii.1918, *Eyles* 1356 (BM; K; SRGH), but is not at present known to be naturalized.

11. LEUCAENA Benth.

(By J. P. M. Brenan & R. K. Brummitt)

Leucaena Benth. in Hook., Journ. Bot. **4**: 416 (1842).

Trees or shrubs, unarmed. Leaves 2-pinnate; a gland often present at the junction of the lowest pair of pinnae, petiole and rhachis otherwise eglandular, or rarely with glands between other pairs of pinnae; pinnae each with one to several or many pairs of leaflets. Inflorescences of rounded heads, pedunculate, axillary, 1–3 together, often racemosely aggregated. Flowers ☿, sessile. Calyx gamosepalous with 5 teeth. Petals 5, free, pubescent to glabrous outside. Stamens 10, fertile. Anthers eglandular at the apex (except in the extra-African *L. forsteri*). Ovary pubescent or sometimes glabrous. Pods oblong or linear-oblong, compressed, usually thinly subcoriaceous, splitting into 2 non-recurving valves. Seeds lying ± transversely in the pod, compressed, brown, glossy, unwinged, with endosperm.

A genus of c. 50 spp., one widespread in the tropics (*L. leucocephala*), one in the Pacific islands, the rest tropical American.

Leucaena leucocephala (Lam.) De Wit in Taxon, **10**: 54 (1961). TAB. **14**. Type an American plant cultivated in France.

Mimosa leucocephala Lam., Encycl. Méth. Bot. **1**: 12 (1783). Type as above.

Mimosa glauca sensu L., Sp. Pl. ed. 2, **2**: 1504 (1763) pro parte, non L., Sp. Pl. **1**: 520 (1753).

Leucaena glauca sensu auct. mult., e.g. Benth. in Hook. Journ. Bot. **4**: 416 (1842).—Bak. f., Legum. Trop. Afr. **3**: 814 (1930).—Brenan, T.T.C.L.: 345 (1949); F.T.E.A. Legum.-Mimos.: 48 (1959).—Gilbert & Boutique, F.C.B. **3**: 231 (1952).—Torre, C.F.A. **2**: 268 (1956).

Shrub or small tree 0·6–9 m. high; young branchlets densely grey-puberulous. Leaves: petiole 2–4·7 cm. long, often with a gland at the junction of the lowest pair of pinnae, glands otherwise absent; rhachis (2·5)7–15 cm. long; pinnae (2)3–8 pairs, opposite; leaflets (5)7–17(21) pairs, 7–18 × 1·5–5 mm., obliquely oblong-lanceolate, acute at the apex, puberulous on the margins and sometimes also on the midrib beneath. Peduncles 2·5 cm. long. Heads of flowers white to cream. Calyx 2–3·5 mm. long, puberulous above outside. Petals 4–5·25 mm. long, puberulous above outside. Stamen-filaments 6·5–7·5 mm. long; anthers hairy. Pods 8–18 × (1·4)1·8–2·1 cm., with a stipe up to 3 cm. long. Seeds 7·5–9 × 4–5 mm., elliptic to obovate.

Mozambique. N: Mogovolas, agricultural station, fl. & fr. v.1934, *Ribeiro* (LISC). T: Tete, Boroma, fl. vii.1891, *Menyharth* 217 (K; W). LM: Lourenço Marques, slope of Ponta Vermelha, fl. 22.i.1960, *Lemos & Balsinhas* 20 (COI; K; LISC; LMJ).

Widespread in the tropics and subtropics, probably native only in the New World. Widely cultivated in the Flora area, and sometimes escaping. The specimens quoted above are from localities where there is some suggestion of possible naturalization of the species, as for example *Temos & Balsinhas* 20 which was recorded as growing in secondary bush. It may certainly be expected to become established elsewhere also.

The hairs on the anthers (to see which a lens is necessary) are a most useful diagnostic character of *L. leucocephala*, and indeed distinguish it from all other *Mimosoideae* in our area.

12. ACACIA Mill.

(By J. P. M. Brenan)

Acacia Mill., Gard. Dict., abridg. ed. 4 (1754)

Trees or shrubs, sometimes climbing; the native species in our area almost invariably armed with prickles or spines, the introduced ones usually unarmed.

Leaves 2-pinnate or (in introduced species) often modified to phyllodes (entire leaflike often flattened organs without pinnae or leaflets); pinnae each with one to many pairs of leaflets; gland on the upper side of the petiole usually present; glands also often present at the insertion of the pinnae. Flowers in spikes, spiciform racemes or round heads, ☿ or ♂ and ☿; if in heads then central flowers not enlarged and modified; inflorescences usually axillary, racemose or paniculate. Calyx (in our species) gamosepalous, subtruncate or usually with 4–5 teeth or lobes. Corolla 4–5(7)-lobed. Stamens many, fertile, their filaments free or (in *A. albida* and *A. eriocarpa*) connate into a tube at their extreme base only; anthers (at least some) glandular at the apex, or all eglandular (in all native species glandular except in *A. albida*, in introduced species mostly eglandular). Ovary stipitate to sessile, glabrous to puberulous. Pods very variable, dehiscent or sometimes indehiscent, flat, ± compressed, or sometimes cylindric, straight, curved, spiral or contorted, continuous or moniliform. Seeds unwinged, often with a hard smooth testa, without endosperm.

A genus of c. 750–800 spp., mostly tropical or subtropical; more than half in Australia, many in Africa and America, fewer in Asia.

In some of our species the remarkable structures derived from the stipules and commonly known as ant-galls occur. There is some evidence that these may not be galls at all, but natural outgrowths of the plant itself. Certainly their presence is taxonomically important. In the text I have compromised by keeping the familiar term " ant-gall " but enclosing it by inverted commas. A new and more accurate term may have to be devised.

Root-nodules have been recorded in species Nos. 1, 2, 3, 5–8, 10–12, 15, 25, 27, 32, 37, 39, 41–43, 46–49, 51–53 and 58.

There is a very interesting paper on the pollen-morphology of the S. African species by Coetzee in S. Afr. Journn. Sci. **52**: 23–27 (1955), in which differences in morphology are shown to give clear support to the division of our species into those with capitate and those with spicate inflorescences. Among the former, " *A. pennata* " is the only anomaly, but this group of species occupies an isolated position on other grounds.

Flowers in spikes or spiciform racemes; inflorescences sometimes short and ellipsoid (e.g. *A. mellifera*) but even then the axis clearly elongate - - Group I (below)
Flowers in globose capitula - - - - - - Group II (p. 56)

Group I

Stipules spinescent, straight; no other prickles; leaves without any gland on the petiole, but one on the upper side of the rhachis between each pair of pinnae; pods indehiscent, falcate or coiled, orange; anthers 0·2–0·4 mm. wide, eglandular even in bud; stamen-filaments connate and tubular for about 1 mm. at the base - - 1. *albida*
Stipules not spinescent; prickles (usually present) borne below the stipules, usually ± hooked, deflexed or curved; petiole usually glandular; rhachis eglandular or with glands between some pairs of pinnae (very rarely, in *A. erubescens*, between all pairs); pods dehiscent, straight or slightly curved, rarely indehiscent and straight, never coiled and orange; anthers 0·1–0·25 mm. wide, glandular at least in bud; stamen-filaments free:
Prickles absent (rarely and abnormally):
Pinnae 9–14 pairs; gland on the petiole 0·75–1·5 mm. in diam.; calyx 0·75–1·25 mm. long, red or purplish, as is the corolla - - - 6. *galpinii*
Pinnae of the larger leaves 15–20 or more pairs; gland on the petiole 2–4 mm. long; calyx 1·7–2·25 mm. long, not red or purplish - - - - 8. *polyacantha*
Prickles present:
The prickles irregularly scattered along the internodes, only a few sometimes tending to be irregularly grouped near the nodes:
Leaflets 0·5–1·5 mm. wide, 1–5(7·5) mm. long; calyx 1–2·5 mm. long, glabrous to slightly pubescent; corolla 2·5–3 mm. long; pods glabrous to puberulous; young branchlets glabrous to pubescent:
Young branchlets puberulous to pubescent; racemes of flowers 4–10 cm. long; leaflets usually ciliate on the margins and often ± hairy on the lower surface; ovary pubescent - - - - - - - - - 2. *ataxacantha*
Young branchlets glabrous; spikes of flowers 1·5–4 cm. long; leaflets glabrous; ovary glabrous - - - - - - - - 3. *chariessa*
Leaflets 3–9 mm. wide, (5)7–18 mm. long; calyx 3·5 mm. long, tomentose outside; corolla 4–4·5 mm. long; pods densely long-tomentose; young branchlets tomentose - - - - - - - - - 4. *eriocarpa*

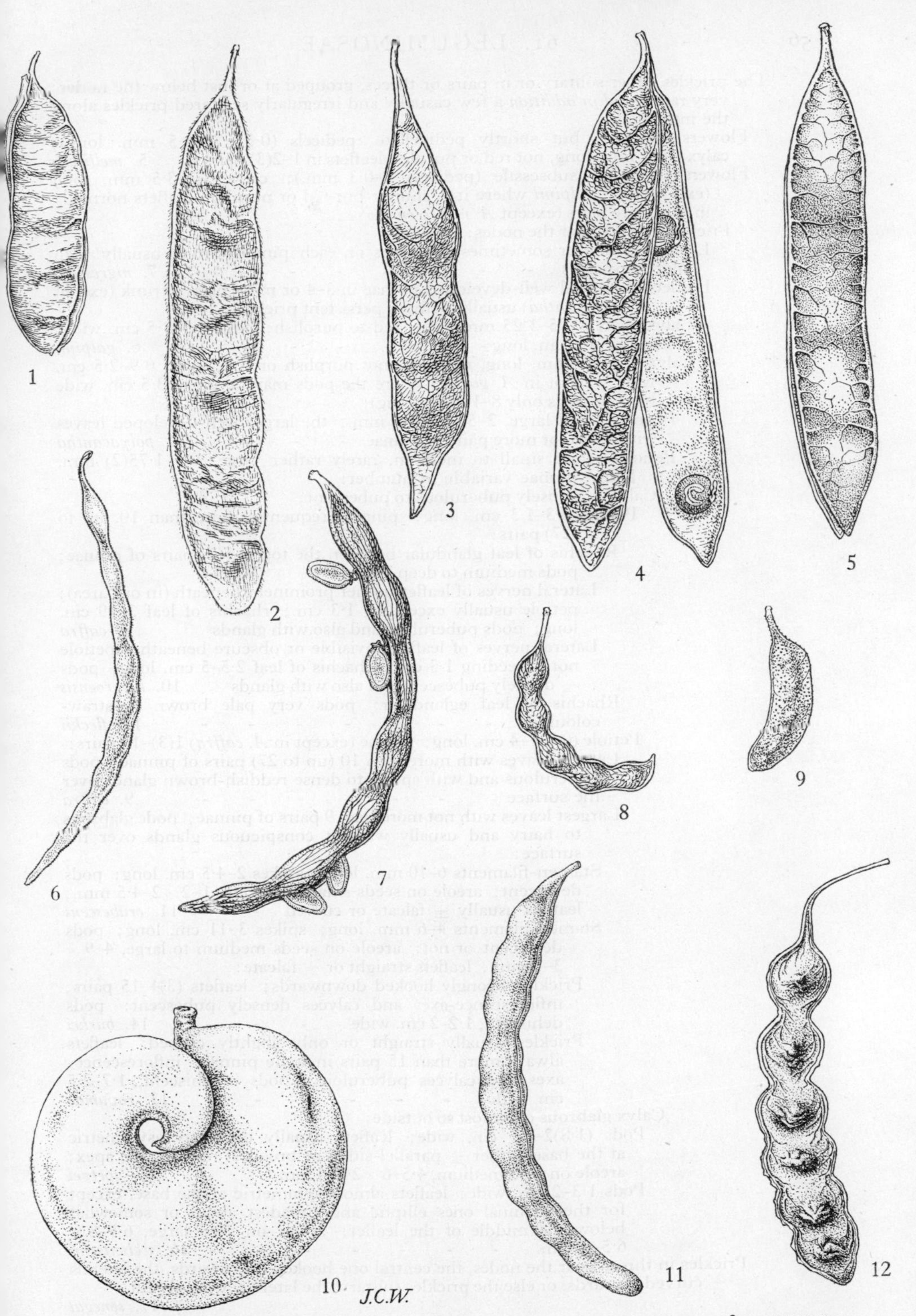

Tab. 15. Pods of ACACIA (all ×⅔). 1. A. ATAXACANTHA; 2. A. ERIOCARPA; 3. A. NIGRESCENS; 4. A. POLYACANTHA SUBSP. CAMPYLACANTHA; 5. A. SENEGAL VAR. LEIORHACHIS; 6. A. HOCKII; 7. A. KARROO; 8. A. BORLEAE; 9. A. NEBROWNII; 10. A. GIRAFFAE; 11. A. HAEMATOXYLON; 12. A. KIRKII SUBSP. KIRKII.

The prickles either solitary or in pairs or threes, grouped at or just below the nodes; very rarely *and in addition* a few casually and irregularly scattered prickles along the internodes:

Flowers distinctly but shortly pedicellate (pedicels (0·5)0·75–1·5 mm. long); calyx 0·6–1 mm. long, not red or purple; leaflets in 1–2(3) pairs 5. *mellifera*

Flowers sessile or subsessile (pedicels 0–0·3 mm.); calyx 1·5–3·5 mm. long (except in *A. galpinii* where it is shorter but red or purple); leaflets normally in 3 or more pairs (except *A. nigrescens*):

Prickles in pairs near the nodes:

Leaflets in one or sometimes two pairs on each pinna; trunk usually with knobby prickles - - - - - - - - - 7. *nigrescens*

Leaflets of normal well-developed pinnae in 3–4 or more pairs; trunk (except in *A. polyacantha*) usually without persistent prickles:

Calyx short, 0·75–1·25 mm. long, red to purplish; pods 2.7–3·5 cm. wide; seeds 12–15 mm. long - - - - - - - - 6. *galpinii*

Calyx 1·5–3·5 mm. long, normally not purplish or red; pods 0·9–2·5 cm. wide (except in *A. goetzei* where the pods may be up to 3·5 cm. wide but the seeds only 8–10 mm. long):

Petiolar gland large, 2–5 × 1·75–3 mm.; the larger well-developed leaves with 15–20 or more pairs of pinnae - - - - 8. *polyacantha*

Petiolar gland small to medium, rarely rather large, 0·5–1·75(2) mm. long; pinnae variable in number:

Calyx ± densely puberulous to pubescent:

Petiole 0·3–1·3 cm. long; pinnae frequently more than 10, up to 20(27) pairs:

Rhachis of leaf glandular between the top 1–5(7) pairs of pinnae; pods medium to deep brown:

Lateral nerves of leaflets rather prominent beneath (in our area); petiole usually exceeding 1·3 cm.; rhachis of leaf 3–19 cm. long; pods puberulous and also with glands 9. *caffra*

Lateral nerves of leaflets invisible or obscure beneath; petiole not exceeding 1·3 cm.; rhachis of leaf 2·5–5 cm. long; pods ± densely pubescent and also with glands 10. *hereroensis*

Rhachis of leaf eglandular; pods very pale brown to straw-coloured - - - - - - - - 12. *fleckii*

Petiole (0·7)1–4 cm. long; pinnae (except in *A. caffra*) 1(3)–10 pairs:

Largest leaves with more than 10 (up to 27) pairs of pinnae; pods puberulous and with sparse to dense reddish-brown glands over the surface - - - - - - - - 9. *caffra*

Largest leaves with not more than 9 pairs of pinnae; pods glabrous to hairy and usually without conspicuous glands over the surface:

Stamen-filaments 6–10 mm. long; spikes 2–4·5 cm. long; pods dehiscent; areole on seeds small and wide, 1–2 × 2–4·5 mm.; leaflets usually ± falcate or curved - - 11. *erubescens*

Stamen-filaments 4–6 mm. long; spikes 3–11 cm. long; pods dehiscent or not; areole on seeds medium to large, 4–9 × 3–5 mm.; leaflets straight or ± falcate:

Prickles strongly hooked downwards; leaflets (3)4–15 pairs; inflorescence-axes and calyces densely pubescent; pods dehiscent, 1·2–2 cm. wide - - - 14. *burkei*

Prickles usually straight or only slightly curved; leaflets always more than 15 pairs in some pinnae; inflorescence-axes and calyces puberulous; pods indehiscent, 1·7–2·5 cm. wide - - - - - - - 13. *rovumae*

Calyx glabrous or almost so outside:

Pods (1·8)2–3·5 cm. wide; leaflets usually distinctly asymmetric at the base, either ± parallel-sided or broadest towards the apex; areole on seed medium, 4·5–6 × 2·5–4 mm. - - 15. *goetzei*

Pods 1·3–2 cm. wide; leaflets almost symmetric at the base, except for the terminal ones elliptic and broadest about or somewhat below the middle of the leaflet; areole on seed large, 6·5–8 × 6·5–8 mm. - - - - - - - - 16. *welwitschii*

Prickles in threes near the nodes, the central one hooked downwards, the laterals ± curved upwards, or else the prickles solitary, the laterals being absent 17. *senegal*

Group II

Plant armed with prickles or spines:

Prickles scattered along the internodes of the stem, not grouped at or near the nodes;

rhachis of leaves often prickly - - - Group II A below
Prickles or spines in pairs at or near the nodes; rhachis of leaves unarmed:
Plant in flower (key hereafter based mainly on floral and vegetative characters):
Flowers bright-yellow or golden- or orange-yellow - - Group II B below
Flowers cream, white, pink or greenish:
Pinnae at least 15 pairs (and usually more than 20 pairs) per leaf on the well-developed leaves of flowering shoots (reduced leaves with fewer pairs of pinnae usually also present) - - - Group II C (p. 59)
Pinnae 1–14 pairs - - - - - Group II D (p. 59)
Plant in fruit (key hereafter based mainly on pod and vegetative characters):
Pods contorted or spirally twisted; spines a mixture of some short and hooked with others long and straight - - - - - 46. *tortilis*
Pods straight, curved or falcate, but not contorted or spirally twisted:
The pods indehiscent and thin-valved (except often for tubercles in the centre of the joints), usually ± moniliform or jointed and breaking up transversely, ± transversely or net-veined, glabrous except usually for sessile glands; bark on trunk usually yellow to green, sometimes grey or brown, powdery or with papery peel:
Bark of twigs grey-brown to plum-coloured, not yellow, of trunk grey to brown or greenish; pinnae of leaves of fruiting shoots 6–14 pairs, but some leaves almost always with 8–9 or more pairs; joints of pod often tubercled in the middle, mostly as wide as or wider than long - - - 39. *kirkii*
Bark of twigs soon becoming pale yellow, of trunk lemon-coloured or greenish-yellow; pinnae of leaves of fruiting shoots 3–6(8) pairs (only on juvenile shoots as many as 10 pairs); joints of pod not tubercled, mostly longer than wide - - - - - - - 40. *xanthophloea*
The pods indehiscent or dehiscent, if indehiscent then the valves markedly thickened, woody or pulpy in texture, not venose and glandular as above:
Valves of pod markedly thickened, woody or pulpy in texture; pods indehiscent or slowly dehiscent - - - - - Group II E (p. 60)
Valves of pod membranous to subcoriaceous or coriaceous, not markedly thickened:
Pods straight or nearly so - - - - - Group II F (p. 60)
Pods ± falcate - - - - - Group II G (p. 60)
Plant altogether unarmed; cultivated species - - - - Group II H (p. 63)

Group II A

Leaflets comparatively large, (2)3–8 mm. wide, (5)9–23 mm. long; prickles all short, up to c. 1·5 mm. long - - - - - - 18. *kraussiana*
Leaflets much smaller, 0·3–2(3) mm. wide, 1–10 mm. long; prickles mostly longer, up to c. 6 mm.:
Petiole 0·5–1·5 cm. long:
Midrib of leaflets excentric at the base; calyx eglandular - - - 19. *brevispica*
Midrib of leaflets almost central at the base; calyx glandular outside 20. *adenocalyx*
Petiole 1·5–6 cm. long:
Stipules (best seen at the base of the peduncles of flower-heads) broadly ovate, 4·5–9 mm. long, 3–4·5 mm. wide, subcordate at the base 21. *latistipulata*
Stipules narrower, linear to lanceolate, oblanceolate or falcate, 0·3–2 mm. wide, not subcordate at the base:
Young branchlets with numerous often minute red-purple glands; pods dehiscent:
Branchlets when young ± densely pubescent with fulvous hairs, dark-brown, later going blackish; leaflets glabrous or sparsely and inconspicuously ciliate; typically in upland forest - - - - - 22. *montigena*
Branchlets ± puberulous to glabrous, olive-green to olivaceous-brown or grey; leaflets variable, sometimes silky-puberulous on the surface; in woodland, thickets and bushland:
Leaflets 0·5–0·8(1) mm. wide, frequently appressed-silky-puberulous on the lower surface, sometimes glabrous; pods glandular and also rather densely puberulous; petiole 0·5–2(2·5) cm. long (in S. Mozambique only) 19. *brevispica*
Leaflets (0·8)1–2(2·5) mm. wide, usually glabrous on the surface beneath, sometimes silky-pubescent; pods glabrous except for glands; petiole (1·5)2·6–6 cm. long (widespread) - - - 23. *schweinfurthii*
Young branchlets usually eglandular, glabrous to sparsely puberulous; pod indehiscent; usually in evergreen forest - - - - 24. *pentagona*

Group II B

Capitula of flowers borne in panicles; leaves large (rhachis, with petiole, 10–37 cm. long) 25. *macrothyrsa*

Capitula of flowers on peduncles which are axillary and clustered or borne singly, sometimes aggregated into " racemes ", but not paniculate:

Leaves 2-pinnate but the leaflets so small close and imbricate that the pinnae resemble single linear crenulate densely grey-tomentellous leaflets; each true leaflet 0·25–0·75 × 0·5 mm.; corolla tomentellous on the lobes outside - - 38. *haematoxylon*

Leaves 2-pinnate but the leaflets larger and distinct from one another, not closely imbricate as above:

Involucel apical; leaflets with lateral nerves visible and somewhat raised beneath; leaflets (6)8–21 pairs; pods indehiscent:

Petiole without a gland on its upper side, but rhachis of leaf glandular at the junction of each pair of pinnae; tree 4·5–22 m. high; spines often stout and long, up to 5(8) cm.; leaflets 1·5–4·5 mm. wide; native species - - 37. *giraffae*

Petiole with a gland on its upper side; rhachis sometimes with a gland near the top pair of pinnae; shrub 1·5–4 m. high; spines usually short, up to 1·8(6) cm.; leaflets 0·75–1·75 mm. wide; only cultivated in our area - - 55. *farnesiana*

Involucel basal or up to 4/5-way up peduncle; leaflets without raised lateral nerves beneath (except in *A. swazica* which has (3)4–6(7) pairs of leaflets); pods dehiscent, not turgid:

Stems clothed with powdery bark:

Peduncles (usually at least) on ± elongate lateral or terminal shoots of the current season, whose leaves are persistent or undeveloped; bark red to yellow or white; involucel firmer and more opaque than that of *A. xanthophloea*; calyx 2–2·5 mm. long; some inflated " ant-galls " always present in our area - - - - - - - - - - 28. *seyal*

Peduncles (usually at least) on abbreviated lateral shoots whose axes do not elongate and are represented by clustered scales; the capitula thus appearing to be in lateral fascicles on older often yellow-barked twigs whose leaves have fallen; bark yellow or greenish-yellow; involucel thinner and more transparent-looking towards the margins than that of *A. seyal*; calyx 1–1·5 mm. long; no " ant-galls " - - - - - - - 40. *xanthophloea*

Stems without powdery bark:

Leaflets with lateral nerves ± prominent and visible beneath, 1·5–5·5 mm. wide; pinnae 1–2(3) pairs - - - - - - - 36. *swazica*

Leaflets with lateral nerves invisible beneath, variable in width; pinnae 1–20 pairs:

Young branchlets rather densely hairy with spreading whitish hairs 0·75–2 mm. long; leaflets ciliate, spinulose-mucronate at the apex 34. *permixta*

Young branchlets except for sessile glands glabrous to puberulous or shortly pubescent or tomentose, but hairs not more than 0·5 mm. long; if branchlets conspicuously and densely pubescent or tomentose, then the leaflets not spinulose-mucronate at the apex:

Involucel at or near the base of the peduncle; pinnae 1(2) pairs 35. *nebrownii*

Involucel 2/5–5/6-way up peduncle; pinnae 1–20 pairs:

Leaflets glandular on the surface and conspicuously crenulate-glandular along the margins which are also (in our area) minutely ciliolate 32. *borleae*

Leaflets eglandular on the surface and also on the margins, or at most with a very few inconspicuous glands on the margins near the apex:

Leaflets spinulose-mucronate at the apex:

Pinnae 7–13(20) pairs; leaflets ciliate on the margins 33. *torrei*

Pinnae 1–6 pairs; leaflets glabrous:

Leaflets 0·9–1·5 mm. wide; rhizomatous shrub 0·5–1·2(2·4) m. high; pods not or only slightly constricted, conspicuously glandular - - - - - - 30. *tenuispina*

Leaflets 1·5–4·5 mm. wide; small tree or sometimes a shrub, 1·5–5 m. high; pods constricted between the seeds, eglandular or nearly so - - - - - - - 31. *exuvialis*

Leaflets not spinulose-mucronate at the apex:

Pinnae (10)14–27 pairs; capitula conspicuously racemose; bark corky, yellowish; branchlets puberulous - - 29. *davyi*

Pinnae usually (1)2–11, rarely to 17 pairs, and then capitula not conspicuously racemose, bark not corky and yellowish, and branchlets ± densely pubescent:

Young branchlets glabrous to puberulous:

Spines mostly short, up to 2(4) cm. long; leaflets up to 5 mm. long; branchlets usually ± puberulous, occasionally glabrous; pods up to 6(8) mm. wide - - - - - 26. *hockii*

Spines usually longer and stouter, up to 7(17) cm. long; leaflets

very commonly more than 5 mm. long; branchlets usually glabrous, rarely sparsely puberulous; pods 6–9(10) mm. wide - 27. *karroo*
Young branchlets ± densely pubescent (in our area) - 41. *nilotica*

Group II C

Leaflets all exceedingly narrow, c. 0·25(0·5) mm. wide (in the Flora area); bark on twigs not peeling off - - - - - - 42. *abyssinica*
Leaflets 0·5 mm. wide or more; bark on twigs usually peeling off:
Involucel apical or in upper half of peduncle:
Calyx shorter than the projecting part of the corolla (corolla 2–4 × calyx); peduncles pubescent and glandular - - - - - - - 51. *arenaria*
Calyx longer than the projecting part of the corolla (corolla $1–1\frac{3}{4}$ × calyx); peduncles variable in indumentum but eglandular - - - - - - 52. *sieberana*
Involucel basal or in lower half of peduncle:
Bark on older branchlets yellow or sometimes greenish; corolla glabrous or sparsely puberulous outside; pinnae 8–16(26) pairs - - - - 44. *pilispina*
Bark on older branchlets rusty-red; corolla ± densely pubescent on the lobes outside; pinnae of well-developed leaves mostly 15–44 pairs:
Flower-heads 2–20 per axil, aggregated into a terminal "raceme" 43. *rehmanniana*
Flower-heads mostly solitary in leaf-axils, not in terminal "racemes" 50. *lasiopetala*

Group II D

Peduncles (at least below the involucel) with ± numerous very small reddish apparently sticky glands (use lens of ×10 or more); other hairs often sparse; similar glands on young branchlets and often elsewhere; involucel mostly at or below the middle of the peduncle, usually 2–3·5 mm. long, conspicuous:
Young branchlets shortly and thinly pubescent, or glabrous:
Bark of twigs grey-brown to plum-coloured, not yellow, of trunk grey to brown or greenish; pinnae of leaves of flowering shoots 6–14 pairs, but some leaves almost always with 8–9 or more pairs; peduncles rather densely (rarely sparsely) pubescent and glandular throughout - - - - - 39. *kirkii*
Bark of twigs soon becoming pale-yellow, of trunk lemon-coloured or greenish-yellow; pinnae of leaves of flowering shoots 3–6(8) pairs (only on juvenile shoots as many as 10 pairs); peduncles sparingly pubescent to subglabrous (very rarely rather densely pubescent), glandular below and sometimes also above the involucel 40. *xanthophloea*
Young branchlets ± densely and coarsely pubescent, often showing a rusty-red colour 49. *gerrardii*

Peduncles eglandular or with very small inconspicuous glands; involucel variable in position, mostly 1–2 mm. long:
Involucel at the apex of or above the middle of the peduncle - - 52. *sieberana*
Involucel at the base of or below the middle of the peduncle (sometimes, in *A. grandicornuta*, at about the middle):
Twigs normally becoming coated with yellowish or greenish powdery bark: young branchlets with spreading often slightly yellowish hairs 0·75–1·5(2) mm. long 44. *pilispina*
Twigs without yellowish or greenish powdery bark:
Peduncles glabrous or almost so:
Branchlets stout; pods 1·8–2·5(3) cm. wide - - 47. *robusta* subsp. *robusta*
Branchlets much more slender; pods 0·6–1·1 cm. wide 48. *grandicornuta*
Peduncles ± densely pubescent or puberulous:
The shorter spines downwardly hooked, the longer ones straight:
Leaf-rhachis short, up to c. 2(2·5) cm. long; longer spines whitish and often ± glossy; pods contorted - - - - - - - - 46. *tortilis*
Leaf-rhachis mostly 2·5 cm. or more long; longer spines not conspicuously whitish; pods straight or falcate, not contorted:
Epidermis of twigs often splitting to expose a rusty-red inner layer; pods falcate - - - - - - - - - 49. *gerrardii**
Epidermis of twigs without a rusty-red inner layer; pods straight:
Indumentum rather short, with the hairs on the peduncle rather shorter than its diameter; leaflets grey-green, often narrower and more parallel-sided than in No. 53; pods flattened, finely puberulous to subglabrous - - - - - - - - - 45. *luederitzii**

* When in flower and without pods these species may be sometimes very difficult to distinguish.

Indumentum rather long, with the hairs on the peduncle as long as or longer than its diameter; leaflets often slightly broadened towards the apex; pods turgid, grey-tomentellous - - - - 53. *hebeclada**
The spines, long and short, all straight or almost so:
Young branchlets glabrous to grey-pubescent or -tomentose; corolla glabrous or only slightly hairy outside:
Branchlets when young glabrous or shortly and not very densely pubescent; epidermis not splitting to expose a rusty-red inner layer 47. *robusta*
Branchlets when young densely and obviously pubescent or tomentose:
Epidermis of twigs often splitting to expose a rusty-red inner layer; pods falcate, narrow - - - - - - - - 49. *gerrardii**
Epidermis of twigs not splitting to expose a rusty-red inner layer; pods straight, broader, turgid - - - - - - 53. *hebeclada**

Group II E

Pods glabrous or sparingly hairy:
The pods very fat, almost round in section, glabrous, finely longitudinally striate; shrub 1·5–4 m. high, cultivated - - - - - 55. *farnesiana*
The pods distinctly compressed or flattened; trees usually 2·5–14 m. high, indigenous:
Pods moniliform or jointed with the valves marked with distinct raised bumps each one corresponding to a seed inside; pods indehiscent - - 41. *nilotica*
Pods not moniliform or jointed, the valves with an irregular but ± smooth and glossy surface, without any bumps corresponding to the seeds inside; pods very slowly dehiscent - - - - - - - - - 52. *sieberana*
Pods ± densely puberulous, pubescent, tomentellous or villous:
Indumentum on pod short, the hairs less than 0·5 mm. long; young branchlets not golden-villous:
Pods pubescent, not densely and continuously grey-tomentellous, flattened, ± moniliform - - - - - - - - 41. *nilotica*
Pods densely and continuously grey-tomentellous:
Pods ± arcuate or falcate, or if straight then asymmetric with one suture straight and the other markedly curved, indehiscent:
Pods 2·5–4·7 cm. wide; leaflets not closely imbricate; tree (4·5)6–16(22) m. high - - - - - - - - 37. *giraffae*
Pods 0·9–1·3 cm. wide; leaflets minute and very closely imbricate, grey; shrub or small tree 1–6 m. high - - - 38. *haematoxylon*
Pods straight and almost symmetric, ultimately dehiscent, erect - 53. *hebeclada*
Indumentum on pods long, the hairs spreading, 2–4 mm. long; young branchlets spreading-golden-villous - - - - - 54. *stuhlmannii*

Group II F

Leaves very large for the genus, 10–20 cm. wide; rhachis together with petiole 10–37 cm. long; pods coriaceous, glossy, glabrous - - - 25. *macrothyrsa*
Leaves much smaller, less than 10 cm. wide, rarely as much as 10 cm. long; pods various:
Twigs normally becoming coated with yellowish or greenish powdery bark; young branchlets with spreading, often slightly yellowish hairs 0·75–1·5(2) mm. long 44. *pilispina*
Twigs not coated with yellowish or greenish powdery bark; indumentum various:
Flower-heads aggregated into terminal " racemes "; young branchlets at first with spreading golden hairs, the epidermis falling off to expose a powdery rusty-red inner layer; pinnae of well-developed leaves 15–44 pairs 43. *rehmanniana*
Flower-heads not aggregated into terminal " racemes "; young branchlets with grey to somewhat yellowish indumentum; no rusty-red bark; pinnae 2–51 pairs:
Pinnae mostly 15–51 pairs; leaflets 0·25–0·4(0·5) mm. wide; all spines straight 42. *abyssinica*
Pinnae 2–9(12) pairs; leaflets 0·5–1·5 mm. wide; shorter spines hooked 45. *luederitzii*

Group II G

Pinnae 15 or more (up to 51) pairs per leaf on well-developed leaves of mature shoots:
Leaflets all exceedingly narrow, c. 0·25–0·4(0·5) mm. wide; usually a large, flat-crowned tree 6–20 m. high; pods at most slightly arcuate, flattened, 1·2–2·8 cm. wide, glabrous to puberulous - - - - - - - - 42. *abyssinica*

* When in flower and without pods these three species may be sometimes very difficult to distinguish.

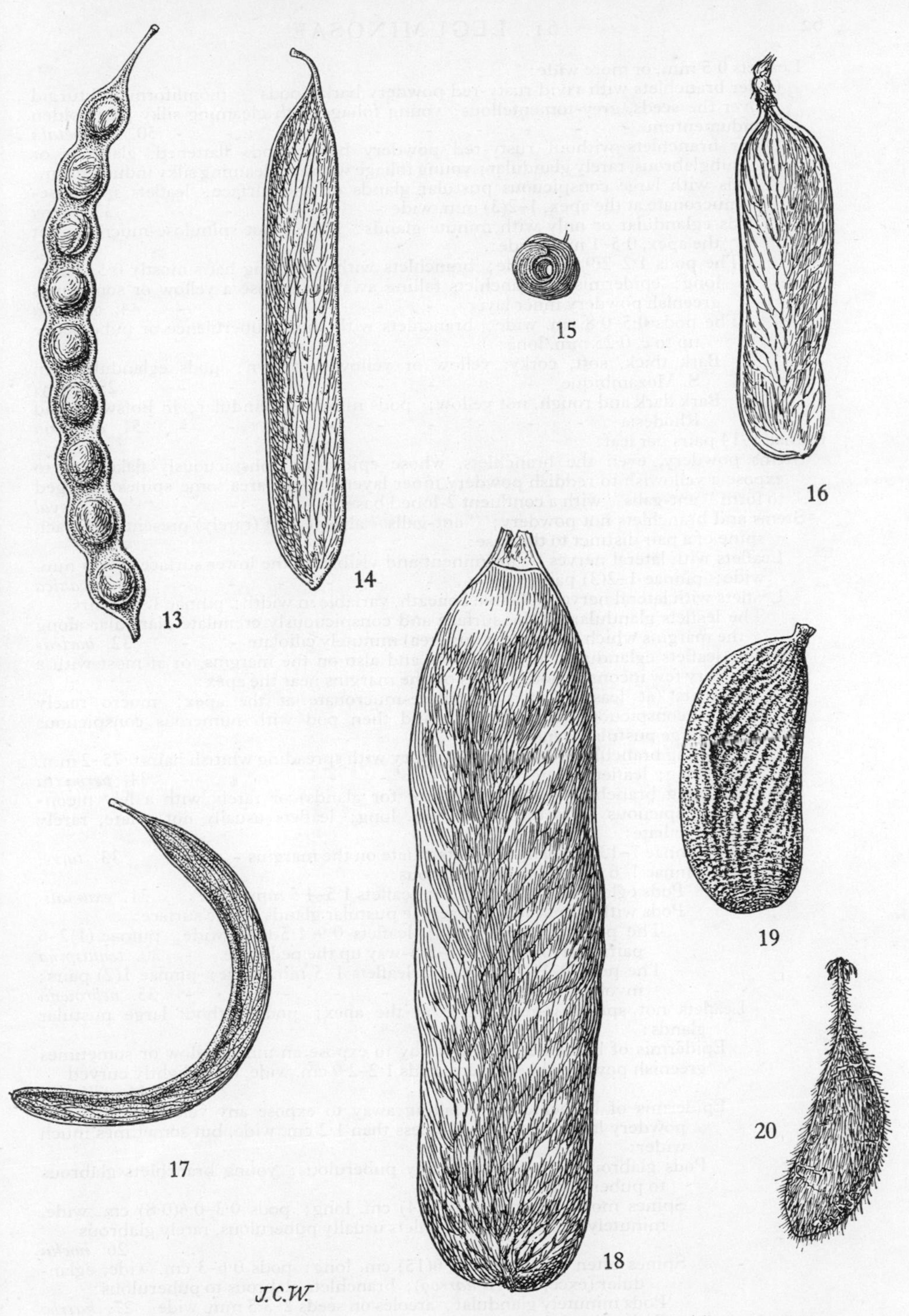

Tab. 16. Pods of ACACIA (all × $\frac{2}{3}$). 13. A. NILOTICA SUBSP. KRAUSSIANA; 14. A. LUEDERITZII VAR. LUEDERITZII; 15. A. TORTILIS SUBSP. HETERACANTHA; 16. A. ROBUSTA SUBSP. ROBUSTA; 17. A. GERRARDII VAR. GERRARDII; 18. A. SIEBERANA VAR. WOODII; 19. A. HEBECLADA; 20. A. STUHLMANNII.

Leaflets 0·5 mm. or more wide:
Older branchlets with vivid rusty-red powdery bark; pods ± moniliform and turgid over the seeds, grey-tomentellous; young foliage with gleaming silky pale-golden indumentum - - - - - - - - - - - 50. *lasiopetala*
Older branchlets without rusty-red powdery bark; pods flattened, glabrous or subglabrous, rarely glandular; young foliage without gleaming silky indumentum:
Pods with large conspicuous pustular glands on the surface; leaflets spinulose-mucronate at the apex, 1–2(3) mm. wide - - - - - - 33. *torrei*
Pods eglandular or only with minute glands; leaflets not spinulose-mucronate at the apex, 0·5–1 mm. wide:
The pods 1·2–2·9 cm. wide; branchlets with spreading hairs mostly 0·5–2 mm. long; epidermis of branchlets falling away to expose a yellow or sometimes greenish powdery inner layer - - - - - - - 44. *pilispina*
The pods 0·5–0·8 cm. wide; branchlets with short puberulence or pubescence up to c. 0·25 mm. long:
Bark thick, soft, corky, yellow or yellowish-brown; pods eglandular; in S. Mozambique - - - - - - - - - - 29. *davyi*
Bark dark and rough, not yellow; pods minutely glandular; in Botswana and Rhodesia - - - - - - - - - - - 51. *arenaria*
Pinnae 1–14 pairs per leaf:
Stems powdery, even the branchlets, whose epidermis conspicuously flakes off to expose a yellowish to reddish powdery inner layer; in our area some spines enlarged to form " ant-galls " with a confluent 2-lobed base - - - - - 28. *seyal*
Stems and branchlets not powdery; " ant-galls " absent or if (rarely) present then each spine of a pair distinct to the base:
Leaflets with lateral nerves ± prominent and visible on the lower surface, 1·5–5 mm. wide; pinnae 1–2(3) pairs - - - - - - - - - 36. *swazica*
Leaflets with lateral nerves invisible beneath, variable in width; pinnae 1–14 pairs:
The leaflets glandular on the surface and conspicuously crenulate-glandular along the margins which are also (in our area) minutely ciliolate - - 32. *borleae*
The leaflets eglandular on the surface and also on the margins, or at most with a very few inconspicuous glands on the margins near the apex:
Leaflets (at least mostly) spinulose-mucronate at the apex; mucro rarely inconspicuous (*A. nebrownii*) and then pod with numerous conspicuous large pustular glands:
Young branchlets rather densely hairy with spreading whitish hairs 0·75–2 mm. long; leaflets ciliate - - - - - - - - - 34. *permixta*
Young branchlets glabrous (except for glands) or rarely with a few inconspicuous hairs up to 0·5 mm. long; leaflets usually not ciliate, rarely ciliate:
Pinnae 7–13(20) pairs; leaflets ciliate on the margins - - 33. *torrei*
Pinnae 1–6 pairs; leaflets glabrous:
Pods eglandular or almost so; leaflets 1·5–4·5 mm. wide 31. *exuvialis*
Pods with numerous dark sessile pustular glands on the surface:
The pods 5–7 mm. wide; leaflets 0·9–1·5 mm. wide; pinnae (1)2–6 pairs; involucel c. 1/2–3/4-way up the peduncle - 30. *tenuispina*
The pods 9–11 mm. wide; leaflets 1–5 mm. wide; pinnae 1(2) pairs; involucel basal - - - - - - - - 35. *nebrownii*
Leaflets not spinulose-mucronate at the apex; pods without large pustular glands:
Epidermis of branchlets falling away to expose an inner yellow or sometimes greenish powdery bark layer; pods 1·2–2·9 cm. wide, only slightly curved 44. *pilispina*
Epidermis of branchlets not falling away to expose any yellow or greenish powdery layer; pods usually less than 1·2 cm. wide, but sometimes much wider:
Pods glabrous or inconspicuously puberulous; young branchlets glabrous to puberulous:
Spines mostly short, up to 2(4) cm. long; pods 0·3–0·6(0·8) cm. wide, minutely glandular; branchlets usually puberulous, rarely glabrous 26. *hockii*
Spines often longer, up to 6(15) cm. long; pods 0·6–3 cm. wide, eglandular (except in *A. karroo*); branchlets glabrous to puberulous:
Pods minutely glandular; areoles on seeds 2–3·5 mm. wide 27. *karroo*
Pods eglandular; areoles on seeds 3·5–6·5 mm. wide:
Leaf-rhachis ± pubescent, or if glabrous then pods 1·8–3 cm. wide or pinnae more than 4 pairs and leaflets more than 15 pairs 47. *robusta*
Leaf-rhachis glabrous; pods 0·6–1·1 cm. wide; pinnae 1–4 pairs;

leaflets 5–15 pairs - - - - - - 48. *grandicornuta*
Pods usually ± densely grey-puberulous or tomentellous; young branchlets ± densely grey-pubescent - - - - - - 49. *gerrardii*

Group II H*

Leaves 2-pinnate:
Pinnae 2–5 pairs per leaf:
Leaflets mostly 2·5–5 cm. long, appressed-puberulous on the surface; young branchlets puberulous or pubescent - - - - - - - - 56. *elata*
Leaflets 0·15–2·3 cm. long, glabrous on the surface (sometimes ciliate on the margins):
Petiole 2–7·5 cm. long; pinnae mostly 5–14 cm. long, not crowded; leaflets 1–2·3 cm. long - - - - - - - - - - - 57. *schinoides*
Petiole very short, c. 2 mm.; pinnae c. 0·8–3 cm. long, crowded; leaflets 0·3–0·8 cm. long - - - - - - - - - - 60. *baileyana*
Pinnae 5–26 pairs per leaf:
Leaf-rhachis with a gland on its upper side at the insertion of each pair of pinnae, and also between the insertions; young shoots golden-yellow-tomentellous; pod constricted between the seeds - - - - - - - - 58. *mearnsii*
Leaf-rhachis with a gland on its upper side at the insertion of each pair (or most pairs) of pinnae, but not between the insertions; young shoots grey- or sometimes yellowish-pubescent; pods not or only slightly constricted between the seeds 59. *dealbata*
Leaves apparently simple, modified to phyllodes by dilation of the petiole and rhachis:
Young branchlets glabrous:
Phyllodes linear-lanceolate to linear, mostly 6–20 cm. long 61. *cyanophylla*
Phyllodes obliquely obovate-lanceolate to ovate-triangular, 0·8–3 cm. long 63. *cultriformis*
Young branchlets densely grey-pubescent; phyllodes ovate to elliptic or elliptic-oblong, pubescent - - - - - - - - - 62. *podalyriifolia*

1. **Acacia albida** Del., Fl. Égypte Expl. Planches: 286, t. 52 fig. 3 (1813).—Oliv., F.T.A. **2**: 339 (1871).—Burkill in Johnston, Brit. Centr. Afr.: 245 (1897).—Sim, For. Fl. Port. E. Afr.: 54, t. 34 (1909).—Eyles in Trans. Roy. Soc. S. Afr. **5**; 361 (1916).—Bak. f., Legum. Trop. Afr. **3**: 825 (1930).—Burtt Davy, F.P.F.T. **2**: 335 (1932).—Steedman, Trees etc. S. Rhod.: 12 (1933).—Hutch., Botanist in S. Afr.: 392 cum photogr. (1946).—O. B. Mill., B.C.L.: 16 (1948); in Journ. S. Afr. Bot. **18**: 18 (1952).—Brenan, T.T.C.L.: 330 (1949); F.T.E.A. Legum-Mimos.: 78, fig. 14/1 (1959).—Codd, Trees & Shrubs Kruger Nat. Park: 38, fig. 32 (1951).—Pardy in Rhod. Agric. Journ. **50**: 325 cum photogr. (1953).—Wild, Guide Fl. Vict. Falls: 148 (1953); S. Rhod. Bot. Dict.: 46 (1953).—Young in Candollea, **15**: 89 (1955).—Torre, C.F.A. **2**: 272 (1956).—Burtt Davy & Hoyle, rev. Topham, N.C.L., ed. 2: 63 (1958).—Karschon in La-Yaaran, **11**: 4 (1961).—Palmer & Pitman, Trees of S. Afr.: 148 cum fig., t. 5, photogr. 31 (1961).—Fanshawe, Fifty Common Trees N. Rhod.: 4 cum tab. (1962).—F. White, F.F.N.R.: 82, fig. 17 B–C (1962).—Mitchell in Puku, **1**: 103 (1963).—Boughey in Journ. S. Afr. Bot. **30**: 157 (1964).—Gomes e Sousa, Dendrol. Moçamb. Estudo Geral, **1**: 232, t. 36 (1966). TAB. **17**. Type from Egypt.
Acacia mossambicensis Bolle in Peters, Reise Mossamb. Bot. **1**: 5 (1861). Type: Mozambique, Rios de Sena and R. Chimazo W. of Tete, *Peters* (B, ? syntypes †).
Prosopis ? kirkii Oliv., F.T.A. **2**: 332 (1871). Type: Malawi, Shire R., *Kirk* (K, holotype).
Faidherbia albida (Del.) A. Chev., Rev. Bot. Appl. **14**: 876 (1934).—Gilbert & Boutique, F.C.B. **3**: 169 (1952). Type as for *A. albida*.

Tree 6–30 m. high, with rough dark-brown or greenish-grey bark and spreading branches; young branchlets ashen to whitish. Stipules spinescent, up to 1·3(2·3) cm. long, straight, never enlarged and inflated; no prickles below the stipules. Leaves: rhachis with a single conspicuous gland at the junction of each of the (2)3–10 pairs of pinnae; no gland on the petiole; leaflets 6–23 pairs, (2·5)3·5–9(14) × 0·7–3(5) mm., rounded to subacute and mucronate at the apex. Flowers cream, sessile or to 0·5(2) mm. pedicellate, in inflorescences 3·5–14 cm. long on peduncles 1·3–3·5 cm. long. Calyx 1–1·7(2·5) mm. long. Corolla 3–3·5(4·5) mm. long, with 5 lobes 1·5–2·5 mm. long. Stamen-filaments 4–6 mm. long, connate for c. 1 mm. at the base; anthers 0·2–0·4 mm. across, eglandular even in bud. Pods bright orange, thick, indehiscent, 6–25 × (1·5)2–3·5(5) cm., glabrous or very rarely

* This group consists wholly of introduced species from Australia. Additional species are very likely to occur.

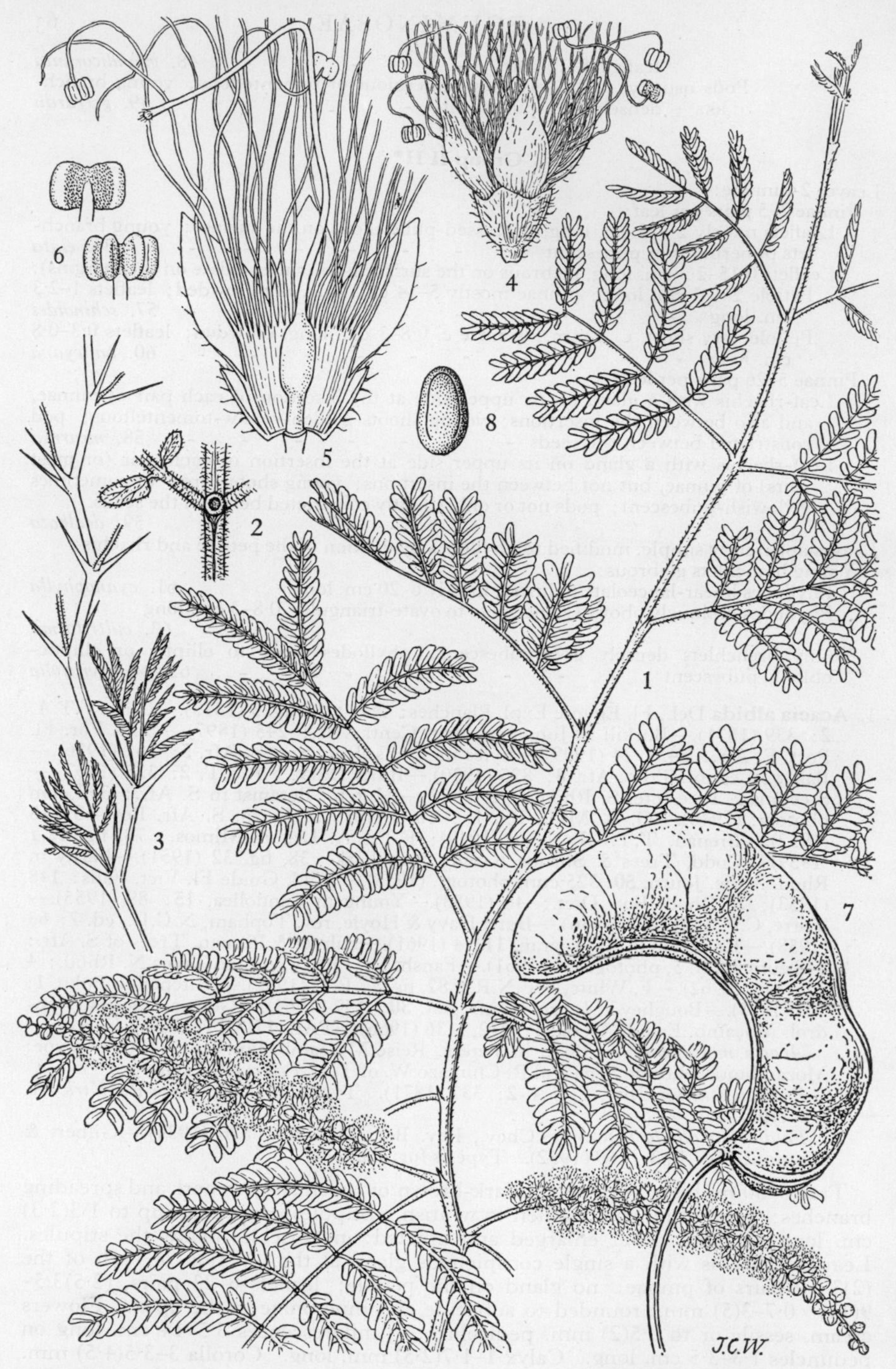

Tab. 17. ACACIA ALBIDA. 1, flowering branch ($\times\frac{2}{3}$); 2, part of leaf-rhachis showing gland ($\times 4$), both from *Robinson* 269; 3, juvenile shoot ($\times\frac{2}{3}$) *Lovemore* 68; 4, flower ($\times 6$); 5, flower opened out to show ovary ($\times 8$); 6, anthers ($\times 20$), all from *Robinson* 269; 7, pod ($\times\frac{2}{3}$) *Meikle* s.n.; 8, seed ($\times 1$) *Kirk* s.n.

puberulous, falcate or curled into a circular coil. Seeds 9–11 × 6–8 mm., elliptic-lenticular; central areole large, 7–9 × 4–6 mm.

Caprivi Strip. Lisikili, 25 km. E. of Katima Mulilo, fl. 17.vii.1952, *Codd* 7100 (K). **Botswana.** N: Chobe, fr. 2.viii.1950, *Robertson & Elffers* 100 (K; PRE; SRGH). SE: Mochudi, fl. i–iv.1914, *Harbor* in *Rogers* 6610 (K; PRE). **Zambia.** B: Sesheke, fl. & fr. viii, *Macaulay* 75 (K). N: shore of Lake Tanganyika at Sumbu, fr. 6.vii.1957, *Savory* 200 (K; SRGH). C: Mt. Makulu, st. without date, *Cole* 16 (K). S: Mapanza NE., fl. & fr. 24.v. & 10.x.1953, *Robinson* 269 (K). **Rhodesia.** N: Sebungwe, fl. & fr. 26.vi.1951, *Lovemore* 68 (K; LISC; SRGH). W: Bulawayo, Kennedy Siding, fl. 15.v.1953, *Hodgson* 2/53 (SRGH). S: Sabi-Lundi Junction, Chitsa's Kraal, fl. 6.vi.1950, *Wild* 3369 (K; LISC; SRGH). **Malawi.** N: Kondowe to Karonga, fl. vii.1896, *Whyte* (K). S: Shire R., fl. 6.vi.1938, *Pole Evans & Erens* 576 (K; PRE). **Mozambique.** N: Macondes, between Moeda and Nairoto, fr. 20.ix.1948, *Barbosa* in *Mendonça* 2227 (BM; K; LISC). Z: Morrumbala, near Morire, fl. 21.v.1943, *Torre* 5349 (LISC; LM). T: Mutarara, between Ancuaze and Pandocane, fl. & fr. 19.vi.1949, *Andrada* 1608 (COI; LISC). SS: between Guijá and Mapai, fl. 6.v.1944, *Torre* 6581 (K; LISC). LM: Sábiè, Moamba, fl. & fr. 7.vi.1948, *Torre* 7950 (LISC).

Widespread in tropical and subtropical Africa from Egypt, Senegal and the Gambia southwards to Botswana, the Transvaal and Natal; also in Syria, Palestine and (?native) Cyprus. Forest, woodland and wooded grassland on alluvial soil by rivers and sometimes by lakes; 40–1070 m. The ecology of *A. albida* in general has been recently investigated by Radwanski & Wickens in Journ. Appl. Ecol. **4**: 569 (1967).

The leaves without any gland on the petiole but with a gland on the rhachis between each of the pairs of pinnae, the large anthers eglandular at the apex even in bud, and the very distinctive pods, whose appearance when ripe has inspired the popular name of "Apple-Ring Acacia", are apparently unique among the species of *Acacia* with spiciform inflorescences occurring in the Flora Zambesiaca area. The stamen-filaments shortly connate at the base are also very unusual, otherwise occurring only in *A. eriocarpa* Brenan among our species.

The pods of *A. albida* are relished by game, including elephants. Along the Zambezi the large boles are used to make canoes (Pardy, loc. cit.). The shoots of young plants of *A. albida* are very different in appearance from the mature ones, having whitish stems with short internodes, usually larger and more conspicuous spines, and smaller closely-set foliage.

In the Flora Zambesiaca area *A. albida* appears always to have the leaflets ± pubescent on the surface and ± pubescent young branchlets, inflorescence-axes, calyces and (often) corollas, thus corresponding to Race B as defined in F.T.E.A. Legum.-Mimos.: 79 (1959). Race A is not known in our area.

2. **Acacia ataxacantha** DC., Prodr. **2**: 459 (1825).—Oliv., F.T.A. **2**: 343 (1871).—Bak. f., Legum. Trop. Afr. **3**: 834 (1930).—Brenan, T.T.C.L.: 332 (1949); F.T.E.A. Legum.-Mimos.: 82, fig. 14/5 (1959).—Codd, Trees & Shrubs Kruger Nat. Park: 40, fig. 33a (1951).—Gilbert & Boutique, F.C.B. **3**: 153 (1952).—Wild, S. Rhod. Bot. Dict.: 46 (1953).—Pardy in Rhod. Agric. Journ. **53**: 615 cum photogr. (1956).—Torre, C.F.A. **2**: 278 (1956).—Mitchell in Puku, **1**: 103 (1963).—Boughey in Journ. S. Afr. Bot. **30**: 157 (1964).—J. Ross in Webbia, **21**: 629 (1966). TAB. **15** fig. 1. Syntypes from Senegal.

Acacia eriadenia Benth. in Hook., Lond. Journ. Bot. **5**: 98 (1846).—O. B. Mill., B.C.L.: 18 (1948). Type from the Transvaal.

Acacia lugardiae N.E.Br. in Kew Bull. **1909**: 107 (1909).—Bak. f., loc. cit.—O. B. Mill., tom. cit.: 20 (1948); in Journ. S. Afr. Bot. **18**: 23 (1952). Type: Botswana, Kwebe Hills, *Mrs. Lugard* 195 (K, holotype).

Acacia ataxacantha var. *australis* Burtt Davy in Kew Bull. **1922**: 324 (1922); F.P.F.T. **2**: 335 (1932).—Bak. f., loc. cit.—O. B. Mill., B.C.L.: 16 (1948); in Journ. S. Afr. Bot. **18**: 19 (1952).—Young in Candollea, **15**: 84 (1955).—F. White, F.F.N.R.: 82, fig. 17 A (1962).—Boughey in Journ. S. Afr. Bot. **30**: 157 (1964). Type from the Transvaal.

Acacia senegal sensu Wild, Guide Fl. Vict. Falls: 149 (1953) quoad specim. *Rogers* 5544.

Scandent shrub up to 15 m. high or a straggling non-climbing shrub or small tree 2–10 m. high. Young branchlets puberulous to densely pubescent and often glandular. Stipules not spinescent, obliquely ovate to linear. Prickles scattered along the internodes, ± hooked or deflexed, often broad-based, up to 7(15) mm. long. Leaves: rhachis mostly 5–13 cm. long, prickly or unarmed; usually with a gland on the petiole and between the uppermost 1–3(5) pairs of pinnae; pinnae (4)6–25(29) pairs; leaflets 14–62 pairs, 2–5(7·5) × 0·5–1·5 mm., ± ciliate (rarely with almost glabrous margins), otherwise glabrous or ± appressed-hairy on the surface beneath, apex obtuse to subacute, lateral nerves usually invisible or faintly apparent. Flowers cream to white, 0·25–0·4 mm. pedicellate, or appearing sessile, in spiciform racemes usually 4–10 cm. long on peduncles 0·5–2·5 cm. long; axis ± densely puberulous or pubescent. Calyx 1–1·7(2·5) mm. long, glabrous or slightly pubescent, rarely more than ⅓–½ as long as the corolla. Corolla 2·5–3 mm. long, with 5 lobes 0·5–0·8 mm. long. Stamen-filaments 3–6 mm. long, free; anthers 0·15 mm. across, with a caducous gland. Ovary pubescent, on a stipe longer than itself. Pods purple-brown to brown, dehiscent, 5–20 × 1–2·4 cm., linear-oblong, straight, very acuminate at both ends or sometimes merely subacute at the apex, puberulous or almost glabrous. Seeds 6–9 mm. in diam., subcircular-lenticular; central areole small, obscure, 2·5–3 × 2·5–3 mm.

Botswana. N: Xudum R., fl. 16.iii.1961, *Richards* 14745 (K; SRGH). SW: 80 km. N. of Kang, fl. 18.ii.1960, *Wild* 5068 (K; SRGH). SE: Tamasetzi, fr. 5.iii.1876, *Holub* (K). **Zambia.** B: Sesheke, fl. 26.i.1952, *White* 1971 (K). S: Zeze, Sinazongwe, fl. 29.xii.1958, *Robson* 990 (BM; K; LISC; SRGH). **Rhodesia.** N: between Macuti and Kariba, fl. iv.1960, *Brewer* 6891 (K; SRGH). W: Wankie Game Reserve, Dett road, fl. 22.ii.1956, *Wild* 4788 (K; SRGH). C: Chilimanzi Distr., Mtao, fl. 3.iii.1951, *Greenhow* 32/51 (K; SRGH). S: Nuanetsi, fr. 25.iv.1962, *Drummond* 7715 (SRGH). **Mozambique.** N: mouth of Messalo R., fl. ii.1912, *Allen* 132 (K). T: Tete, fr. 5.v.1948, *Mendonça* 4098 (LISC). SS: Massinga, fr. iv.1936, *Gomes e Sousa* 1725 (COI; K). LM: Namaacha, Libombo Mts., fr. 24.iv.1947, *Pedro & Pedrógão* 694 (LMJ; PRE).

From Senegal to the Sudan Republic in the N., extending southwards to SW. Africa, Swaziland, Natal and the Cape Province. Woodland, wooded grassland and thicket, sometimes by rivers, sometimes not; 100–1520 m.

Although many of the specimens from our area have the leaflets ± appressed-hairy both on the surface and margins and also the denser pubescence on the young stems, leaf-rhachides and inflorescence-axes characteristic of var. *australis* Burtt Davy, typical *A. ataxacantha* also occurs together with numerous intermediates which make almost impossible any clear subdivision of the species. The strongly developed indumentum of var. *australis* appears to occur only in southern Africa.

3. **Acacia chariessa** Milne-Redh. in Kew Bull. **1933**: 143 (1933).—Steedman, Trees etc. S. Rhod.: 15 (1933).—Wild, S. Rhod. Bot. Dict.: 47 (1953).—Boughey in Journ. S. Afr. Bot. **30**: 157 (1964). Type: Rhodesia, Bulawayo, *Borle* 13 (K, holotype).

Shrub 1–3 m. high, sometimes sprawling and forming thickets; young branchlets slender, glabrous, purplish. Stipules not spinescent, subulate-triangular. Prickles mostly scattered and single along the stems, rather small, to c. 6 mm. long, slightly arcuate-recurved; occasionally a few prickles tending to be grouped irregularly in pairs near the nodes. Leaves: rhachis 0·5–4·5 cm. long, unarmed or sometimes armed; gland on petiole often absent, sometimes a small gland towards the apex; glands also between the top 1–4 pairs of pinnae; pinnae 2–10 pairs, short, up to 1·9 cm. long; leaflets 14–32 pairs, imbricate, 1–3 × 0·5–0·7(1) mm., glabrous, rounded to subacute at the apex; midrib and lateral nerves obscure or only slightly prominent. Flowers white to yellow, sessile, in spikes 1·5–4 cm. long on peduncles (0·7)1·5–2·5(3) cm. long; axis glandular, otherwise glabrous or almost so. Calyx 1·5–2·25 mm. long, glabrous. Corolla 2·5–3 mm. long, glabrous. Stamen-filaments 5–6 mm. long; anthers 0·15–0·2 mm. across, with a caducous gland. Ovary glabrous, shortly stipitate. Pods brown or purplish, dehiscent, 3–6·5 (excluding stipe) × 1–1·6 cm., glabrous except for a few small glands (and occasionally some puberulence) towards the base, oblong to linear-oblong, straight, rounded to acuminate at the apex, attenuate at the base into a slender stipe 0·5–1·5 cm. long; margins usually ± irregularly sinuate or constricted. Seeds 5–8 mm. in diam., flattened; central areole small, 1·5–2 × 0·75–1·25 mm.

Rhodesia. W: Matopos Dam, fl. xii.1948, *Miller* B/817 (K; PRE). C: Charter Distr., Mhlaba Hills near Windsor Chrome Mine, fl. 16.i.1962, *Wild* 5602 (K; SRGH). S: Shabani Distr., near Mashaba, fl. 21.xi.1962, *Loveridge* 491 (K; SRGH).

Apparently confined to Rhodesia. With *Colophospermum mopane*, also *Acacia* and *Combretum*, almost always on serpentine soils; 1070–1520 m.

4. **Acacia eriocarpa** Brenan in Kew Bull. **12**: 360 (1957).—F. White, F.F.N.R.: 432 (1962). —Boughey in Journ. S. Afr. Bot. **30**: 157 (1964). TAB. **15** fig. 2. Type: Rhodesia, Chirundu, *Goodier* 81 (K, holotype; LISC; SRGH).

Shrub or small to medium tree c. 3–6 m. high, sometimes many-stemmed; young branchlets tomentose, soon glabrescent. Stipules not spinescent, 5–7 × 4–6 mm., broadly and obliquely ovate. Prickles scattered along the internodes, ± hooked or deflexed, up to c. 5 mm. long. Leaves: rhachis 4–19 cm. long, prickly or unarmed; petiole with a prominent gland 1–2·5 mm. long on the upper side near the base; rhachis glandular below the top 1–2 pairs of pinnae; pinnae 4–9 pairs; leaflets 4–15 pairs, (5)7–18 × 3–9 mm., obliquely oblong-elliptic or -lanceolate (or the terminal ones obovate), ± densely pubescent especially beneath when young, ± glabrescent at maturity; lateral nerves and usually venation also clearly visible beneath. Flowers sessile, in spikes 4–5 (? more) cm. long on peduncles 0·5–1·3 cm. long; axis tomentose. Calyx 3·5 mm. long, tomentose. Corolla 4–4·5 mm. long, with 5 lobes 1–1·25 mm. long which are tomentose outside. Stamen-filaments 5–6 mm. long, irregularly connate towards the base; anthers 0·2 mm. across, with a caducous gland. Ovary tomentose, on an almost equally long stipe. Pods dehiscent, 7–14 × 1·8–2·4 cm., with dense often ± matted long brownish tomentum outside, linear-oblong, straight, obtuse rarely subacute at the apex. Seeds subcircular, 6–10 mm. in diam.; central areole small, 1·5–3 × 2·5 mm.

Zambia. E: Luangwa R., fr. 5.vi.1958, *Fanshawe* 4533 (K; LISC). **Rhodesia.** N: Urungwe Distr., 16 km. S. of Chirundu, fr. 26.iii.1956, *Goodier* 63 (K; SRGH). W: Wankie, fl. 11.xii.1934, *Eyles* 8287 (K; SRGH). **Mozambique.** T: Mágué, between Mágué and R. Zambeze, fr. 30.iv.1964, *Wild* 6538 (K; SRGH).
Known only from our area. Woodland and thicket.

A. eriocarpa is unique among our spicate species in having tomentose pods.

5. **Acacia mellifera** (Vahl) Benth. in Hook., Lond. Journ. Bot. **1**: 507 (1842).—Oliv., F.T.A. **2**: 340 (1871).—Bak. f., Legum. Trop. Afr. **3**: 828 (1930).—Brenan, T.T.C.L.: 329 (1949); in Kew Bull. **11**: 191 (1956); F.T.E.A. Legum.-Mimos.: 84, fig. 14/8 (1959).—Torre, C.F.A. **2**: 273, t. 52 A (1956). Type from Arabia.
Mimosa mellifera Vahl, Symb. Bot. **2**: 103 (1791). Type as above.
Acacia senegal subsp. *mellifera* (Vahl) Roberty in Candollea, **11**: 153 (1948). Type as above.

Shrub or small tree 1–6(9) m. high; young branchlets pubescent or glabrous, grey-brown to purplish-black. Stipules not spinescent. Prickles in pairs just below each node, deep brown to blackish, hooked, 2·5–5(6) mm. long. Leaves: petiole usually glandular; rhachis glabrous to pubescent, frequently with a gland between the top 1–2 pairs of pinnae; pinnae 2–3, very rarely 4 pairs; leaflets 1–2 (very rarely 3) pairs, 3·5–22 × 2·5–16 mm., obliquely obovate to obovate-elliptic or -oblong, glabrous to pubescent, venose, rounded to emarginate or subacute and often apiculate at the apex. Flowers cream to white, on pedicels (0·5)0·75–1·5 mm. long in subglobose to ± elongate racemes; axis 0·15–3·5 cm. long, glabrous or sometimes pubescent; peduncle 0·4–1·3 cm. long. Calyx 0·6–1 mm. long, glabrous. Corolla 2·5–3·5 mm. long, 5-lobed. Stamen-filaments 4–6 mm. long, free; anthers 0·15–0·25 mm. across, with a caducous gland. Ovary glabrous; stipe very short. Pods pale brown to straw-coloured, (2·5)3·5–8(9) × 1·5–2·5(2·8) cm., dehiscent, glabrous, oblong, straight, venose, rounded to shortly and abruptly acuminate at the apex. Seeds 9–10 × 8 mm., subcircular-lenticular; central areole small, 2–3 × 2·5–3 mm., slightly impressed.

Typical subsp. *mellifera* has normally two pairs of pinnae per leaf and ± elongate racemes whose peduncles are 0·4–1·3 cm. long and usually shorter than the 0·5–3·5 cm. long inflorescence-axis. Subsp. *mellifera* is found in Arabia, in NE. Africa from Egypt to Tanzania, and in Angola.

Subsp. **detinens** (Burch.) Brenan in Kew Bull. **11**: 191 (1956); F.T.E.A. Legum.-Mimos.: 85 (1959).—Palmer & Pitman, Trees of S. Afr.: 159 cum fig. (1959).—F. White, F.F.N.R.: 82, fig. 17 D (1962).—Boughey in Journ. S. Afr. Bot. **30**: 158 (1964). Type from S. Africa (Prieska Division).
Acacia detinens Burch., Trav. Int. S. Afr. **1**: 310 (1822).—Warb., Kunene-Samb. Exped. Baum: 243 (1903).—Bak. f., Legum. Trop. Afr. **3**: 828 (1830).—Burtt Davy, F.P.F.T. **2**: 345 (1932).—Hutch., Botanist in S. Afr.: 175, 179, 543, 631 (1946).—O. B. Mill., B.C.L.: 17 (1948); in Journ. S. Afr. Bot. **18**: 20 (1952).—Wild, S. Rhod. Bot. Dict.: 47 (1953).—Torre, C.F.A. **2**: 273, t. 52 B (1956).—Story, in Mem. Bot. Surv. S. Afr. **30**: 22 (1958). Type as above.

Pinnae normally in 3, rarely 4 pairs. Racemes very short or subglobose, their peduncles 0·4–1·1 cm. long and normally exceeding the very short (1·6–6·5 mm. long) axis.

Caprivi Strip. Linyanti, fr. 28.xii.1958, *Killick & Leistner* 3165 (K; PRE). **Botswana.** N: Ngamiland, fl. 10.viii.1897, fr. 22.ix.1897, leaf 1.i.1898, *Lugard* 13 (K). SW: 32 km. N. of Wenda, st. 26.ii.1960, *Wild* 5164 (K: SRGH). SE: c. 10 km. E. of Pharing, fr. 14.xi.1948, *Hillary & Robertson* 535 (K; PRE). **Zambia.** B: Machili, fr. 6.x.1960, *Fanshawe* 5826 (K). N: Mwunyamadzi R., fr. 5.x.1933, *Michelmore* 628 (K). S: between Katombora and Kasungula, fl. 25.viii.1947, *Brenan & Greenway* 7749 (K). **Rhodesia.** N: Lake Kariba, fr. viii.1960, *Goldsmith* 96/60 (K; SRGH). W: Nyamandhlovu, fl. 12.ix.1953, *Plowes* 1628 (K; SRGH). **Mozambique.** T: Tete, Boroma, fl. & fr. 25.vii.1950, *Chase* 2806 (BM; K; LISC; SRGH).

Also in Tanzania, Angola, Transvaal and SW. Africa. In dry bushland or bush-savanna, sometimes marginal to mopane; 210–1370 m.

6. **Acacia galpinii** Burtt Davy in Kew Bull. **1922**: 326 (1922); F.P.F.T. **2**: 337 (1932).—Steedman, Trees etc. S. Rhod.: 13 (1933).—O. B. Mill., B.C.L.: 18 (1948); in Journ. S. Afr. Bot. **18**: 20 (1952).—Pardy in Rhod. Agric. Journ. **49**: 12 cum photogr. (1952).—Wild, S. Rhod. Bot. Dict.: 47 (1953).—Young in Candollea, **15**: 97 (1955).—Palgrave, Trees of Central Afr.: 239, cum tab. et photogr. (1956).—Burtt Davy & Hoyle, rev. Topham, N.C.L. ed. 2: 64 (1958).—Brenan, F.T.E.A. Legum.-Mimos.: 87 (1959).—Palmer & Pitman, Trees of S. Afr.: 151, cum fig. et photogr. 16, 30 (1961).—F. White, F.F.N.R.: 83, fig. 17 F (1962).—Mitchell in Puku, **1**: 104 (1963).—Boughey in Journ. S. Afr. Bot. **30**: 157 (1964).—de Winter, de Winter & Killick, Sixty-six Transvaal Trees: 46 cum photogr. (1966). Type from the Transvaal.

Acacia caffra sensu Oliv., F.T.A. **2**: 345 (1871) tantum quoad specim. *McCabe* pro parte.—Bak. f., Legum. Trop. Afr. **3**: 833 (1930) etiam pro parte ut praec.

Acacia senegal sensu O. B. Mill., B.C.L.: 21 (1948), vide O. B. Mill. in Journ. S. Afr. Bot. **18**: 20 (1952).

Tree 8–25 m. high; bark rough, corky, longitudinally furrowed with fibrous strips coming away here and there; young branchlets subglabrous to ± densely short-pubescent. Stipules not spinescent. Prickles in pairs just below the nodes, straight or recurved, up to c. 1 cm. long, on some twigs few or even apparently absent. Leaves: petiole often glandular (gland 0·75–1·5 mm. in diam.); rhachis subglabrous to ± puberulous or pubescent, glandular (sometimes interruptedly) between the top 1–4 pairs of pinnae, sometimes with glands at some basal pairs also; pinnae (4)9–14 (? more) pairs; leaflets (8)13–35(45) pairs, (2)4–11(15) × (0·5)1–3(4) mm., narrowly oblong to linear-oblong, slightly ciliolate on the margins and at the base, otherwise glabrous, or wholly glabrous, obtuse to subacute at the apex, with lateral nerves almost or quite invisible beneath. Spikes (4)5–11 cm. long, often clustered or several together on short lateral leafless shoots from twigs of the previous year, sometimes opening when the tree is leafless; peduncle 0·3–1·5 cm. long. Flowers sessile. Calyx purple or reddish-purple, 0·75–1·25 mm. long, cupular, ± puberulous outside. Corolla coloured as the calyx, 2 mm. long, 5-lobed, ± puberulous outside. Stamen-filaments 4–5 mm. long; anthers 0·15–0·2 mm. across, glandular at the apex. Ovary glabrous, on a stipe half its length. Pods purplish-brown, dehiscent, 11·5–28 × 2·7–3·5 cm., straight, glabrous or almost so; valves thinly woody. Seeds 12–15 × 10–12 mm.; central areole large, 7–8 × 3·5–5 mm.

Botswana. N: 10 km. NE. of Maun, fr. vi.1946, *Miller* B/441 (PRE). SE: Serowe, at Kgotla, fr. *Miller* B/278 (PRE). **Zambia.** B: Masese, fl., *Fanshawe* 6694 (K; LISC). C: Great North Road, Chisamba–Broken Hill, S. of Mumbwa turn-off, fl. 13.ix.1947, *Trapnell* in *Brenan* 7857 (K). S: Namwala, fl. 18.x.1959, *Fanshawe* 5247 (K). **Rhodesia.** N: Mazoe Distr., Chipoli, fl. 20.ix.1958, *Moubray* 17 (SRGH). W: Bulawayo, fl. & fr. x.1930, *Eyles* 6622 (K; SRGH). C: Salisbury, fl. & fr. 4.x.1936, *Eyles* 8791 (K; SRGH). E: Umtali Commonage, Lynwoods Farm, fl. ix.1948, *Chase* 983 (K; LISC; SRGH). S: Bikita Distr., Marangaranga, fl. 5.x.1955, *Chase* 5725 (BM; COI; K; LISC; SRGH). **Malawi.** N: Rumpi Distr., Deep Bay, Lake Nyasa, immat. fr., *Chapman* 2819 (SRGH). C: Lilongwe, fr. 5.vii.1960, *Chapman* 800 (BM; SRGH). S: Cholo, Sankulwani Station, fr. vii.1943, *Hornby* 2886 (PRE). **Mozambique.** Z: Massingire (Morrumbala), near M'bôbo Hospital, st. 4.viii.1942, *Torre* 4496 (BM; K; LISC). T: R. Mudzi, 16 km. from Rhodesian border, fl. 26.ix.1948, *Wild* 2640 (K; SRGH). MS: Matondo, fr. 3.vii.1947, *Simão* 1343 (LISC; LM).

Also in Tanzania and the Transvaal. Woodland and wooded grassland, particularly, though not always, by rivers; 360–1490 m.

The purple calyces and corollas are unusual among the African species of *Acacia*. As pointed out in F.T.E.A. Legum.-Mimos.: 87 (1959), *A. galpinii* is more clearly and closely related to *A. persicifolia* Pax, a species occurring from Kenya and Uganda northward, than to any other species in southern Africa.

When leafless, flowering *A. galpinii* may be hard to separate from *A. nigrescens* in the same condition. The differences are given under the latter species (p. 71).

7. **Acacia nigrescens** Oliv., F.T.A. **2**: 340 (1871).—Burkill in Johnston, Brit. Centr. Afr.: 245 (1897).—Sim, For. Fl. Port. E. Afr.: 54, t. 33 (1909).—Eyles in Trans. Roy. Soc. S. Afr. **5**: 362 (1916).—Bak. f., Legum. Trop. Afr. **3**: 829 (1930).—Steedman, Trees etc. S. Rhod.: 13 (1933).—O. B. Mill., B.C.L.: 20 (1948); in Journ. S. Afr. Bot. **18**: 23 (1952).—Brenan, T.T.C.L.: 329 (1949); in Mem. N.Y. Bot. Gard. **8**: 429 (1954); F.T.E.A. Legum.-Mimos.: 85, fig. 14/9 (1959).—Codd, Trees & Shrubs Kruger Nat. Park: 47, figs. 40–43 (1951).—Wild, Guide Fl. Vict. Falls: 148 (1953); S. Rhod. Bot. Dict.: 48 (1953).—Pardy in Rhod. Agric. Journ. **51**: 173, cum photogr. (1954).—Young in Candollea, **15**: 119 (1955).—Palgrave, Trees & Shrubs of Central Afr.: 253, cum tab. et photogr. (1956).—Torre, C.F.A. **2**: 274 (1956).—Burtt Davy & Hoyle, rev. Topham, N.C.L., ed. 2: 64 (1958).—Cardoso in Moçamb. Publ., sér. A, **4**: 1–43 (1960).—Palmer & Pitman, Trees of S. Afr.: 161, cum fig. (1961).—F. White, F.F.N.R.: 82, fig. 17 E (1962).—Mitchell in Puku, **1**: 104 (1963).—Boughey in Journ. S. Afr. Bot. **30**: 158 (1964).—de Winter, de Winter & Killick, Sixty-six Transvaal Trees: 50 cum photogr. (1966).—Gomes e Sousa, Dendrol. Moçamb. Estudo Geral, **1**: 233, t. 37 (1966).—J. Ross in Bol. Soc. Brot., Sér. 2, **42**: 11 (1968). TAB. **15** fig. 3. Type: Malawi, near Mitonda, *Kirk* (K, holotype).

Acacia caffra sensu Oliv., F.T.A. **2**: 345 (1871) pro parte, tantum quoad specim. *McCabe* pro parte.—Bak. f., Legum. Trop. Afr. **3**: 833 (1930) etiam pro parte ut praec.

Acacia nigrescens var. *pallens* Benth. in Trans. Linn. Soc. **30**: 517 (1875).—Young in Candollea, **15**: 119 (1955). Type: Mozambique, near Sena, *Kirk* 201 (K, holotype).

Acacia perrotii Warb. in Notizbl. Bot. Gart. Berl. **2**: 249 (1898). Type from Tanzania (Lindi).

Acacia passargei Harms in Passarge, Kalahari: 789 (1904); in Engl., Pflanzenw. Afr. **3**, 1: 384 (1915). Type presumably from Botswana and collected by Passarge, but details uncertain (? B, holotype †).

Acacia pallens (Benth.) Rolfe in Kew Bull. **1907**: 361 (1907).—Bak. f., Legum. Trop. Afr. **3**: 829 (1930).—Burtt Davy, F.P.F.T. **2**: 339, fig. 57 (1932).—Gomes e Sousa, Dendrol. Moçamb. **2**: 49, cum tab. (1949). Type as for *A. nigrescens* var. *pallens*.

Albizia lugardii N.E.Br. in Kew Bull. **1909**: 169 (1909). Type: Botswana, Ngamiland, Okavango Valley, *Lugard* 246 (K, holotype).

Acacia nigrescens var. *pallida* Eyles in Trans. Roy. Soc. S. Afr. **5**: 362 (1916) *nom. nud*. Probably a mistake for var. *pallens* (vide supra).

Acacia schliebenii Harms in Notizbl. Bot. Gart. Berl. **12**: 507 (1935).—Brenan, T.T.C.L.: 329 (1949). Type from Tanzania (Lindi Distr.).

Acacia nigrescens var. *nigrescens*.—Young in Candollea, **15**: 119 (1955).

Tree 3–30 m. high; trunk usually ± beset with knobby prickles; young branchlets glabrous to sometimes pubescent. Stipules not spinescent. Prickles in pairs just below each node, hooked, blackish, persistent, 2·5–7 mm. long (on branchlets). Leaves: petiole glandular or not; rhachis glabrous to pubescent, sometimes with a gland between the top 1–2 pairs of pinnae; pinnae 2–4 pairs; leaflets normally 1–2 pairs, (6·5)10–35(50) × (5·3)7–30(49·8) mm., obliquely obovate-orbicular to broadly obovate-elliptic, glabrous to sometimes pubescent, venose, subcoriaceous, apex rounded and often emarginate. Flowers white or cream, sessile, in ± aggregate or solitary spikes 1–10(12) cm. long on peduncles 0·3–2·4 cm. long; axis glabrous except for minute sessile glands, or sometimes pubescent. Calyx 1·5–2 mm. long, glabrous. Corolla 2–2·5 mm. long, 5-lobed. Stamen-filaments 3·5–6 mm. long, free: anthers 0·1 mm. across, with a caducous gland. Ovary glabrous, very shortly stipitate. Pods darkish brown, dehiscent, 6·1–17·8 × 1·4–2·4(2·7) cm., glabrous, oblong, straight or nearly so, hardly venose, acuminate at the apex. Seeds 12–13 mm. in diam., subcircular-lenticular; central areole large, 6–8 × 6–8 mm., somewhat impressed.

Caprivi Strip. E. of the Cuando R., st. x.1945, *Curson* 1182 (PRE). **Botswana.** N: Chobe Concessions, st. 28.vii.1950, *Robertson & Elffers* 64 (K; PRE). SE: Mahalapye,

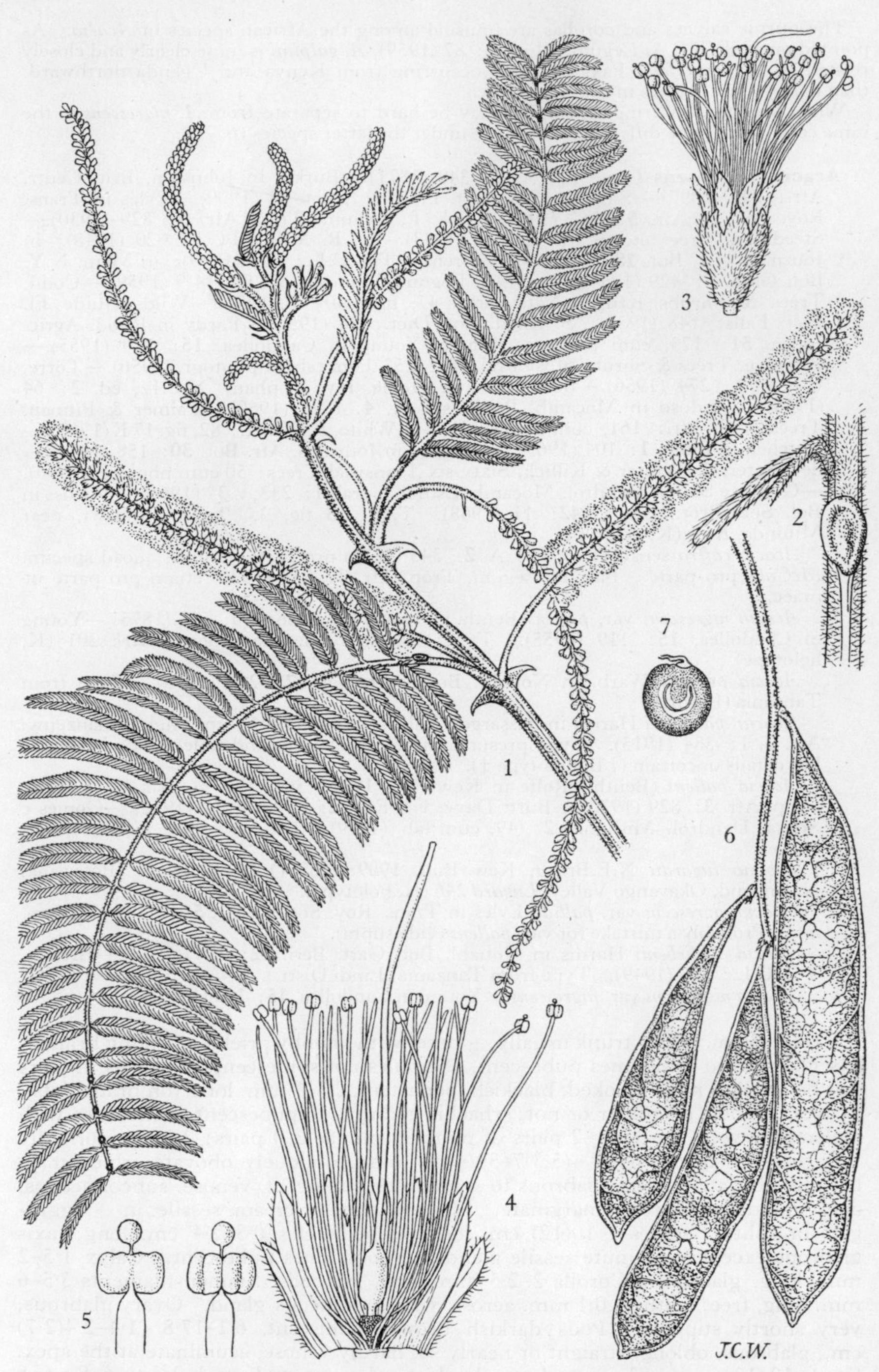

Tab. 18. ACACIA POLYACANTHA SUBSP. CAMPYLACANTHA. 1, flowering branch ($\times\frac{2}{3}$); 2, gland on petiole ($\times 4$); 3, flower ($\times 6$); 4, flower, opened out to show ovary ($\times 6$); 5, anthers ($\times 46$), all from *Lusaka Natural History Club* 172; 6, pods ($\times\frac{2}{3}$); 7, seed ($\times\frac{2}{3}$), all from *Gilliland* 199.

fl. 15.x.1959, *de Beer* 782 (K; SRGH). **Zambia.** B: Masese, fr. 24.v.1962, *Fanshawe* 6839 (K). S: Monze, fl. 3.x.1963, *Van Rensburg* K.B.S. 2525 (K). **Rhodesia.** N: Darwin Distr., Chimanda Reserve near Winda Pools, fl. 4.ix.1958, *Phipps* 1299 (K; SRGH). W: Nyamandhlovu, fl. 25.ix.1953, *Plowes* 1632 (K; SRGH). E: Chipinga, fl. x.1959, *Soane* 66 (SRGH). S: Beitbridge, between Customs Post and Limpopo R., fr. 25.iii.1959, *Drummond* 6010 (K; LM; SRGH). **Malawi.** N: Nyungwe, fl. 10.x.1930, *Migeod* 975 (BM). C: Chitala to Domera Bay, fl. 29.x.1941, *Greenway* 6377 (K; PRE). S: Chikwawa Distr., Lower Mwanza R., fl. 4.x.1946, *Brass* 17951 (K; PRE; SRGH). **Mozambique.** N: Nacala, between Nacala and Fernão Veloso, fl. 14.x.1948, *Barbosa* 2406 (LISC; LM). Z: Mocuba, Namagoa, fl. ix.1945, *Faulkner* Pretoria No. 326 (K; LM; PRE; SRGH). T: Tete, Boroma, 11 km. from Msusa on Tete road, fl. 22.vii.1950, *Chase* 2693 (BM; COI; K; LISC; SRGH). MS: between Quicuaca and Machimeja, fl. 2.ix.1942, *Mendonça* 131 (LISC). SS: between Vilanculos and Mabote, fl. 1.ix.1944, *Mendonça* 1941 (BM; K; LISC). LM: Maputo, fr. 11.iii.1947, *Hornby* 2593 (K; PRE; SRGH).

From Tanzania southwards to Natal. In woodland and wooded grassland, particularly on alluvial soils by rivers and lakes; 40–1610 m.

This easily recognized species shows comparatively little variation. It is generally glabrous, but is occasionally puberulous or even quite densely pubescent. The number of leaflets per pinna is usually only two, but some specimens show an inconstant tendency, though quite probably genetically controlled, to produce four. The characteristic raised knobs on the trunk are evidently variable in their occurrence.

A. nigrescens frequently flowers when leafless and may then be difficult to distinguish from *A. galpinii*, which often does the same. *A. galpinii* may then be told by its shorter (0·75–1·25 mm. long) ± puberulous calyx and by the corolla-lobes being ± puberulous outside. In *A. nigrescens* the calyx is 1·5–2 mm. long and glabrous outside, as are the corolla-lobes.

The wood of *A. nigrescens* is hard, heavy and durable.

8. **Acacia polyacantha** Willd. in L., Sp. Pl. ed. 4, **4**: 1079 (1806).—Brenan in Kew Bull. **11**: 195 (1956). Type from India.

Tree up to 20(25) m. high, rarely shrubby; trunk with fissured bark and knobby persistent prickles; young branchlets pubescent or puberulous, rarely subglabrous, grey to brown. Stipules not spinescent. Prickles in pairs just below each node, straw-coloured to brown or blackish, 4–12 mm. long. Leaves: petiole glandular (gland usually 2–5 × 1·75–3 mm.); rhachis pubescent or puberulous, rarely subglabrous, glandular between the top 3–17 pairs of pinnae; pinnae (6)13–40(60) pairs: leaflets (15)26–66 pairs, 2–5(6) × 0·4–0·75(1·25) mm., linear to linear-triangular, pubescent usually only on the margins, only the midrib (and sometimes some very small basal nerves) visible, subacute to narrowly obtuse at the apex. Flowers cream or white, sessile or nearly so, in spikes (3·5)6–12·5 cm. long, produced with the new leaves; axis densely pubescent or tomentellous; peduncle (0·5)1·2–2 cm. long. Calyx 1·7–2·25 mm. long, pubescent or puberulous, rarely puberulous on the lobes only or subglabrous. Corolla 2–3 mm. long, 5-lobed, usually $1\frac{1}{3}$ times or more as long as the calyx. Stamen-filaments 4·5–6 mm. long; anthers 0·1 mm. across, with a caducous gland. Ovary glabrous; stipe very short. Pods brown, dehiscent, 7–18 × 1–2·1 cm., oblong, straight, venose, usually acuminate at the apex, glabrous or nearly so, rarely ± pubescent. Seeds subcircular to elliptic-lenticular, 8–9 × 7–8 mm.; central areole medium to small, 3–4 × 3–3·5 mm., not impressed.

Typical subsp. *polyacantha*, with prickles straight or almost so, is known only from India and (probably) Ceylon.

Subsp. **campylacantha** (Hochst. ex A. Rich.) Brenan in Kew Bull. **11**: 195 (1956); F.T.E.A. Legum.-Mimos.: 88, fig. 14/12 (1959).—Mitchell in Puku, **1**: 104 (1963). —Boughey in Journ. S. Afr. Bot. **30**: 158 (1964). TAB. **15**, fig. 4, **18.** Syntypes from Ethiopia.

Acacia campylacantha Hochst. ex A. Rich., Tent. Fl. Abyss. **1**: 242 (1847).—R.E.Fr., Wiss. Ergebn. Schwed. Rhod.-Kongo-Exped. **1**: 64 (1914).—Bak. f., Legum. Trop. Afr. **3**: 831 (1930).—Burtt Davy, F.P.F.T. **2**: 337 (1932).—O. B. Mill., B.C.L.: 17 (1948); in Journ. S. Afr. Bot. **18**: 19 (1952).—Brenan, T.T.C.L.: 331 (1949); in Mem. N.Y. Bot. Gard. **8**: 429 (1954).—Codd, Trees & Shrubs Kruger Nat. Park: 42, fig. 35, 37 e–f (1951).—Pardy in Rhod. Agric. Journ. **48**: 404 cum photogr. (1951).—Wild, Guide Fl. Vict. Falls: 148 (1953); S. Rhod. Bot. Dict.: 46 (1953).—Young in Candollea, **15**: 99 (1955).—Palgrave, Trees of Central Afr.: 235, cum tab. et photogr. (1956).—Torre, C.F.A. **2**: 276 (1956).—Burtt Davy & Hoyle, rev. Topham, N.C.L. ed. 2: 64 (1958).—Fanshawe, Fifty Common Trees N.

Rhod.: 6, cum tab. (1962).—F. White, F.F.N.R.: 83, fig. 17 H (1962). Syntypes as above.

Acacia catechu sensu Oliv., F.T.A. **2**: 344 (1871).—Harms in Warb., Kunene-Samb.-Exped. Baum: 243 (1903).—Sim, For. Fl. Port. E. Afr.: 56 (1909).

Acacia caffra sensu Oliv., tom. cit.: 345 (1871) pro parte quoad specim. angol.—Bak. f. in Journ. Linn. Soc., Bot. **40**: 64 (1911); Legum. Trop. Afr. **3**: 833 (1930) quoad specim. angol.—Eyles in Trans. Roy. Soc. S. Afr. **5**: 361 (1916).

Acacia suma sensu Eyles, tom. cit.: 362 (1916).

Acacia caffra var. *tomentosa* sensu Bak. f., Legum. Trop. Afr. **3**: 833 (1930) pro parte quoad specim. *Swynnerton* 61.

Acacia pallens sensu Steedman, Trees etc. S. Rhod.: 14, t. 8 (1933).

Acacia catechu subsp. *suma* var. *campylacantha* (Hochst. ex A. Rich.) Roberty in Candollea, **11**: 157 (1948). Types as above.

Acacia caffra var. *campylacantha* (Hochst. ex A. Rich.) Aubrév., Fl. Forest. Soud.-Guin.: 272 (1950).—Gilbert & Boutique, F.C.B. **3**: 150 (1952). Types as above.

Bark whitish to yellowish or grey. Prickles ± hooked.

Botswana. N: Ngamiland, fl. xii.1930, *Curson* 548 (PRE). **Zambia.** B: Zambezi R. at Shingi, 11 km. N. of Chavuma, fl. 14.x.1952, *Angus* 635 (BM; K; PRE). N: Mutinondo R., fr. 7.vi.1957, *Savory* 185 (SRGH). W: Solwezi Distr., Mehela R., fr. 21.vii.1930, *Milne-Redhead* 743 (K). C: Mt. Makulu, fl. 17.xi.1956, *Angus* 1443 (BM; K; PRE). E: Petauke Distr., between Changwe and Luangwa, fl. 16.xii.1958, *Robson* 963 (BM; K; LISC; SRGH). S: Katombora, by Zambezi R., fl. & fr. 25.xi.1949, *Wild* 3212 (K; LISC; SRGH). **Rhodesia.** N: Shamva, Walwyn Farm, fr. 21.ii.1956, *Guy* 3/56 (K; LM; SRGH). W: Shangani Reserve, fl. iii.1949, *Davies* 1 (SRGH). C: Salisbury, fl. & fr. 16.xii.1921, *Eyles* 3233 (K; SRGH). E: Umtali, fl. 4.xi.1954, *Chase* 5320 (BM; K; LISC; SRGH). S: Chitsa's Kraal, fr. 8.vi.1950, *Wild* 3381 (K; LISC; SRGH). **Malawi.** N: c. 13 km. S. of Rukuru R., fr. 9.vi.1938, *Pole Evans & Erens* 680 (K; PRE). C: Kota Kota Distr., Nchisi, fr. viii.1946, *Brass* 17116 (BM; K; PRE; SRGH). S: Mlanje Mt., fl. 18.x.1957, *Chapman* 476 (BM; K). **Mozambique.** N: Amaramba, between Missão de Mepanhira and Mecanhelas, fl. 20.x.1943, *Andrada* 1429 (COI; LISC). Z: Mocuba, fl. 1945, *Faulkner* PRE 374 (K; LM; PRE; SRGH). T: Mutarara, between Ancuaze and Doa, fr. 21.vi.1949, *Barbosa & Carvalho* in *Barbosa* 3201 (K; LMJ). MS: Báruè, between Mungári and Vila Gouveia, fl. 30.x.1941, *Torre* 3722 (BM; K; LISC).

Widespread in tropical Africa from the Gambia and Ethiopia (Eritrea) to the Transvaal. In wooded grassland and woodland, usually on colluvial or alluvial clays and loams by rivers and streams; 50–1460 m.

Although the pinnae range from 6–60 pairs per leaf, the larger leaves on all specimens are likely to have 15–20 or more pairs. A variant lacking prickles has been recorded in East Africa and should be looked for in our area.

9. **Acacia caffra** (Thunb.) Willd. in L., Sp. Pl. ed. 4, **4**: 1078 (1806).—Burtt Davy, F.P.F.T. **2**: 337, fig. 55 (1932).—O. B. Mill., B.C.L.: 17 (1948); in Journ. S. Afr. Bot. **18**: 19 (1952).—Codd, Trees & Shrubs Kruger Nat. Park: 42 (1951).—Young in Candollea, **15**: 102 (1955).—Brenan in Kew Bull. **11**: 193 (1956).—Palmer & Pitman, Trees of S. Afr.: 150 cum fig., t. 4 (1961).—Boughey in Journ. S. Afr. Bot. **30**: 157 (1964).—de Winter, de Winter & Killick, Sixty-six Transvaal Trees: 44 cum photogr. (1966). Type from S. Africa.

Mimosa caffra Thunb., Prodr. Pl. Cap. **2**: 92 (1800). Type as above.

Acacia caffra var. *rupestris* Sim, For. Fl. Port. E. Afr.: 56 (1909). Type: Mozambique, Marracuene and Lourenço Marques, *Sim* 6235 (not seen, location doubtful).

Acacia caffra var. *tomentosa* Glover in Ann. Bolus Herb. **1**: 146 (1915).—Young in Candollea, **15**: 107 (1955). Syntypes from S. Africa (Natal and Transvaal).

Acacia mellei sensu O. B. Mill. in Journ. S. Afr. Bot. **18**: 23 (1952) pro parte saltem quoad specim. *Miller* B/950.

A shrub or tree 2–10 m. high, not climbing; bark rough, scaly, dark-grey to brown; young branchlets varying from subglabrous with a little puberulence to densely spreading-pubescent or even (though not in our area) tomentose. Stipules not spinescent, subulate-linear. Prickles in pairs just below the nodes, patent to ± hooked, up to 7 mm. long, often sparse; occasionally with a few additional prickles scattered elsewhere on the stem. Leaves: petiole (1)1·3–3·5 cm. long; rhachis 3–19 cm. long, prickly or unarmed; gland on petiole present or absent, 0·75–1·75 mm. long, also with glands between top 1–4 pairs of pinnae; pinnae (4)6–27 pairs; leaflets 13–57 pairs, 2–7(10) × 0·6–1·5(2·5) mm., glabrous or with some appressed or spreading cilia on the margins or (but not in our area) densely hairy on the margins and ± so on the lower surface, rounded to obtuse at the apex, with midrib and often also lateral nerves usually rather prominent beneath. Flowers creamy-yellow to white, sessile, in spikes (2)4–10 cm. long on peduncles (0·5)1–3 cm. long; axis ± densely puberulous to pubescent or tomentose, also glandular. Calyx 2–3 mm. long, rather densely puberulous to pubescent, $\frac{2}{3}$–$\frac{5}{6}$ as long as the corolla. Corolla 2·5–3·5 mm. long, ± puberulous to pubescent on the outside of the lobes. Stamen-filaments 6–7 mm. long; anthers 0·15 mm. across, with a caducous gland. Ovary glabrous, shortly stipitate. Pods brown, dehiscent, 6–17 × 0·9–1·5(1·9) cm., puberulous, rarely pubescent, and with sparse to dense reddish-brown glands over the surface, linear, straight to ± curved, acute to attenuate at the base and apex. Seeds 8–10 × 7–8 mm., flattened; central areole small, 2–3 × 2–2·7 mm.

Botswana. SE: Kgatla Distr., Sikwani, fl. 13.x.1955, *Reyneke* 420 (K; PRE). **Mozambique.** LM: between Boane and Impamputo, fl. 8.xi.1961, *Lemos & Balsinhas* in *Lemos* 213 (BM; COI; K; LISC; LMJ; PRE; SRGH).

Also in S. Africa and Swaziland. Ecology uncertain in our area: recorded from woodland and dry rocky hills at c. 1070–1220 m.

A. caffra only just reaches the southernmost parts of our area. In S. Africa it becomes very variable, particularly in indumentum but also in number of pinnae and size of leaflets. Some varieties have been described but they do not appear to correspond with any clearly defined taxonomic entities.

A. caffra is normally clearly separated from *A. ataxacantha* by having paired not scattered prickles, but when, as occasionally happens, a few scattered prickles are additionally present, confusion is possible, though *A. caffra* can always be recognized by the glabrous not pubescent ovary and by the calyx being considerably larger in proportion to the corolla.

10. **Acacia hereroensis** Engl., Bot. Jahrb. **10**: 20 (1888).—Bak. f., Legum. Trop. Afr. **3**: 835 (1930).—O. B. Mill., B.C.L.: 19 (1948).—J. Ross in Journ. S. Afr. Bot. **31**: 220 (1965). Type from SW. Africa.
Acacia mellei Verdoorn in Fl. Pl. S. Afr.: t. 860 (1942).—O. B. Mill., B.C.L.: 20 (1948): in Journ. S. Afr. Bot. **18**: 23 (1952) pro parte saltem excl. specim. *Miller* B 950.—Young in Candollea, **15**: 109 (1955).—Brenan in Kew Bull. **11**: 197 (1956). Type from the Transvaal.

A shrub or tree 1–10 m. high, not climbing; young branchlets densely short- to long-pubescent and with many red-brown glands. Stipules not spinescent. Prickles in pairs just below the nodes, ± downwardly hooked, up to 5(7) mm. long. Leaves: petiole 3–13 mm. long; rhachis 2·5–5 cm. long, unarmed or rarely sparsely prickly; gland on petiole 0·5–1 mm. long, sometimes absent; glands also between top 1–5(7) pairs of pinnae; pinnae 8–20 pairs; leaflets 15–32 pairs, 1–4 × 0·25–1 mm., pubescent on the margins and lower surface or near the margins only, obtuse to rounded at the apex, with midrib and lateral nerves invisible or obscure beneath. Flowers white to cream, sessile, in spikes 3–7 cm. long on peduncles 0·5–2·5 cm. long; axis densely pubescent, also glandular. Calyx 2–2·75 mm. long, densely short-pubescent outside. Corolla 2·5–3·5 mm. long, appressed-pubescent on the outside of the lobes. Stamen-filaments 7–7·5 mm. long; anthers 0·15 mm. across, with a caducous gland. Ovary glabrous, shortly stipitate. Pods brown, dehiscent, 5(6)–14 × (1)1·4–2·3 cm., ± densely pubescent and with ± numerous brown glands, linear-oblong, straight, ± acuminate to acute at the base and apex. Seeds 7–8·5 × 5–8·5 mm., flattened; central areole small, 1–1·5 × 1·5 mm.

Botswana. SE: S. of Lobatsi, fr. 10.iv.1931, *Pole Evans* 3152 (2) (K; PRE). **Rhodesia.** W: Matopos Research Station, fl. 25.i.1952, *Plowes* 1405 (SRGH).

Also in S. Africa and SW. Africa. Ecology uncertain, probably in dry habitats.

This is extremely close to *A. caffra*, differing in the usually shorter petiole and in having the lateral nerves invisible or obscure on the lower surface of the leaflet.

11. **Acacia erubescens** Welw. ex. Oliv., F.T.A. **2**: 343 (1871).—Bak. f., Legum. Trop. Afr. **3**: 830 (1930).—O. B. Mill. in Journ. S. Afr. Bot. **18**: 20 (1952).—Wild, S. Rhod. Bot. Dict.: 47 (1953).—Young in Candollea, **15**: 111 (1955).—Torre, C.F.A. **2**: 276, t. 53 B (1956).—Burtt Davy & Hoyle, rev. Topham, N.C.L. ed. 2: 64 (1958). —Brenan, F.T.E.A. Legum.-Mimos.: 88, fig. 14/13 (1959).—F. White, F.F.N.R.: 83, fig. 17 G (1962).—Mitchell in Puku, **1**: 103 (1963).—Boughey in Journ. S. Afr. Bot. **30**: 157 (1964). Type from SW. Africa.

Acacia dulcis Marl. & Engl. in Engl., Bot. Jahrb. **10**: 24 (1888).—Bak. f., Legum. Trop. Afr. **3**: 830 (1930).—Burtt Davy, F.P.F.T. **2**: 337 (1932).—O. B. Mill., B.C.L.: 17 (1948); in Journ. S. Afr. Bot. **18**: 20 (1952).—Codd, Trees & Shrubs Kruger Nat. Park: 42, fig. 37a–b (1951).—Gilbert & Boutique, F.C.B. **3**: 151 (1952).—Story in Mem. Bot. Surv. S. Afr. **30**: 22 (1958). Type from SW. Africa.

Acacia kwebensis N.E.Br. in Kew Bull. **1909**: 108 (1909). Type: Botswana, Kwebe Hills, *Mrs. Lugard* 24 (K, holotype).

Shrub or tree 2–10 m. high; bark grey to yellowish or whitish, with papery flaking or peeling outer layer; young branchlets ± pubescent. Stipules not spinescent. Prickles in pairs just below the nodes, brown or grey, up to 4(6) mm. long, hooked. Leaves: petiole (0·7)1·3–2·5(3) cm. long, glandular or not, gland small, 0·4–1 mm. long; rhachis pubescent; glands variable, either between each pair of pinnae, or absent from some, or between the top pair only; pinnae 3–7 pairs; leaflets 10–27 pairs, 3–8(10) × (0·75)1–2(3) mm., obliquely oblong, often slightly falcate or the upper somewhat obovate, slightly pubescent especially on the margins, or becoming glabrous, veins somewhat prominent at first beneath, becoming obscure as the leaves age, apex usually oblique, acute or subacute, occasionally obtuse. Flowers white or cream with pink tinge, sessile, in spikes 2–4·5 cm. long on peduncles 0·7–2·4 cm. long; axis pubescent. Calyx 2·25–4·5 mm. long, densely pubescent. Corolla 2·5–6·5 mm. long, 5-lobed, ± densely appressed-pubescent on the lobes outside. Stamen-filaments 6–10 mm. long; anthers 0·2–0·25 mm. across, glandular at the apex. Ovary glabrous; stipe very short. Pods brown or deep brown, dehiscent, 3–13 × 1·2–1·9 cm., subglabrous except for pubescent margins and stipe, linear-oblong, straight, venose, coriaceous, rounded to acute, rarely acuminate at the apex; glands usually few and inconspicuous. Seeds 7–10 × 8–11 mm., often slightly wider than long; areole 1–2 × 2–4·5 mm., small and wide.

Botswana. N: Toteng, NE. tip of Lake Ngami, fl. 12.ix.1954, *Story* 4659 (K; PRE). SE: Metsimaklaba, fr. 11.iv.1931, *Pole Evans* 3154 (4) (K; PRE). **Zambia.** N: Lamyas village, Mwunyamadzi R., fl. & fr. 30.ix.1933, *Michelmore* 623 (K). S: Namwala, fr. 30.iii.1962, *Mitchell* 13/67 (K; SRGH). **Rhodesia.** N: Kariba, fl. 26.viii.1956, *Armitage* 228/56 (SRGH). W: Bulalima Mangwe Distr., Brunapeg near Mphoengs, fr. 3.vii.1962, *Wild* 5839 (K; SRGH). E: Lower Sabi, Mtema, fr. 28.i.1948, *Wild* 2404 (K). S: Gwanda, fl. & fr. 6.x.1959, *Kennan* in *M.R.S.H.* 3019 (K; LISC; SRGH). **Malawi.** N: 21 km. NE. of Rumpi, fr. 2.vii.1953, *Langdale-Brown* 91 (EA). **Mozambique.** SS: Guijá, between Mucatine and Munhamane, fr. 25.vii.1945, *Pedro & Pedrógão* 2054 (LMJ; PRE).

Also in the Congo, Tanzania, Angola, SW. Africa and the Transvaal. In drier types of woodland, often with *Colophospermum mopane*; 240–1070 m.

A. erubescens is closely related to *A. fleckii* Schinz and misidentifications can easily occur. Probably the easiest distinguishing character is the petiole-length. In *A. erubescens* the petiole is normally 1·3–2·5 cm. long, and only occasionally more or less, these occasional extremes usually occurring on the same shoot as petioles of more usual length; in *A. fleckii* the petiole is normally 0·5–1 cm. long, occasionally as long as 1·3 cm. The leaf-rhachis is ± glandular in *A. erubescens*, but eglandular in *A. fleckii*. The leaflets are usually markedly curved-falcate in *A. erubescens* but straight in *A. fleckii*, and the corolla-lobes ± densely appressed-pubescent outside in *A. erubescens* but only slightly puberulous to subglabrous in *A. fleckii*. The gland on the petiole is small in *A. erubescens*, 0·4–1 mm. long, while in *A. fleckii* it is 0·75–2 mm. long. This character is less useful than the others owing to the amount of overlapping.

12. **Acacia fleckii** Schinz in Mém. Herb. Boiss. **1**: 108 (1900).—Bak. f., Legum. Trop. Afr. **3**: 832 (1930).—O. B. Mill., B.C.L.: 18 (1948).—Brenan in Kew Bull. **11**: 197 (1956).—Torre, C.F.A. **2**: 277, t. 54 (1956).—Story in Mem. Bot. Surv. S. Afr. **30**: 22 (1958).—F. White, F.F.N.R.: 84, fig. 17 I (1962).—Mitchell in Puku, **1**: 104 (1963). —Boughey in Journ. S. Afr. Bot. **30**: 157 (1964). Type from SW. Africa.

Acacia cinerea Schinz in Verh. Bot. Verein. Brand. **30**: 240 (1888) non Spreng. (1826).—Bak. f., loc. cit.—O. B. Mill., tom. cit.: 17 (1948); in Journ. S. Afr. Bot. **18**: 19 (1952). Type from SW. Africa.

Acacia caffra var. *tomentosa* sensu Bak. f., tom. cit.: 833 (1930) pro parte quoad specim. *Lugard* 93.—O. B. Mill., B.C.L.: 17 (1948); in Journ. S. Afr. Bot. **18**: 19 (1952) saltem quoad specim. *Curson* 173.

Shrub or small round-crowned tree 1·5–10 m. high; bark on trunk grey to cream, peeling in small flakes; young branchlets densely grey-pubescent with many small reddish glands intermixed, later becoming rather smooth (except for prickles), pale grey to grey-brown. Stipules subulate, not spinescent. Prickles in pairs just below the nodes, often many and strong, brown to grey or blackish, strongly hooked, up to 8 mm. long, usually broad-based. Leaves: petiole 0·5–1(1·3) cm. long, with a gland 0·75–2 mm. long; rhachis pubescent, eglandular; pinnae 6–20 pairs; leaflets (9)12–30 pairs, 2–5 × 0·3–1(1·2) mm., linear-oblong, straight or almost so, ± ciliate on the margins and sometimes pubescent on the lower surface also; veins somewhat prominent at first, becoming obscure as the leaves age, apex not oblique, rounded to obtuse or sometimes subacute. Flowers white, sessile, in spikes 3–6·5 cm. long on peduncles 1–2 cm. long; axis densely short-pubescent. Calyx 2–3 mm. long, shortly pubescent or puberulous or sometimes subglabrous. Corolla 3–4 mm. long, 5-lobed, glabrous to slightly puberulous on the outside of the lobes. Stamen-filaments 7–9 mm. long; anthers 0·2 mm. across, glandular at the apex. Ovary puberulous, stipitate. Pods pale-brown to straw-coloured, dehiscent, 6–12·5 (including stipe) × 1·3–2(2·3) cm., sparsely pubescent to subglabrous, with numerous small reddish glands, linear-oblong, straight, finely reticulately and transversely venose, rounded to acute at the apex. Seeds 8–12 mm. in diam., subcircular; areole small, 1–4 × 1–2·5 mm.

Caprivi Strip. Katima Mulilo, on banks of Zambezi, fl. 24.xii.1958, *Killick & Leistner* 3090 (K; SRGH). **Botswana.** N: Makalamebedi, fr. 21.iv.1931, *Pole Evans* 3316 (12) (K; PRE). SW: 58 km. N. of Kan on road to Ghanzi, fl. 17.ii.1960, *de Winter* 7358 (BM; K; PRE; SRGH). SE: 6 km. N. of Mahalapye, fl. 14.i.1960, *Leach & Noel* 63 (K; SRGH). **Zambia.** B: Sesheke, fl. 26.xii.1952, *Angus* 1033 (BM; K). S: Namwala, fr. 11.vi.1949, *Hornby* 3009 (K; SRGH). **Rhodesia.** W: Main Camp, Wankie Game Reserve, fl. x–xi.1960, *Tapping* 16/60 (K; SRGH). Gatooma Distr., Umsweswe R. 16 km. S. of Gatooma, fr. 15.vi.1968, *Rushworth* 1187 (K). S: Shabani, fl. xii.1957, *Miller* 4880 (SRGH).

Also in Angola, Transvaal and SW. Africa. In the drier types of deciduous woodland (e.g. with *Baikiaea*), thicket and bushland. Frequent on Kalahari Sand; 850–1370 m.

The differences between this species and *A. erubescens* are given under the latter (p. 74).

13. **Acacia rovumae** Oliv., F.T.A. **2**: 353 (1871).—Sim, For. Fl. Port. E. Afr.: 58 (1909).—Bak. f., Legum. Trop. Afr. **3**: 831 (1930).—Brenan, T.T.C.L.: 331 (1949); F.T.E.A. Legum.-Mimos.: 90, fig. 14/15 (1959). Type: Tanzania or Mozambique, Rovuma Bay, *Kirk* (K, holotype).

Tree 10–15 m. high, with openly branched flat crown and rough or smooth dark-grey or grey-green bark; young branchlets puberulous or very shortly pubescent with short curved hairs that are yellowish, at least when dry. Stipules not spinescent. Prickles in pairs just below the nodes, deep-grey to blackish, up to 4–6 mm. long, spreading or pointing a little upwards, usually straight or only slightly curved. Leaves: petiole (10)18–40 mm. long, with a small gland 0·4–0·7 mm. in diam.; rhachis puberulous, glandular between the top 1–4 pairs of pinnae; pinnae 6–9 pairs; leaflets (9)13–31 pairs, 4–8 × 1·5–2(3·5) mm., oblong, oblique at the base and the subacute to obtuse apex, pale-glaucescent beneath, puberulous especially beneath, lateral nerves visible when young, becoming obscure. Flowers sessile or nearly so, in spikes 6–10 cm. long on 1·5–3 cm. long peduncles, produced with the leaves; axis puberulous. Calyx 1·5–2 mm. long, puberulous. Corolla 2–3 mm. long, glabrous or slightly puberulous on the lobes outside, 5-lobed, exceeding the calyx. Stamen-filaments 4–5 mm. long; anthers 0·1 mm. across, with a caducous gland. Ovary glabrous, very shortly stipitate. Pods probably not dehiscent, irregularly breaking up, 7–15 × 1·7–2·5 cm., glabrous, oblong, straight, and smooth or nearly so, dark-brown when dry, green when living, rather thick and turgid, rounded or acute at the apex. Seeds 10–13 × 7–9 mm., oblong-elliptic-lenticular, hard-walled; central areole 7–9 × 4·5–5 mm., large, not impressed.

? **Mozambique.** N: Palma, Rovuma Bay, fr. iii.1862 & x.1862, *Kirk* (K).

Also in Kenya, Tanzania and Madagascar. Ecology uncertain in our area; elsewhere in riverine forest and saline-water swamp-forest near sea-level or at low altitudes.

Apparently closely related to *A. burkei* Benth., under which species the differences are discussed.

The appearance of the pods of *A. rovumae* suggests that they are indehiscent and perhaps water-borne in their dispersal. If this is confirmed, and it requires further observation on the spot, then it is a very unusual feature in any *Acacia*. *A. rovumae* is also outstanding among its nearest relatives by usually having its prickles not or scarcely hooked.

14. **Acacia burkei** Benth. in Hook., Lond. Journ. Bot. **5**: 98 (1846).—Sim, For. Fl. Port. E. Afr.: 56 (1909).—Burtt Davy, F.P.F.T. **2**: 337, fig. 56 (1932).—O. B. Mill., B.C.L.: 17 (1948) pro parte excl. syn. *A. mossambicensis*; in Journ. S. Afr. Bot. **18**: 19 (1952) pro parte ut praec.—Codd, Trees & Shrubs Kruger Nat. Park: 41 (1951).—Young in Candollea, **15**: 115 (1955).—Palmer & Pitman, Trees of S. Afr.: 150 (1961).—de Winter, de Winter & Killick, Sixty-six Transvaal Trees: 42 cum photogr. (1966).—J. Ross in Bol. Soc. Brot., Sér. 2, **42**: 275 (1968). Type from the Transvaal.

Tree 3–27 m. high with smooth or scaly greyish-yellow to brownish or almost black bark; young branchlets densely pubescent. Stipules not spinescent. Prickles in pairs just below the nodes, brown when young, then grey to blackish, up to 3–8 mm. long, strongly hooked downwards. Leaves: petiole (4)13–35 mm. long, with a small gland 0·5–0·8 mm. in diam. (rarely the gland placed between the lowest pair of pinnae); rhachis ± densely pubescent, eglandular or with a gland between the top pair of pinnae only; pinnae (1)3–13 pairs; leaflets variable in number, size and shape, (1)4–19 pairs, 1·2–20·2 × 0·3–13·1 mm., usually ± obovate but varying to oblong or sometimes ± elliptic, rounded to subacute or acute at the apex, usually markedly asymmetric at the base, varying from pubescent all over on both sides to bearded only on the midrib beneath near the base and with a few hairs on the margin: venation usually ± prominent beneath. Flowers sessile or almost so (pedicel to 0·25 mm. long), white, in spikes 3–6(14·6) cm. long on peduncles (0·5)1–3 cm. long, produced with the leaves; axis densely pubescent. Calyx 1·7–2·5 mm. long, ± densely pubescent. Corolla 2–3 mm. long, glabrous or shortly pubescent on the outside of the lobes, 5-lobed. Stamen-filaments 4·5–6 mm. long; anthers c. 0·15 mm. across, with a caducous gland. Ovary glabrous, on a stipe about half its length. Pods dehiscent, 4·1–16·9 × 0·9–2·4 cm., glabrous to hairy particularly near the margins base and apex, linear-oblong, straight, obscurely venose, purplish-brown, ± acuminate to mucronate at the apex. Seeds c. 6–13 × 6–11 mm.; central areole 4–8 × 3–8 mm.

Botswana. SE: Kanye Distr., Pharing, fl. & fr. x.1946, *Miller* B/492 (PRE). **Rhodesia.** S: Nuanetsi Distr., road from Chikwedza Camp to Litshani's kraal, fl. & fr. xi.1956, *Davies* 2203 (K; LISC; SRGH). **Mozambique.** N: between Namapa and Lúrio, fl. 11.x.1948, *Barbosa* in *Mendonça* 2366 (BM; K; LISC; LM). MS: Chimoio, between Vila Pery and Vanduzi, near Tembe, fr. 1.iii.1948, *Garcia* in *Mendonça* 434 (BM; K; LISC). SS: between Inharrime and Panda, fl. & fr. 26.ix.1947, *Pedro & Pedrógão* 1888 (LMJ; PRE). LM: Matola, fl. 11.x.1940, *Torre* 1752 (BM; K; LISC).

Also in S. Africa. In woodland and bushland.

A. burkei is almost as variable in its foliage as *A. goetzei* Harms, but I do not think it possible to recognize any subspecies or varieties. It differs from *A. goetzei* by the ± densely hairy calyx; the pods of *A. burkei* are generally narrower also.

A. burkei is close to *A. rovumae* Oliv. and *A. tanganyikensis* Brenan. It differs from the first in the strongly hooked prickles, the less numerous leaflets (always more than 15 pairs in some pinnae in *A. rovumae*), the pubescent not puberulous inflorescence-axes and calyces, and the narrower dehiscent pods; and from the second in the broader less numerous leaflets.

A few specimens have been collected in Mozambique with pods wider than usual, to c. 2·5 cm.; examples are *Garcia* in *Mendonça* 434, cited above, and *Torre* 2241 from LM: between Moamba and Sábiè, fr. 4.xii.1940 (LISC). These may be variants of *A. burkei*, or represent occasional hybrids with other species. This possibility is also indicated by *Torre* 2055 (LISC), from Mozambique, LM: Goba, fl. 16.xi.1940, which is very like some forms of *A. burkei*, except that there are only 3–4 pairs of broad leaflets per pinna and the calyx is only very sparingly pubescent.

A. schlechteri Harms in Engl., Bot. Jahrb. **51**: 367 (1914) may well represent a similar plant. It was based on *Schlechter* 11901, from Mozambique, LM: Ressano Garcia, xi.1897 (B, holotype †), of which I have seen no material. See also Young in Candollea, **15**: 123 (1955).

Further observation in the field of these aberrant plants would be valuable.

15. **Acacia goetzei** Harms in Engl., Bot. Jahrb. **28**: 395 (1900).—Bak. f., Legum. Trop. Afr. **3**: 830 (1930).—Brenan, T.T.C.L.: 329 (1949); in Kew Bull. **11**: 198 (1956); F.T.E.A. Legum.-Mimos.: 91, fig. 14/16 (1959).—Torre, C.F.A. **2**: 276, t. 53 A (1956).—F. White, F.F.N.R.: 83 (1962).—Mitchell in Puku, **1**: 104 (1963). Type from Tanzania (Kilosa Distr.).

Tree 3–20 m. high, with rounded crown and rough grey or brown bark; young branchlets glabrous to pubescent. Stipules not spinescent. Prickles in pairs just below the nodes, pale then dark-brown or grey, up to 7 mm. long, hooked downwards. Leaves: petiole 1·3–5 cm. long, with or rarely without a small gland; rhachis glabrous to pubescent, usually glandular between the top 1–3(5) pairs of pinnae (and sometimes the basal pair as well); pinnae 3–10 pairs; leaflets (2)5–20(25) pairs, (2)3–17(25) × (0·75)1–7(14) mm., rounded to mucronate or subacute at the apex, usually distinctly asymmetric at the base, glabrous to pubescent; venation somewhat prominent beneath. Flowers sessile or nearly so, white or slightly yellowish, in spikes (2)3–12 cm. long on 0·4–4·5 cm. long peduncles, produced with the leaves; axis glabrous to pubescent. Calyx 1·5–2·75 mm. long, glabrous. Corolla 2–3·75 mm. long, 5-lobed, glabrous, exceeding the calyx. Stamen-filaments 4·5–6 mm. long; anthers 0·2–0·25 mm. across, with a caducous gland. Ovary glabrous, very shortly stipitate. Pods dehiscent, (5)8–18 × (1·8)2–3·5 cm., glabrous or nearly so, oblong or irregularly constricted, straight or nearly so, venose, red- to purplish-brown, acuminate or apiculate at the apex. Seeds 8–11 × 8–10 mm., subcircular-lenticular; central areole medium, 4·5–6 × 2·5–4 mm.

Subsp. **goetzei** Brenan in Kew Bull. **11**: 204 (1956); F.T.E.A. Legum.-Mimos.: 91 (1959).—F. White, F.F.N.R.: 83 (1962).—Boughey in Journ. S. Afr. Bot. **30**: 157 (1964).

Acacia mossambicensis sensu Bak. f., Legum. Trop. Afr. **3**: 831 (1930).—Wild, S. Rhod. Bot. Dict.: 48 (1953).—Burtt Davy & Hoyle, rev. Topham, N.C.L., ed. 2: 64 (1958).

Acacia welwitschii sensu Eyles in Trans. Roy. Soc. S. Afr. **5**: 363 (1916).—Wild, Guide Fl. Vict. Falls: 149 (1953).—Burtt Davy & Hoyle, tom. cit.: 65 (1958) synon. dub.

Leaflets (of leaves on mature flowering shoots) nearly all more than 3 mm. wide (range 2·5–14 mm.), usually wider towards the apex and thus obovate, obovate-oblong or oblanceolate-oblong, often in comparatively few ((2)5–11(14, very rarely to 18)) pairs; rhachis of leaf frequently (by no means always) unarmed and with a gland between the topmost pair of pinnae only.

Zambia. C: Chingombe, fl. 26.ix.1957, *Fanshawe* 3730 (K). E: Nsadzu Bridge, fl. 27.xi.1958, *Robson* 742 (BM; K; LISC). S: 64 km. E. of Mazabuka, fl. 8.x.1930, *Milne-Redhead* 1230 (K). **Rhodesia.** N: Mazoe Distr., Chipoli, fl. & fr. 17.x.1958, *Moubray* 37 (SRGH). W: Shangani Distr., Leopard Mine, fl. immat. ix.1910, *Nobbs & Sim* 959 (SRGH). C: Gwelo Distr., Nahla, fl. immat. 26.x.1924, *Steedman* 99 (SRGH). E: Umtali, Plantation Drive, fl. & fr. 4.x.1962, *Chase* 7824 (K; LISC; SRGH). **Malawi.** C: Lilongwe, fr. vi.1944, *Leech* 1 (PRE). S: Mlanje Distr., Mchese Mt., fr. 22.vii.1958, *Chapman* W/616 (K; SRGH). **Mozambique.** N: Malema, Lioma road, 7 km. from Mutuáli, fl. 8.x.1953, *Gomes e Sousa* 4142 (COI; K; LISC; PRE; SRGH). Z: between Mocuba and Munhamade, fl. & fr. 23.x.1941, *Torre* 3455 (BM; K; LISC; LM). T: Moatize, between Zóbuè and Muatize, fl. 27.viii.1943, *Torre* 5809 (BM; K; LISC). MS: Gorongosa, fl. 14.x.1946, *Simão* 1082 (LM; PRE). LM: Namaacha, Goba, fl. 16.xi.1960, *Torre* 2055 (LM).

Also in the Congo, Tanzania and Angola. Woodland of various types, wooded grassland and thicket, 610–1400 m.

Vegetatively, especially in indumentum and leaflets, extremely variable, but constant in the characters of the individual flowers and fruits. The armature of the leaf-rhachis is also variable, the type of subsp. *goetzei* having a prickly rhachis which is, however, less common than an unarmed one in this subspecies.

It is possible that the variability may be due to hybridization between subsp. *microphylla* and species such as *A. nigrescens* Oliv. having given rise to part at least of the gamut of variation at present included in subsp. *goetzei*. This is at present only speculation and needs experimental testing.

Subsp. **microphylla** Brenan in Kew Bull. **11**: 204 (1956); F.T.E.A. Legum.-Mimos.: 91 (1959).—F. White, F.F.N.R.: 83 (1962).—Boughey in Journ. S. Afr. Bot. **30**: 157 (1964). Type: Malawi, Mombera Distr., Njakwa to Fort Hill, *Greenway* 6393 (EA; K, holotype).

Acacia ulugurensis Harms in Engl., Bot. Jahrb. **28**: 396 (1900).—Bak. f., Legum. Trop. Afr. **3**: 831 (1930).—Brenan, T.T.C.L.: 332 (1949).—Burtt Davy & Hoyle, rev. Topham, N.C.L., ed. 2: 64 (1958). Type from Tanzania (Ulugurus).

Acacia joachimii Harms in Notizbl. Bot. Gart. Berl. **12**: 507 (1935).—Brenan, T.T.C.L.: 331 (1949). Type from Tanzania (Lindi Distr.).

Acacia van-meelii Gilbert & Boutique in Bull. Jard. Bot. Brux. **22**: 177 (1952); F.C.B. **3**: 149 (1952). Type from the Congo.

Leaflets (of leaves on mature flowering shoots) nearly all narrower than 3 mm. (range 0·75–3 mm.), usually (except terminal pairs) not wider towards apex, and thus oblong or linear-oblong, often in more numerous (8–23) pairs than in subsp. *goetzei*; rhachis of leaf frequently (by no means always) ± prickly and with glands between the topmost 1–3(–5) pairs of pinnae.

Zambia. N: Karindi Valley, Lutikila Basin, E. of Lake Bangweulu, st. 22.ix.1933, *Michelmore* 587 (K). C: Lusaka, fr. 20.iii.1960, *Fanshawe* 5556 (K; SRGH). E: Jumbe-Machinje Hills, fl. 11.x.1958, *Robson* 37 (BM; K; LISC; SRGH). S: Monze, fr. iii.1934, *Trapnell* 1443 (K). **Rhodesia.** N: Karoi, fl. 4.x.1946, *Wild* 1284 (K; SRGH). E: Umtali Distr., Commonage below Darlington, fl. 21.x.1958, *Chase* 7005 (K; SRGH). **Malawi.** N: Mwanemba Mt., fl. ii–iii.1903, *McClounie* 152 (K). C: Chitala to Dowa, fl. & fr. 28.x.1941, *Greenway* 6367 (K). **Mozambique.** N: Nampula, fl. 14.xi.1948, *Andrada* 1463 (BM; COI; K; LISC; LMJ). Z: Pebane, between Pebane and Mualama, fl. 30.x.1942, *Torre* 4725 (BM; K; LISC). T: Macanga, between Furancungo and Vila Gamito, fl. 20.x.1943, *Torre* 6070 (LISC).

Aso in the Congo, Ethiopia, Kenya and Tanzania. Woodland of various types; 600–1400 m.

Robson 173, from Zambia, is said to have smooth bark.

There is no hard and fast line separating subsp. *goetzei* and subsp. *microphylla*, although the extremes are very different. Some specimens have been collected in Zambia and Rhodesia which can only be regarded as intermediate between the two subspecies.

A. goetzei is extremely close to *A. welwitschii* subsp. *delagoensis* Harms and *A. burkei* Benth. The differences are discussed under those species.

16. **Acacia welwitschii** Oliv., F.T.A. **2**: 341 (1871).—Bak. f., Legum. Trop. Afr. **3**: 829 (1930).—Torre, C.F.A. **2**: 274 (1956). Syntypes from Angola and Mozambique: Zambeze, below Tete, *Kirk* (K, syntype).

Tree 3–15 m. high with rough bark and broad crown; young branchlets glabrous. Stipules not spinescent. Prickles in pairs just below the nodes, grey or blackish, up to c. 5(7) mm. long, hooked downwards. Leaves: petiole with a small to rather large gland; rhachis glabrous, eglandular, or with a gland between the top pair of pinnae; pinnae 2–4(5) pairs; leaflets (2)3–8 pairs, 4–20 × 2·5–13 mm., elliptic or broadly elliptic, sometimes somewhat ovate, except for terminal ones broadest at or near or sometimes below the middle, rounded and often slightly emarginate at the apex, nearly symmetrical at the base, glabrous, with venation rather prominent beneath. Flowers sessile or almost so, white, in spikes 3–13 cm. long on peduncles 0–3 cm. long, produced with the leaves; axis glabrous to slightly pubescent. Calyx 1·5–1·75 mm. long, glabrous. Corolla 2·25–3 mm. long, glabrous, 5-lobed. Stamen-filaments 4·5–5·5 mm. long, free; anthers c. 0·15 mm. across, with a caducous gland. Ovary glabrous, on a stipe about half its length. Pods dehiscent, 5·5–16·5 × 1·3–2 cm., glabrous, linear-oblong, usually straight, obscurely venose, blackish to grey-brown, rounded to ± acuminate at the apex. Seeds 9–12 × 10–14 mm., irregularly subcircular to subreniform; central areole large, 6·5–8 × 6·5–8 mm., almost circular, impressed.

Subsp. *welwitschii*, with inflorescences 6·5–13 cm. long and leaflets 10–20 × 4·5–13 mm., is known only from Angola.

Subsp. **delagoensis** (Harms) J. Ross & Brenan in Kew Bull. **21**: 67 (1967).

Acacia welwitschii Oliv., F.T.A. **2**: 341 (1871) pro parte quoad specim. *Kirk*.—Sim, For. Fl. Port. E. Afr.: 55, t. 37 (1909).—Gomes e Sousa, Pl. Menyharth.: 69 (1936).

Acacia delagoensis [Sims, tom. cit.: 56 (1909) *nom. nud.*] Harms in Engl., Bot. Jahrb. **51**: 367 (1914).—Burtt Davy, F.P.F.T. **2**: 337 (1932).—Codd, Trees & Shrubs Kruger Nat. Park: 42, fig. 37 c–d (1951).—Young in Candollea, **15**: 117 (1955).—Boughey in Journ. S. Afr. Bot. **30**: 157 (1964).—Gomes e Sousa, Dendrol. Moçamb. Estudo Geral, **1**: 234, t. 38 (1966). Type: Mozambique, Umbeluzi, *Schlechter* 11718 (B, holotype †; BM).

Inflorescence 4–6·5 cm. long; leaflets usually smaller than in subsp. *welwitschii*, 5–15 × 3–11 mm.

Rhodesia. E: Lower Sabi, E. bank st. 28.i.1948, *Wild* 2368 (K; SRGH). S: Nuanetsi Dist.r, Fishans, Lundi R., fr. 23.iv.1962, *Drummond* 7698 (K; LISC; SRGH)

Mozambique. N: Macondes, Nangororo, left bank of R. Ridi, fr. 28.x.1959, *Gomes e Sousa* 4490 (COI; K; PRE). Z: Morrumbala, between Murire and Chantengo, st. 6.ix.1949, *Barbosa & Carvalho* in *Barbosa* 3973 (K; LM). T: below Tete, fl. 3.xii.1860, *Kirk* (K). MS: Chemba, Nhamaco longo, st. 6.vii.1947, *Simão* 1387 (LM; PRE). SS: Guijá, Caniçado, fr. 10.vi.1947, *Pedrógão* 282 (COI; LMJ; PRE). LM: Sábiè, between Moamba and Ressano Garcia, fl. 19.xii.1944, *Torre* 6875 (K; LISC).
Also in the Transvaal. Woodland; 45–460 m.

The geographical ranges of subsp. *welwitschii* and subsp. *delagoensis* are widely separated, and specimens of each have a markedly different " look ", yet the only satisfactory difference is in the size of the inflorescence.

Wild 2348, cited above, is rather doubtful, and this is the only specimen so far seen from this division of Rhodesia.

A. welwitschii is very similar indeed to *A. goetzei* subsp. *goetzei*, and they are maintained here as separate species only after considerable hesitation. *A. welwitschii* differs in the leaflets being elliptic and almost symmetric at the base, the consistently narrow pods and the larger areoles on the seed.

Some specimens from Rhodesia, Nuanetsi District (*Müller* 645, 638 (SRGH)) resemble *A. welwitschii* subsp. *delagoensis* but have wider pods than usual, up to 2·2–2·4 cm. wide and rather larger leaflets. The status of these plants is uncertain, but the differences may well be due to hybridization with *A. goetzei* Harms.

17. **Acacia senegal** (L.) Willd., Sp. Pl. ed. 4, **4**: 1077 (1806).—Bak. f., Legum. Trop. Afr. **3**: 827 (1930).—Burtt Davy, F.P.F.T. **2**: 337 (1932).—Brenan, T.T.C.L.: 330 (1949); F.T.E.A. Legum.-Mimos.: 92, fig. 14/17 (1959).—Codd, Trees & Shrubs Kruger Nat. Park: 50 (1951).—Gilbert & Boutique, F.C.B. **3**: 149 (1952).—Young in Candollea, **15**: 93 (1955).—Torre, C.F.A. **2**: 273 (1956).—J. Ross in Bol. Soc. Brot., Sér. 2, **42**: 207 (1968). Type uncertain, presumably from Senegal, probably not from Arabia as stated by Linnaeus.
Mimosa senegal L., Sp. Pl. **1**: 521 (1753). Type as above.

Shrub or tree up to 13 m. high; bark grey to brown or blackish, scaly, rough; young branchlets densely to sparsely pubescent, soon glabrescent. Stipules not spinescent. Prickles just below the nodes, either in threes, up to 7 mm. long, the central one hooked downwards, the laterals ± curved upwards, or else solitary, the laterals being absent. Leaves: petiole glandular or not (gland c. 0·5–0·75 mm. in diam.); rhachis ± pubescent, glandular between the top 1–5 pairs of pinnae, prickly or not; pinnae (2)3–6(12) pairs, 0·5–1·5(2·4, very rarely to 4 or more) cm. long; leaflets 7–25 pairs, 1–4(9) × 0·5–1·75(–3) mm., linear - to elliptic-oblong, ciliate on the margins only or ± hairy on the surface, or wholly subglabrous, lateral nerves not visible or sometimes somewhat prominent beneath, apex obtuse to subacute. Flowers white or cream, fragrant, sessile, in spikes 1·5–10 cm. long on peduncles 0·7–2 cm. long, normally produced with the leaves; axis pubescent to glabrous. Calyx 2–2·75(3·5) mm. long, glabrous to somewhat pubescent. Corolla 2·75–4 mm. long, exceeding the calyx, 5-lobed, glabrous outside. Stamen-filaments 4·5–7 mm. long, free; anthers 0·2–0·25 mm. across, with a caducous gland. Ovary glabrous, very shortly stipitate. Pods usually grey-brown, sometimes pale- or dark-brown, dehiscent, (1·8)4–19 × (1·2)2–3·4 cm., densely to sparsely appressed-pubescent to -puberulous, oblong, straight, venose, rounded to acuminate at the apex. Seeds ± subcircular-lenticular, 8–12 mm. in diam.; central areole small to medium, 2·5–6 × 2·5–5 mm., markedly impressed.

Var. **senegal.**—Brenan, F.T.E.A. Legum.-Mimos.: 93 (1959).

A tree with a single central stem and a usually dense flat-topped crown; bark without any papery peel, rough, grey to brown. Inflorescence-axis normally pubescent, very rarely glabrous (and not so far in our area). Pods variable in size, usually rounded to somewhat pointed but not rostrate or acuminate at the apex.

Mozambique. N: Montepuez, between Balama and Alide, fr. 1.ix.1948, *Barbosa* 1952 (LISC; LM).
Widespread in tropical Africa to the N. of our region. Ecology uncertain.

Var. **rostrata** Brenan in Kew Bull. **8**: 99 (1953).—Young in Candollea, **15**: 96 (1955).—Boughey in Journ. S. Afr. Bot. **30**: 158 (1964).—J. Ross in Bol. Soc. Brot., Sér. 2, **42**: 233 (1968). Type from Transvaal.
Acacia spinosa Marl. & Engl. in Engl., Bot. Jahrb. **10**: 20 (1888) non E. Mey. (1836).—O. B. Mill. in Journ. S. Afr. Bot. **18**: 24 (1952). Type from SW. Africa.

Acacia rostrata Sim, For. Fl. Port. E. Afr.: 55, t. 37 (1909) non Humb. & Bonpl. ex Willd. (1806).—Bak. f., Legum. Trop. Afr. **3**: 827 (1930). Type: Mozambique, Maputo, *Sim* 6263 (not found).
Acacia senegal var.—Codd, Trees & Shrubs Kruger Nat. Park: 51 (1951).
Acacia senegal sensu Wild, S. Rhod. Bot. Dict.: 49 (1953).

Either a shrub branching at or close to the base or a small tree with a single central stem, 1–4(6) m. high; crown dense, flattened; bark normally with a flaking papery peel, creamy-yellow to yellow-green or grey-brown. Inflorescence-axis always pubescent. Pods 2–3½ times as long as wide, very rarely more, 2·3–8·5 × 1·6–2·9 cm., usually ± rostrate or acuminate at the apex.

Botswana. N: N. of Lake Dow, 40 km. ESE. of Rakops, fr. 22.iii.1965, *Wild & Drummond* 7223 (SRGH). SE: Dikhatlon Ranch, fr. 14.iv.1931, *Pole Evans* 3189 (39) (K; PRE). **Rhodesia.** S: Lower Sabi, Birchenough Bridge, fr. 30.i.1948, *Wild* 2458 (K; SRGH); Shashi-Limpopo confluence, fr. 22.iii.1959, *Drummond* 5936 (K; SRGH). **Mozambique.** N: Monte Icoculo, st. 31.viii.1948, *Pedro & Pedrógão* 4997 (LMJ). SS: Guijá, Caniçado, Chamusca, fr. 19.vi.1947, *Pedrógão* 352 (COI; K; LMJ; PRE). LM: Sábiè, Moamba, 5 km. from Sábiè, fr. 7.vi.1948, *Torre* 7968 (K; LISC).
Also in S. Africa (the Transvaal and Natal), Swaziland and SW. Africa. In woodland, wooded grassland and bushland, penetrating Rhodesia along the alluvium of river-valleys; near sea-level–c. 600 m.

It is possible that this variety has a wider range than here given. A few specimens from Kenya and Somalia included in F.T.E.A. Legum.-Mimos.: 93 (1959) under the rather heterogeneous concept of var. *kerensis* Schweinf. have pods similar to those of var. *rostrata*. The situation needs clarifying in eastern and north-eastern tropical Africa.

Var. **leiorhachis** Brenan in Kew Bull. **8**: 98 (1953).—Young in Candollea, **15**: 95 (1955). —Boughey in Journ. S. Afr. Bot. **30**: 158 (1964).—J. Ross in Bol. Soc. Brot., Sér. 2, **42**: 231 (1968). TAB. **15** fig. 5. Type from Tanzania (Pare Distr.).
Acacia circummarginata Chiov. in Ann. di Bot. **13**: 394 (1915).—Bak. f., Legum. Trop. Afr. **3**: 834 (1930).—Brenan, F.T.E.A. Legum.-Mimos.: 94, fig. 14/18 (1959). Type from Ethiopia (Ogaden).
Acacia senegal var. *senegal* sensu Brenan, F.T.E.A. Legum.-Mimos.: 93 (1959) pro parte quoad syn. *A. senegal* var. *leiorhachis* necnon *A. thomasii* et *A. somalensis* sensu T.T.C.L. (1949).
Acacia sp. 1.—F. White, F.F.N.R.: 82 (1962).

Always a tree with a central stem and a crown either rounded or (with us) irregular with straggling branches; bark with conspicuous yellow papery peel. Inflorescence-axis always glabrous.

Zambia. W: Sinazongwe–Memba road, c. 24 km. from Sinazongwe, fl. 10.vi.1963, *Bainbridge* 838 (SRGH). S: Lusitu, fl. 22.v.1961, *Fanshawe* 6602 (K). **Rhodesia.** N: Sipolilo Distr., Hunyani R., 10 km. S. of Dande Mission, fl. 15.v.1962, *Wild* 5755 (K; SRGH). W: Wankie, fr. ix.1960, *Davies* 2810 (K; SRGH). S: Nuanetsi Distr., below Lundi escarpment, fl. 12.viii.1956, *Mowbray* 113 (K; SRGH). **Mozambique.** T: between Tete and Chicoa, fl. & fr. 25.vi.1949, *Andrada* 1641 (COI; LISC). MS: Mossurize, Machaze-Pande, fl. 26.vii.1949, *Pedro & Pedrógão* 7777 (LMJ).
Also in Ethiopia, Kenya and Tanzania. In woodland and bushland, often with *Colophospermum mopane*; 460–910 m.

18. **Acacia kraussiana** Meisn. ex Benth. in Hook., Lond. Journ. Bot. **1**: 515 (1842).—Brenan & Exell in Bol. Soc. Brot., Sér. 2, **31**: 103, t. I fig. C (1957).—Burtt Davy & Hoyle, rev. Topham, N.C.L. ed. 2; 64 (1958).—Mogg in Macnae & Kalk, Nat. Hist. Inhaca I.: 9 (1958).—Palmer & Pitman, Trees S. Afr.: 89, t. 48 (1961). TAB. **19** fig. C. Type from Natal.

Shrub 1–12 m. high, often climbing; young branchlets puberulous and inconspicuously glandular, or rarely pubescent, turning grey-brown. Prickles scattered on the stems, petioles and rhachides, ± deflexed, short, less than 2 mm. long, arising from longitudinal bands along the stem which are concolorous with the intervening lenticellate bands. Leaves: petiole (1·4)1·7–3·5 cm. long: pinnae 3–6 pairs, 2–8 cm. long, leaflets 5–15 pairs, (5)9–23 × 2–8 mm., elliptic-oblong to obovate, glabrous or nearly so, rarely ± puberulous or pubescent on the surface; midrib excentric particularly towards the base. Flowers white, in paniculately arranged heads 8–13 mm. in diam. Stipules at the base of the peduncles small, c. 1 mm. wide, inconspicuous, soon caducous, not subcordate at the base. Calyx eglandular outside. Pods brown, slowly dehiscent, 6·5–16 × 1·4–2·5 cm., linear-oblong, subcoriaceous, flat, margins not strongly thickened. Seeds blackish or chestnut-brown, 6–9 × 4·5–6 mm., ellipsoid, somewhat compressed; areole 4–7 × 2·5–4 mm.

Mozambique. SS: Chibuto, between Chibuto and Gomes da Costa, fl. 14.xi.1957, *Barbosa & Lemos* 8110 (COI; K; LISC). LM: Lourenço Marques, Polana, fl. 20.i.1960, *Lemos & Balsinhas* 14 (COI; K; LISC; SRGH).

Also in Natal. In coastal woodland bushland and thicket, often near the shore, 0–110 m.

A very distinct species. The armature is variable, the prickles on the stem being sometimes numerous, sometimes very few and inconspicuous. The hairiness of the leaflets is also variable. Usually they are glabrous or nearly so, as in *Lemos & Balsinhas* 14, cited above; rarely they are thinly puberulous, as in *Barbosa & Lemos* 8110, also cited above; in *Barbosa & Lemos* 8106 (COI; K; LISC), from the same locality as 8110, the leaflets have a dense puberulence over their surface.

19. **Acacia brevispica** Harms in Notizbl. Bot. Gart. Berl. **8**: 370 (1923).—Bak. f., Legum. Trop. Afr. **3**: 853 (1930).—Brenan, T.T.C.L.: 332 (1949); F.T.E.A. Legum.-Mimos.: 96, fig. 15/22 (1959); in Kew Bull. **21**: 477 (1968).—Brenan & Exell, C.F.A. **2**: 287 (1956); in Bol. Soc. Brot., Sér. 2, **31**: 108, t. I fig. B (1959). TAB. **19** fig. B. Type from Tanzania.
Acacia brevispica var. *brevispica*.—Ross & Gordon-Gray in Brittonia, **18**: 62 (1966).

Subsp. **dregeana** (Benth.) Brenan in Kew Bull. **21**: 479 (1968). Type from Natal.
Acacia pennata var. *dregeana* Benth. in Hook., Lond. Journ. Bot. **1**: 516 (1842). Type as above.
Acacia brevispica sensu Brenan & Exell in Bol. Soc. Brot., Sér. 2, **31**: 114 (1957) quoad pl. mozamb. etc.
Acacia brevispica var. *dregeana* (Benth.) Ross & Gordon-Gray in Brittonia, **18**: 63 (1966). Type as above.

Shrub or small tree 1·2–8 m. high, usually scrambling or scandent; young branchlets very shortly puberulous or rarely glabrous, and also with numerous minute reddish glands. Prickles scattered, recurved or spreading, arising from longitudinal bands along the stem which are usually paler than the intervening lenticellate bands. Leaves: petiole often 0·8–1·5 cm. long, but some normally and characteristically longer than 1·5 cm. (up to 2(2·5) cm.); pinnae 7–18 pairs, 1·5–3·5(4·5) cm. long; leaflets 0·5–0·8(1) mm. wide, very numerous, linear-oblong; midrib nearer one margin at the base; lower surface frequently appressed-silky-puberulous, sometimes glabrous; margins inconspicuously appressed-puberulous or glabrous. Flowers white, in heads 10–15 mm. in diam., racemosely arranged or aggregated into a rather irregular terminal panicle; main axes of inflorescence of heads characteristically zigzag. Stipules at base of peduncles small, 0·75–1 mm. across, inconspicuous, soon caducous, not subcordate at the base. Calyx eglandular outside, puberulous to almost glabrous. Pods 6·5–15 × 2–2·8 cm., subcoriaceous to coriaceous, oblong to linear-oblong, rather densely puberulous and with many minute reddish glands. Seeds olive-brown, 11–12 × 8–10 mm., elliptic, compressed; areole 7–8 × 3–4 mm.

Mozambique. LM: Namaacha, Estação de Goba, fl. & fr. 24.x.1940, *Torre* 1854 (BM; LISC; LM).

Also in S. Africa (the Transvaal, Natal and eastern Cape Province) and Swaziland. Bushland and woodland.

A. brevispica subsp. *dregeana* is difficult to place taxonomically. In general it is closest to *A. brevispica* subsp. *brevispica*, but in one important character, the length of the petiole, it very frequently recalls *A. schweinfurthii* Brenan & Exell.

Typical *A. brevispica* is widespread in tropical Africa, but does not occur anywhere closer to our area than Angola and central Tanzania. Subsp. *dregeana* differs from this in:

(1) Indumentum on the young stem and leaf-rhachides very short and ± appressed, shorter than the glands (in typical *brevispica* the indumentum almost always spreading and longer than the glands).

(2) *Some* petioles on each plant almost always 15 mm. or more long (in typical *brevispica* all up to 13(15) mm. long).

(3) Petiolar gland sometimes 0, but up to 0·5–1·5 mm. long (in typical *brevispica* normally up to 1·5–3 mm., sometimes 0).

(4) Leaflets with marginal cilia very short and inconspicuous (in typical *brevispica* cilia nearly always conspicuous).

(5) Lower surface of leaflet usually ± appressed-pubescent (in typical *brevispica* glabrous or almost so).

None of these differences is perhaps completely free from exceptions on one side or the other, but in combination they work well and this fact, as well as the marked geographical disjunction involved, makes subsp. *dregeana* a generally easily recognizable entity.

A. schweinfurthii differs from *A. brevispica* subsp. *dregeana* in:

(1) Leaflets, except for marginal ciliation, glabrous or practically so on the surface beneath, not appressed-pubescent.

(2) Leaflets usually larger, 1 mm. or more wide.

(3) Pods glabrous or almost so except for glands, not with dense puberulence as well as glands as in subsp. *dregeana*.

20. **Acacia adenocalyx** Brenan & Exell in Bol. Soc. Brot., Sér. 2, **31**: 115, t. I fig. D (1957).—Brenan, F.T.E.A. Legum.-Mimos.: 97, fig. 15/23 (1959). TAB. **19** fig. D. Type from Tanzania.

Compact shrub or small tree 1–5 m. high, sometimes low and spreading, or even scandent; young branchlets puberulous and with very many minute brown glands. Prickles scattered, deflexed, arising from longitudinal bands along the wholly blackish-brown stems. Leaves: petiole 0·5–1·2 cm. long; pinnae 10–23 pairs, 0·6–3·5 cm. long; leaflets very numerous and neat, 0·3–0·75 mm. wide, linear-oblong; midrib subcentral at the base. Flowers white, in heads 8–10 mm. in diam., often irregularly paniculate. Stipules at base of peduncles small, 0·3–0·5 mm. wide, inconspicuous, soon caducous, not subcordate at the base. Calyx-lobes with many minute brown glands outside (use ×20 lens). Pods dehiscent, 6·5–15 × 1·6–3·6 cm., subcoriaceous or stiffly papery, oblong, puberulous or glabrous and with very many minute brown glands. Seeds black, 8–9 × 5·5–6 mm., smooth, elliptic, compressed; areole 5–6 × 2·5–3 mm.

Mozambique. N: Nacala, near Fernão Veloso, fl. & fr. 15.x.1948, *Barbosa* 2417 (BM; EA; K; LISC; LM). MS: Mossurize, Maringa, R. Save, fr. 27.vi.1950, *Chase* 2451 (SRGH).

Extending along the east African coastal belt through Tanzania to southern Kenya. In bushland; c. 10–180 m.

The short petioles enables this and typical *A. brevispica* Harms to be separated from *A. pentagona* (Schumach.) Hook f. and *A. latistipulata* Harms. *A. adenocalyx* differs from all of them in having small brown glands on the outside of the calyx-lobes, and the midrib of the leaflets subcentral at the base and not to one side. In addition the wholly blackish-brown twigs will readily distinguish *A. adenocalyx* from *A. brevispica*.

21. **Acacia latistipulata** Harms in Engl., Bot. Jahrb. **51**: 367 (1914).—Bak. f., Legum. Trop. Afr. **3**: 853 (1930).—Brenan, T.T.C.L.: 332 (1949); F.T.E.A. Legum.-Mimos.: 97, fig. 15/24 (1959).—Brenan & Exell in Bol. Soc. Brot., Sér. 2, **31**: 119, t. I fig. F (1957). TAB. **19** fig. F. Syntypes from Tanzania.

Arborescent or scandent shrub up to 6 m. high; young branchlets densely pubescent or puberulous, eglandular; older ones pale-grey. Prickles scattered, recurved. Leaves mostly large: petiole c. 1·8–5 cm. long; pinnae 10–31 pairs, c. 3–8·5 cm. long; leaflets very numerous, 0·8–2(3) mm. wide, linear to linear-oblong; midrib nearer one margin at the base. Heads of flowers c. 8 mm. in diam., in an ample terminal panicle. Stipules at base of peduncles comparatively large and conspicuous, 4·5–9 × 3–4·5 mm., ovate, acute at the apex, subcordate at the base, pubescent or puberulous. Calyx puberulous, eglandular. Pods dehiscent, 5–20 × 2·5–4·2 cm., subcoriaceous, oblong, glabrous except for some glands, umbonate over the seeds. Seeds dark-brown, 9–11 × 6–7 mm., smooth, elliptic, compressed; areole 5–6 × 2·5–3 mm.

Mozambique. N: Macondes, fr. 18.ix.1948, *Barbosa* 2217 (BM; K; LISC; LM); Mogincual, fl. 30.iii.1964, *Torre & Paiva* 11496 (LISC).

Also in southern Tanzania. Woodland; 50–220 m.

A very distinct species with unusually broad stipules and large foliage.

Torre & Paiva 11543 (LISC) from Mozambique, N: Mogovolas, fr. immat. 1.iv.1964, is near to *A. latistipulata* but the branchlets are glabrous or finely puberulous, scarcely armed, and the leaflets are mostly 2–3 mm. wide. More material of this is needed.

22. **Acacia montigena** Brenan & Exell in Kew Bull. **21**: 480 (1968). Type from Uganda.

Acacia monticola Brenan & Exell in Bol. Soc. Brot., Sér. 2, **31**: 125 (1957) non

J. M. Black (1937).—Brenan, F.T.E.A. Legum.-Mimos.: 98, fig. 15/26 (1959). Type as above.

Scandent shrub to 30 m. high; young branchlets ± densely pubescent with fulvous hairs and many red-purple glands mixed, dark-brown, later going blackish. Prickles deflexed, scattered, arising from longitudinal bands usually darker than the intervening ones. Leaves: petiole 1·5–3·5 cm. long; pinnae 7–19 pairs, 3–4·5(5·5) cm. long; leaflets 0·5–1·25 mm. wide, linear-oblong, glabrous or margins sparsely and inconspicuously ciliolate; midrib nearer one margin at the base. Flowers cream or white, in heads 10–15 mm. in diam. usually in pyramidal panicles. Stipules at base of peduncles small, 1–1·5 mm. wide, inconspicuous, soon caducous, not subcordate at the base. Calyx puberulous and eglandular outside. Corolla puberulous outside. Pods dark-brown, dehiscent, 8–18 × 3–4·5 cm., subcoriaceous, oblong, with margins 1–1·5 mm. wide and not very thickened. Seeds brown or black, 9–12 × 6–7 mm., smooth, elliptic, compressed; areole small, 5–7 × 2·5–3·5 mm.

Zambia. W: Kabompo R., below " boma ", fl. 23.xi.1952, *Holmes* 1017 (K). **Malawi.** N: Misuku Hills, Mugesse Forest, fr. immat. x.1953, *Chapman* 166 (K).
Also from the Congo, Uganda, Kenya and Tanzania.

The two specimens cited above are the only evidence for the occurrence of this species in our area, and the identity of neither specimen is beyond doubt. More material is needed.

A. montigena until recently would have been included under " *A. pennata* ". *A. montigena* is most closely related to *A. pentagona* (Schumach.) Hook. f., differing in the dense indumentum of hairs and also the glands clothing the young branchlets. The leaflets of *A. montigena* are usually narrower than those of *A. pentagona*, and the ovary is always pubescent while in *A. pentagona* it is frequently glabrous. Very important differences are found in the pod, which is dehiscent in *A. montigena* and with much less thickened margins than those of the indehiscent pods of *A. pentagona*; in addition the areole on the seed is small, not large. *A. pentagona* is typical of lowland rain- and swamp-forest, occurring only doubtfully in upland rain-forest, while *A. montigena* is typical of the latter, in spite of the Kabompo occurrence.

23. **Acacia schweinfurthii** Brenan & Exell in Bol. Soc. Brot., Sér. 2, **31**: 128, t. I fig. E (1957).—Brenan, F.T.E.A. Legum.-Mimos.: 99, fig. 15/27 (1959); in Kew Bull. **21**: 477 (1968).—White, F.F.N.R.: 84 (1962).—Mitchell in Puku, **1**: 105 (1963). TAB. **19** fig. E. Type from the Sudan.
Acacia brevispica var. *schweinfurthii* (Brenan & Exell) Ross & Gordon-Gray in Brittonia **18**: 62 (1966). Type as above.

Scandent shrub to 12 m., or sprawling, or a small spreading tree; young branchlets olive-green or pale-brown, later olive-brown, puberulous and glandular. Prickles deflexed, scattered, arising from brownish longitudinal bands darker than the intervening yellowish to grey ones. Leaves: petiole (1·5)2·6–6·5 cm. long, with a gibbous gland 1–2 × 0·5–1 mm.; pinnae 9–17 pairs, 3·5–7 cm. long; glands on rhachis between the top 1–3 pairs; leaflets numerous, (0·8)1–2(2·5) mm. wide, linear or linear-oblong, margins ciliolate with whitish appressed hairs; midrib nearer one margin at the base. Flowers white or palest yellow, in heads 8–12 mm. in diam. in ± pyramidal panicles. Stipules at base of peduncles small, 0·3–1·2 mm. wide, inconspicuous, soon caducous, not subcordate at the base. Calyx eglandular outside. Pods 9·5–19 × 1–2·9 cm., coriaceous or subcoriaceous, oblong, ± transversely plicate and umbonate over the seeds, margins not strongly thickened. Seeds blackish or dark-brown, 9–11 × 6·5–8 mm., smooth, elliptic; areole largish, 6–8 × 3–5 mm.

Var. **schweinfurthii.**—Brenan & Exell, tom. cit.: 130 (1957).—Brenan, loc. cit.

Leaflets glabrous beneath except for the ciliolate margins.

Botswana. N: Ngamiland, fl. xii.1930, *Curson* 443 (PRE). SE: Palapye Distr., Ratlolo, fl. 7.ii.1958, *de Beer* 699 (K; LISC; SRGH). **Zambia.** B: Sesheke, fl. 27.xii.1952, *Angus* 1047 (K). N: Lake Mweru, Chienge sandbank, fr. 4.vi.1933, *Michelmore* 384 (K). C: Chiwefwe, fr. 1.v.1957, *Fanshawe* 3227 (K). E: Sasare to Petauke, fl. 9.xii.1958, *Robson* 883 (K; LISC; SRGH). S: Mapanza E., fl. & fr. 12.xii.1953, *Robinson* 399 (K). **Rhodesia.** N: Urungwe Reserve, Mlelechi R., fl. 18.xii.1952, *Lovemore* 337 (SRGH). W: Bulawayo, Hillside Dam, fr. iii.1958, *Miller* 5140 (SRGH).

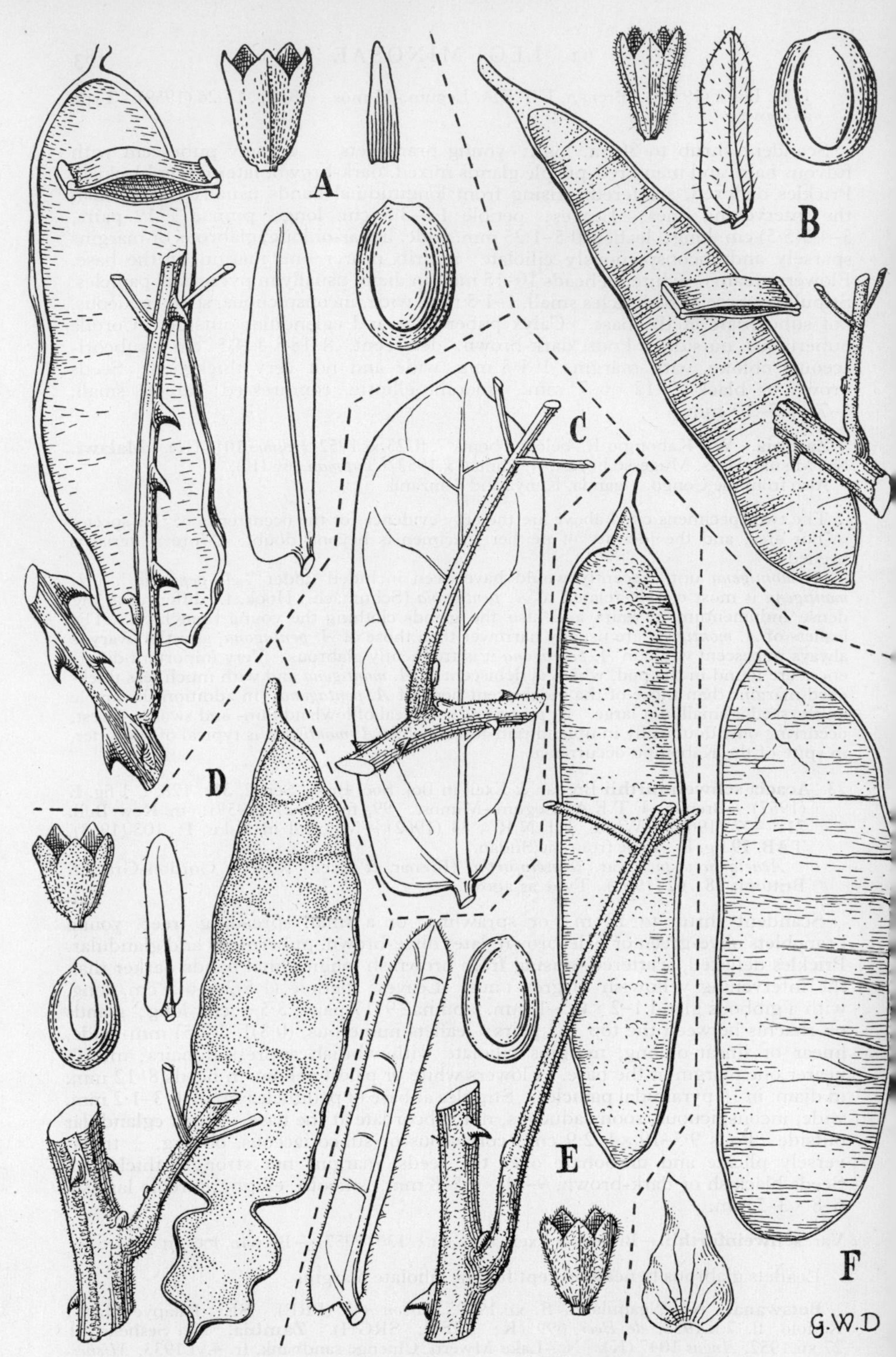

Tab. 19. A.—ACACIA PENTAGONA. Part of branchlet with petiole, showinggland ($\times 2$) leaflet; ($\times 8$); stipule ($\times 3$); calyx ($\times 8$); pod, with transverse section ($\times \frac{2}{3}$); seed ($\times 2$). B.—ACACIA BREVISPICA. Part of branchlet with petiole showing gland ($\times 2$); leaflet ($\times 8$); calyx ($\times 8$); pod, with transverse section ($\times \frac{2}{3}$); seed ($\times 2$). C.—ACACIA KRAUSSIANA. Part of branchlet with petiole showing gland ($\times 2$); leaflet ($\times 8$); pod ($\times \frac{2}{3}$); seed ($\times 2$). D.—ACACIA ADENOCALYX. Part of branchlet with petiole showing gland ($\times 2$); leaflet ($\times 8$); pod ($\times \frac{2}{3}$); seed ($\times 2$). E.—ACACIA SCHWEINFURTHII. Part of branchlet with petiole showing gland ($\times 2$); leaflet ($\times 8$); pod ($\times \frac{2}{3}$) seed ($\times 2$). F.—ACACIA LATISTIPULATA. Pod ($\times \frac{2}{3}$); stipule ($\times 3$). From Bol. Soc. Brot., Sér. 2, 31.

C: Salisbury, fl. xii.1924, *Eyles* 4044 (K; SRGH). E: Christmas Pass, fl. 10.xii.1961, *Chase* 7599 (K; LISC; SRGH). S: Simlala R., Tuli Circle, fr. v.1959, *Thompson* T/13/59 (K; SRGH). **Malawi.** C: Bua R. drift, Kasungu–Kota Kota road, fl. & fr. 13.i.1959, *Robson & Jackson* in *Robson* 1142 (K; LISC; SRGH). S: Shire–Zambezi, Chiloma, fl. xii, *Scott Elliot* 8699 (K). **Mozambique.** N: Porto Amélia, fl. 20.xii.1963, *Torre & Paiva* 9630 (LISC). MS: Cheringoma, between Inhaminga and Inhamitanga, fl. 18.ii.1948, *Andrada* 1055 (LISC). SS: Bilene, Macia, fr. 9.vii.1947, *Pedro &Pedrógão* 1396 (COI; LM; SRGH). LM: Maputo, Salamanga-Maputo, fl. 21.xi.1940, *Torre* 2136 (BM; K; LISC; LM).

From the Sudan Republic southwards to the Transvaal and Natal. Often forming thickets in woodland and forest fringing rivers and springs; sometimes in woodland away from rivers or on termite mounds; 30–1460 m.

Var. **sericea** Brenan & Exell in Bol. Soc. Brot., Sér. 2, **31**: 131 (1957).—Brenan, F.T.E.A. Legum.-Mimos.: 99 (1959). Type from Tanzania.

Leaflets ± appressed-silky-pubescent beneath.

Zambia. N: Chiengi Distr., near Lake Mweru, fr. 27.v.1961, *Astle* 734 (K). **Rhodesia.** E: Chipinga Distr., Sabi Valley Experimental Station, fl. xii.1959, *Soane* 208 (K; LISC; SRGH). **Mozambique.** Z: Mopeia, Chamo, fr. iii.1859, *Kirk* (K). T: Temangan, R. Mazoe, fr. 17.v.1966, *Neves Rosa* 264 (LM). MS: below Chigogo, fr. 16.iv.1860, *Kirk* (K).

Also in Tanzania. Habitat similar to that of var. *schweinfurthii*.

There are numerous references in the literature about our area to the occurrence of "*A. pennata* (L.) Willd.". True *A. pennata* does not occur in Africa, but until recently the name was used in a wide sense to cover species Nos. 19–24 inclusive. It is usually difficult or impossible to decide the precise application of these references, and for that reason they have not been cited here. Often *A. schweinfurthii* is likely to be the species referred to, but there is usually no certainty.

24. **Acacia pentagona** (Schumach.) Hook. f. in Hook., Niger Fl.: 331 (1849).—Brenan & Exell in Bol. Soc. Brot., Sér. 2, **31**: 134, t. I fig. A (1957).—Brenan, F.T.E.A. Legum.-Mimos.: 100, fig. 15/28 (1959).—F. White, F.F.N.R.: 432 (1962). TAB. **19** fig. A. Type from Ghana.

Mimosa pentagona Schumach., Beskr. Guin. Pl.: 324 (1827). Type as above.

Acacia pentaptera Welw. in Ann. Conselho Ultram. **1858**: 584 (1859).—Brenan & Exell, C.F.A. **2**: 287 (1956). Type from Angola.

Acacia silvicola Gilbert & Boutique in Bull. Jard. Bot. Brux. **22**: 179 (1952) pro parte quoad typum; F.C.B. **3**: 155 (1952). Type from the Congo.

An often tall liane; young branchlets sparsely puberulous to glabrous and eglandular, very rarely with inconspicuous sessile glands, red-brown to deep purplish. Prickles deflexed, scattered, arising from longitudinal bands usually darker than the intervening ones. Leaves: petiole (1·5)2–6 cm. long; pinnae 8–15 pairs, 2·5–9 cm. long; leaflets 0·7–1·8(2) mm. wide, linear or linear-oblong, glabrous or nearly so; midrib nearer one margin at the base. Flowers white, in heads 8–10(12) mm. in diam. usually in ample panicles. Stipules at base of peduncles small, 0·75–1·5 mm. wide, inconspicuous, soon caducous, not subcordate at the base. Calyx eglandular outside. Corolla glabrous or rarely sparingly puberulous outside. Pods thick, hard, dark-brown, indehiscent, 7·5–16 × 1·8–3·5 cm., oblong, with markedly thickened margins 2–4 mm. wide. Seeds black, 10–13 × 6–8 mm., smooth, ellipsoid, thick but somewhat compressed; areole large, 7–10 × 4·5–6 mm.

Zambia. N: Lake Mweru, fr. 6.viii.1958, *Fanshawe* 4655 (K). **Rhodesia.** E: Chirinda, fr. 20.x.1947, *Chase* in GHS 18029 (COI; K; SRGH). **Mozambique.** MS: Mossurize, Espungabera, fl. & fr. 15.xi.1943, *Torre* 6200 (K; LISC).

Widespread in the rain-forest regions of tropical Africa. In rain-forest and evergreen fringing forest; 450–1220 m.

A. pentagona is typically a liane of evergreen forest. From all the other species of the "*A. pennata*" complex occurring in our area it is separable by its thick normally indehiscent pods. It is distinct from *A. schweinfurthii* Brenan & Exell also in the darker twigs and larger leaflets. For the distinction between *A. pentagona* and *A. montigena* Brenan & Exell, see under the latter species.

25. **Acacia macrothyrsa** Harms in Engl., Bot. Jahrb. **28**: 396 (1900).—Bak. f., Legum. Trop. Afr. **3**: 851 (1930).—Brenan, T.T.C.L.: 336 (1949); F.T.E.A. Legum.-Mimos.: 101, fig. 15/30 (1959).—Gilbert & Boutique, F.C.B. **3**: 157 (1952).—Wild, S. Rhod. Dict. Bot.: 48 (1953).—Palgrave, Trees of Central Afr.: 246 (1956).—Burtt Davy & Hoyle, rev. Topham, N.C.L., ed. 2: 64 (1958).—F. White, F.F.N.R.: 86,

fig. 19 (1962).—Boughey in Journ. S. Afr. Bot. **30**: 157 (1964). Type from Tanzania.
Acacia buchananii Harms in Engl., Bot. Jahrb. **30**: 76 (1901).—Bak. f., tom. cit.: 852 (1930).—Steedman, Trees etc. S. Rhod.: 13 (1933). Type: Malawi, *Buchanan* 256 (B, holotype †; BM).

Small or medium tree 2–15 m. high; bark rough, fissured, grey (or brown *fide* F.C.B.). Stipules spinescent, up to 1·6(5·5) cm. long, stout, brown, glossy, compressed. Leaves large, 10–20 cm. wide; rhachis (with petiole) 10–37 cm. long; pinnae mostly 9–16(27) pairs; leaflets 12–70 pairs, 4(6)–11(20) × 1–3·5(6) mm., rather stiff and glossy above, glabrous, or ciliolate on the margins. Flowers orange or yellow, strongly and sweetly scented, in heads 8–13 mm. in diam., in a panicle up to c. 45 × 30 cm. whose branches (and usually also the main axis) are leafless. Pods blackish, blackish-purple or brown, 7–20 × 1·5–2·5 cm., coriaceous, glossy, glabrous, oblong, straight. Seeds olive- or deep-brown, 8–10 × 7–9 mm., smooth, elliptic to subcircular, compressed; areole 4–5 × 3–3·5 mm.

Zambia. B: Balovale, fr. vii.1933, *Trapnell* 1217 (K). N: Mporokoso Distr., Kabwe Plain, Mweru Wantipa, fl. immat. 15.xii.1960, *Richards* 13708 (K; SRGH). W: Matonchi Farm, fl. 15.ii.1938, *Milne-Redhead* 4590 (K). C: Great North Road 17 km. S. of Lusaka, fl. 20.iii.1952, *White* 2303 (FHO). E: between Nyimba and Fort Jameson, fr. 21.viii.1946, *Gouveia & Pedro* 1771 (LMJ; PRE). **Rhodesia.** N: Sipolilo, fl. 15.ii.1956, *Cleghorn* 139 (SRGH). C: Salisbury Commonage, fl. 13.ii.1956, *Orpen* 1/56 (K; SRGH). E: Umtali Commonage, fl. 30.i.1956, *Chase* 5693 (BM; K; SRGH). **Malawi.** N: Mzimba, fr. 8.vi.1938, *Pole Evans & Erens* 651 (K; PRE). C: Dedza Distr., Chongoni Forest, between Kanjole and Cibentu Hills, fl. 26.iii.1961, *Chapman* 1182 (SRGH). S: Blantyre Distr., Naperi house, 12.iii.1938, *Lawrence* 566 (K). **Mozambique.** N: Vila Cabral, Metónia, fl. iii.1933, *Gomes e Sousa* 1330 (BM; COI; K; LM). Z: Lugela, between Mocuba and Muobede, 24 km. from Mocuba, fl. 23.v.1949, *Barbosa & Carvalho* in *Barbosa* 2811 (K; LM; SRGH). T: Macanga, between Cásula and Chiuta, 3·6 km. from Casula, fl. 8.vii.1949, *Barbosa & Carvalho* in *Barbosa* 3495 (K; LMJ). MS: Manica, between Macequece and the frontier, fl. 25.iii.1948, *Mendonça* 3864 (BM; K; LISC).

Also in Ghana, Nigeria, the Congo, Sudan, Uganda, Kenya and Tanzania. Deciduous woodland and wooded grassland; 30–1500 m.

The young branchlets, inflorescence-axes, leaf-rhachides etc. are usually glabrous in East Africa. Elsewhere they may be puberulous, as in all West African material (including the type of *A. dalzielii* Craib) and in some specimens from the Congo, Mozambique and Zambia.

The leaflets vary much in size and number, but the variation does not seem to follow any clear pattern of geography, though specimens with particularly small and numerous leaflets have been collected in the Barotse and Western Provinces of Zambia. The large leaves coupled with the robust panicles of yellow or orange flower-heads make *A. macrothyrsa* easy to recognize.

26. **Acacia hockii** De Wild. in Fedde, Repert. **11**: 502 (1913).—Bak. f., Legum. Trop. Afr. **3**: 849 (1930).—Brenan, F.T.E.A. Legum.-Mimos.: 104, fig. 15/33 (1959).—F. White, F.F.N.R.: 85, fig. 18 c (1962). TAB. **15** fig. 6. Type from the Congo (Katanga).
Acacia stenocarpa sensu auct. mult. e.g.—Oliv., F.T.A. **2**: 351 (1871).—Bak. f., tom. cit.: 845 (1930).
Acacia seyal var. *multijuga* Schweinf. ex Bak. f., tom. cit.: 844 (1930).—O. B. Mill., B.C.L.: 21 (1948); in Journ. S. Afr. Bot. **18**: 24 (1952).—Brenan, T.T.C.L.: 338 (1949); in Mem. N.Y. Bot. Gard. **8**: 429 (1954).—Torre, C.F.A. **2**: 284 (1956). Syntypes from the Sudan.

Shrub or tree (1)2–6(12) m. high; bark not powdery, red-brown to greenish or rarely pale-yellow, peeling off in papery layers when not burned; young branchlets ± densely puberulous, rarely glabrous, with ± numerous reddish sessile glands, usually elongate and slender with reddish or brownish bark which does not peel to expose a powdery layer as in *A. seyal*. Stipules spinescent, up to 2(rarely 4) cm. long, mostly short, straight, suberect or spreading, subulate or flattened on the upper side; " ant-galls " and other prickles absent. Leaves often with a gland on the petiole and between the top 1(3) pairs of pinnae; pinnae (1)2–11 pairs; leaflets 9–29 pairs, 2–6·5 × 0·5–1(1·25) mm., usually ± densely ciliolate, sometimes glabrous, obtuse to acute but not spinulose-mucronate at the apex; lateral nerves invisible beneath. Flowers bright yellow, in axillary pedunculate heads 5–12 mm. in diam. borne along shoots of the current season; involucel $\frac{1}{3}$–$\frac{2}{3}$-way up the peduncle, 1·5–3 mm. long. Apex of bracteoles rounded to rhombic, sometimes pointed. Calyx (1)1·5–2 mm. long, glabrous except in upper part. Corolla 2·5–3·5 mm. long, glabrous outside. Pods (4)5–14 × 0·3–0·6(0·8) cm., as in *A. seyal* except for being often ± puberulous. Seeds olive-brown, 5–7 × 3–4 mm., smooth, elliptic, compressed; areole 3·5–4·5 × 2–2·5 mm.

Zambia. N: Mkupa Katandula, fl. & fr. 12.vi.1950, *Bullock* 2941 (K; SRGH). W: Mwinilunga Distr., slope E. of Matonchi R., fl. & fr. 15.ii.1938, *Milne-Redhead* 4587 (K). E: Luangwa Valley, Kapitomoyo's village, Mwunyamazi R., fr. 28.ix.1933, *Michelmore* 600 (K). **Malawi.** N: Katete, S. Mzimba, fl. & fr. immat. 18.vi.1954, *Jackson* 1356 (BM; K). C: Kasungu, fl. & fr. 24.viii.1946, *Brass* 17405 (BM; K; PRE; SRGH). S: Ncheu, fl. 22.vi.1962, *Robinson* 5405 (K; SRGH). **Mozambique.** Z: 72 km. N. of R. Limboé, fl. 2.vi.1938, *Pole Evans & Erens* 523 (K; PRE). T: Moatize, 50 km. SW. of Zóbuè, fl. 13.v.1961, *Leach & Rutherford-Smith* 10827 (K; LISC; SRGH).

Widespread in tropical Africa from Guinée to the Sudan southwards to our area and Angola. Woodland, wooded grassland, thicket and scrub; 460–1670 m.

A. hockii occupies a wide range both of habitat and altitude, and is also widespread in tropical Africa. The plant is correspondingly variable, and it may later be possible to divide it into infraspecific taxa.

In the past it has been confused with *A. seyal*, but, although the two are closely related, it seems preferable to maintain them as distinct species. *A. hockii* differs from *A. seyal* Del. primarily by having non-powdery bark. The twigs are usually (not always) more elongate and slender, with reddish or brownish bark which does not peel to expose the inner layer so characteristic of *A. seyal*. The young branchlets are usually clothed with a ± dense puberulence which is not found in *A. seyal*; rarely, however, the branchlets are glabrous except for sessile glands. The spines are never " ant-galled " and usually short, but may occasionally be up to 4 cm. long and whitish even on the flowering twigs.

The internode-length varies considerably in different gatherings, often resulting in great differences in general appearance, which may however be mainly or entirely due to the various habitats. There is evidence also that the bark may vary in colour, perhaps in the way that it does in *A. seyal*; but careful observation of this in the field is still wanted.

A strange tendency of *A. hockii*, shared also by *A. seyal* and *A. nilotica* (L.) Willd. ex Del. and some other species with bright yellow flowers, is for a few flowers to arise in the involucel on the peduncle, sometimes giving the appearance of a smaller secondary capitulum below the main one.

The records of *A. seyal* var. *multijuga* by O. B. Miller (ll. cc., supra) from Botswana require confirmation.

27. **Acacia karroo** Hayne, Darstell. u. Beschreib. Arzneyk. Gewächse, **10**: t. 33 (1827).—Bak. f., Legum. Trop. Afr. **3**: 843 (1930).—Burtt Davy, F.P.F.T. **2**: 346 (1932).—Steedman, Trees etc. S. Rhod.: 13 (1933).—Hutch., Botanist in S. Afr.: 138, 260, 411, 472, 512, 543, 547, 550, 552, 664 (1946).—O. B. Mill., B.C.L.: 19 (1948); in Journ. S. Afr. Bot. **18**: 22 (1952).—Codd, Trees & Shrubs Kruger Nat. Park: 44, fig. 38 h, i (1951).—Pardy in Rhod. Agric. Journ. **49**: 317 cum photogr. (1952).—Wild, S. Rhod. Bot. Dict.: 48 (1953).—Verdoorn in Bothalia, **6**: 409–13 (1954); in Fl. Pl. S. Afr. **31**: t. 1220 (1956).—Palgrave, Trees of Central Afr.: 242 cum tab. et photogr. (1956).—Burtt Davy & Hoyle, rev. Topham, N.C.L., ed. 2: 64 (1958).—Mogg in Macnae & Kalk, Nat. Hist. Inhaca I.: 10, 145 (1958).—Palmer & Pitman, Trees S. Afr.: 157 cum fig. et photogr. (1961).—F. White, F.F.N.R.; 85, fig. 18 D (1962).—Boughey in Journ. S. Afr. Bot. **30**: 157 (1964).—de Winter, de Winter & Killick, Sixty-six Transvaal Trees: 50 cum photogr. (1966). TAB. **15** fig. 7. Type from S. Africa (Cape Prov.).

Acacia natalitia E. Mey., Comm. Pl. Afr. Austr. **1**, 1: 167 (1836).—Sim, For. Fl. Port. E. Afr.: 57 (1909).—Burtt Davy, F.P.F.T. **2**: 347 (1932).—Burtt Davy & Hoyle, rev. Topham, N.C.L., ed. 2: 64 (1958). Type from Natal.

Acacia horrida sensu Sim, loc. cit.—Hutch., Botanist in S. Afr.: 541, 542 (1946).

Acacia horrida var. *transvaalensis* Burtt Davy in Kew Bull. **1908**: 158 (1908).—Sim, loc. cit. Type from the Transvaal.

Acacia seyal sensu Sim, loc. cit.

Acacia karroo var. *transvaalensis* (Burtt Davy) Burtt Davy in Kew Bull. **1922**: 328 (1922); F.P.F.T. **2**: 347 (1932). Type as for *A. horrida* var. *transvaalensis*.

Tree (1·5)3–15 m. high, rarely shrubby; bark on trunk dark-red-brown to blackish; young branchlets glabrous or rarely sparsely and inconspicuously puberulous, also with small inconspicuous pale to reddish sessile glands; epidermis flaking off to expose a dark-rusty-red not powdery under-bark, sometimes grey to brown and persistent. Stipules spinescent, up to 7(17) cm. long, rather robust, whitish, often ± deflexed, sometimes fusiform-inflated, up to 1 cm. (? or more) thick, but remaining distinct to the base and not confluent; other prickles absent. Leaves with a small to large (sometimes paired) gland at the junction of each pinna-pair, rarely lacking at the basal 1–2 pairs; sometimes a large gland on the upper side of the petiole; pinnae (1)2–7(9) pairs; leaflets 5–15(27) pairs, 4–7(12) × 1–3(5·5) mm., glabrous or rarely with minutely ciliolate margins, eglandular, obtuse to subacute but not spinulose-mucronate at the apex; lateral nerves invisible beneath. Flowers deep- or golden-yellow, in axillary pedunculate heads 8–12 mm. in diam. borne along shoots of the current season, sometimes aggregated into ± leafless terminal " racemes "; involucel c. $\frac{1}{3}$–$\frac{3}{4}$-way up the peduncle, c. 2 mm. long. Calyx 1·25–2 mm. long, subglabrous. Corolla c. 2·5–3 mm. long, glabrous or almost so. Pods dehiscent, (4)6–16 × 0·6–0·9(1) cm., linear, ± falcate, usually ± constricted (sometimes not) between the seeds, glabrous except for small usually inconspicuous glands. Seeds olive-green to -brown, 5–8 × 3–5 mm., oblong-elliptic, compressed; areole 4·5–5·5 × 2–3·5 mm.

Botswana. N: Ngamiland, fl. xii.1930, *Curson* 375 (PRE). SE: Mahalapye–Shushong road, Bonopitsi, fl. 14.i.1958, *de Beer* 543 (K; SRGH). **Zambia.** C: Lusaka-Kafue, 6 km., fl., *Angus* 1442 (BM; K; PRE). **Rhodesia.** N: Lomagundi, fr. 19.iv.1929, *Eyles* 6341 (K; SRGH). W: Bulawayo, fl. 9.xii.1920, *Borle* 15 (K; PRE). C: Salisbury, fl. xi.1919, fr. 19.viii.1921, *Eyles* 1908, 3180 (PRE; SRGH). E: Melsetter Distr., near Nyanyadzi settlement, fr. vii.1959, *Davies* 2564 (SRGH). S: 16 km. SE. of Chibi, fr. iv.1953, *Vincent* 215 (SRGH). **Malawi.** C: Dedza, fr. 28.vii.1960, *Chapman* 852 (BM; SRGH). S: Mlanje Mt., Fort Lister Gap, fr. 9.viii.1958, *Chapman* 625 (K; SRGH). **Mozambique.** N: Malema, Serra Meripa, fl. & fr. 5.ii.1964, *Torre & Paiva* 10477 (LISC). Z: NW. of Gúruè, fr., *A. J. Hornby* 2705 (K; PRE). T: Angónia, between Vila Coutinho and Dedze, fr. 17.vii.1949, *Andrada* 1780 (LISC). MS: Mossurize, Espungabera, fl. 15.xi.1943, *Torre* 6202 (BM; K; LISC; PRE). SS: Bilene, Macia, fr. 9.vii.1947, *Pedro & Pedrógão* 1393 (COI; K; LMJ; PRE). LM: Maputo, Bela Vista, Ponta de Ouro, fl. 10.xii.1961, *Lemos & Balsinhas* in *Lemos* 285 (BM; COI; K; LISC; SRGH).

Also in Angola, S. Africa, Swaziland and SW. Africa. Woodland, wooded grassland, coastal scrub, often by rivers and in valleys; near sea-level to 1520 m.

Acacia karroo shows a wide range of variation, but not in such a way as to allow any infraspecific taxa to be recognized. *A. natalitia* E. Mey., which has often been maintained as a separate species, merely corresponds with those forms of *A. karroo* with more numerous and smaller leaflets. These extremes are connected with normal *A. karroo* by innumerable gradations.

A. karroo is closely related to *A. seyal* Del., differing in the bark and (as far as our area is concerned) by the enlarged inflated spines being distinct to the base and not confluent below into a rounded ± 2-lobed structure.

28. **Acacia seyal** Del., Fl. Égypte Expl. Planches: 286, t. 52 fig. 2 (1813).—Oliv., F.T.A. **2**: 351 (1871).—Burkill in Johnston, Brit. Centr. Afr.: 245 (1897).—R.E.Fr., Wiss. Ergebn. Schwed. Rhod.-Kongo-Exped. **1**: 64 (1914).—Bak. f., Legum. Trop. Afr. **3**: 844 (1930).—Hutch., Botanist in S. Afr.: 327 (1946).—O. B. Mill., B.C.L.: 21 (1948); in Journ. S. Afr. Bot. **18**: 24 (1952).—Brenan, T.T.C.L.: 337 (1949); F.T.E.A. Legum.-Mimos.: 103, fig. 15/32 (1959).—Gilbert & Boutique, F.C.B. **3**: 160, fig. 4 A–D (1952).—F. White, F.F.N.R.: 85, fig. 18 B (1962).—Mitchell in Puku, **1**: 105 (1963). Type from Egypt.

Tree (2)3–9(12) m. high; bark on trunk powdery, white to greenish-yellow or orange-red; young branchlets with few sparse hairs to almost glabrous and with numerous reddish sessile glands (rarely, and not in our area, rather densely puberulous); epidermis of twigs becoming reddish and conspicuously flaking off to expose a greyish or ± reddish powdery bark. Stipules spinescent, up to 8 cm. long; " ant-galls " present or not; other prickles absent. Leaves often with a rather large gland on the petiole and between the top 1–2 pairs of pinnae; pinnae (2)3–7(8) pairs; leaflets (7)11–20 pairs, 3–8(10) × 0·75–1·5(3) mm., in our area sparingly ciliolate to glabrous; lateral nerves invisible beneath. Flowers bright-yellow, in axillary pedunculate heads 10–13 mm. in diam. borne on terminal or short lateral shoots of the current season; involucel in the lower half of the peduncle, 2–4 mm. long; apex of bracteoles rounded to elliptic, sometimes pointed. Calyx 2–2·5 mm. long, inconspicuously puberulous in upper part. Corolla 3·5–4 mm. long, glabrous outside. Pods dehiscent, (5)7–20(22) × 0·5–0·9 cm., linear, ± falcate, ± constricted between the seeds, finely longitudinally veined, glabrous except for some sessile glands. Seeds olive to olive-brown,

7–9 × 4·5–5 mm., faintly and minutely wrinkled, elliptic, compressed; areole 5–6 × 2·5–3·5 mm.

Typical var. *seyal*, in which the " ant-galls " are absent, although occurring in Tanzania and widespread in tropical Africa from there northwards, has not so far been recorded from our area.

Var. **fistula** (Schweinf.) Oliv., loc. cit.—Bak. f. loc. cit.—Brenan, T.T.C.L.: 338 (1949); F.T.E.A. Legum.-Mimos.: 103 (1959). Syntypes from the Sudan.
Acacia fistula Schweinf. in Linnaea, **35**: 344 (1867–8). Types as above.

Some pairs of spines are fused at the base into " ant-galls " 0·8–3 cm. in diam. which are greyish or whitish, often marked with sienna-red and with a longitudinal furrow down the centre, making the " galls " ± 2-lobed.

Zambia. C: Kafue, fl. 8.vi.1958, *Fanshawe* 4561 (K). S: Mazabuka to Choma, mile 8, fl. 14.vii.1952, *White* 3008, 3009 (K). **Malawi.** C: Domira Bay, st. 13.vii.1936, *Burtt* 6048 (K). S: Chisawani near Tuchila on Palombe Plain, fl. & fr. 2.ix.1959, *Willan* (1)59 (K; SRGH). **Mozambique.** N: Porto Amélia, between Porto Amélia and Ancuabe, fl. 24.viii.1948, *Barbosa* in *Mendonça* 1870 (BM; LISC; LM).

Also extending northwards through eastern Africa to the Sudan and Somalia. In woodland and wooded grassland, especially on seasonally flooded black-cotton soils along water-courses; 20–1220 m.

For the differences between *A. seyal* and *A. hockii*, see under the latter.

The colour of the bark is variable. In our area it seems most commonly to be yellow or greenish-yellow, but orange-red bark also occurs. The two specimens collected by White cited above represent trees showing the different colours but occurring in the same place.

A. seyal var. *fistula* may occur in Botswana, though I have not seen material. See records by O. B. Miller (ll. cc., supra) under *A. seyal*.

28 × 10. **Acacia seyal** var. **fistula** (Schweinf.) Oliv. × **xanthophloea** Benth.—Brenan, F.T.E.A. Legum.-Mimos.: 109 (1959).

Differs from *A. seyal* var. *fistula* in being a taller tree up to 20 m. high, and in having the inflorescences in lateral fascicles on the twigs of the previous season and the pods mostly shorter than usual (4–11 cm.) and less falcate. Differs from *A. xanthophloea* in having some of the spines showing a tendency to be enlarged below and thus ± fusiform and the pods ± falcate.

Malawi. S: Shirwa, fr. 22.x.1941, *Greenway* 6349 (K; PRE).

Not known elsewhere. In woodland on a black clay loam on a lake flood-plain; 520 m.

The pods of this apparent hybrid are conspicuously irregular, ranging in length from 4–11 cm., and in width from 6–10 mm. They are often apparently ill-formed, with irregular constrictions, but with a distinct tendency to be wider than is usual in *A. seyal*. They are also correspondingly variable in the extent to which they are curved.

29. **Acacia davyi** N.E.Br. in Kew Bull. **1908**: 161 (1908).—Burtt Davy, F.P.F.T. **2**: 346 (1932).—Hutch., Botanist in S. Afr.: 361 (1946). Syntypes from the Transvaal and Swaziland.

Small tree 2–4·5 m. high; bark thick, soft, corky, yellow to yellowish-brown; young branchlets puberulous, later glabrescent, turning grey-brown. Stipules spinescent, short, up to (2)3 cm. long, slender, mostly ascending; " ant-galls " and other prickles absent. Leaves with small sessile glands at the junctions of the top 1–3 (rarely to 7) pairs of pinnae and often at or near the junction of the lowest pair also; pinnae (10)14–27 pairs; leaflets (17)20–44 pairs, 2·5–5·5 × 0·7–1 mm., glabrous, entire, eglandular, obtuse to subacute but not spinulose-mucronate at the apex; lateral nerves invisible beneath. Flowers deep-yellow, in axillary fascicled pedunculate heads 5–7 mm. in diam. borne along shoots of the current year and often in ± elongate terminal " racemes "; involucel $\frac{1}{3}$–$\frac{2}{3}$-way up the peduncle, c. 1·5 mm. long. Calyx 1·25–1·5 mm. long, glabrous except on the lobes. Corolla 1·75 mm. long, glabrous. Pods dehiscent, 5–12 × 0·5–0·8 cm., linear, straight to falcate, not or somewhat constricted between the seeds, sometimes torulose, glabrous, eglandular, inconspicuously venose. Seeds olive-brown, 6–7·5 × 5·5 mm., elliptic to subcircular, compressed: areole 4·5 × 2·5–3·5 mm.

Mozambique. LM: Goba, Namaacha, fl. 22.xii.1944, *Mendonça* 3447 (BM; K; LISC).
Also in the Transvaal, Swaziland and Natal. Ecology insufficiently known. Recorded from *Acacia-Bauhinia* bush; 600–800 m.

A. davyi has on account particularly of its numerous pinnae a superficial resemblance to *A. arenaria* Schinz. The latter has a very different distribution, not being found in Mozambique, and is also more strongly pubescent and has white or pink flowers.

30. **Acacia tenuispina** Verdoorn in Bothalia, **6**: 156, fig. 5 (1951). Type from the Transvaal.
Acacia permixta var. *glabra* Burtt Davy in Kew Bull. **1922**: 330 (1922); F.P.F.T. **2**: 340 (1932).—Hutch., Botanist in S. Afr.: 664 (1946). Type as above.

Rhizomatous shrub 0·5–1·2(2·4) m. high; young branchlets with numerous rather small pale and inconspicuous sessile glands, otherwise glabrous. Stipules spinescent, up to 5 cm. long, slender, whitish, sometimes slightly arcuate-deflexed; " ant-galls " and other prickles absent. Leaves with a cup-shaped to slightly stipitate gland at the junction of the top 1–2 pairs or of the only pair of pinnae; pinnae (1)2–6 pairs; leaflets (3)4–8 pairs, (2)3–5 × 0·9–1·5 mm., glabrous, entire, eglandular, spinulose-mucronate at the apex; lateral nerves invisible beneath. Flowers yellow, in axillary pedunculate heads 7–10 mm. in diam. scattered along shoots of the current, or a few on shoots of the previous, season; involucel c. $\frac{1}{2}$–$\frac{3}{4}$-way up the peduncle, 1–1·5 mm. long. Calyx 1·5–2 mm. long, glabrous. Corolla c. 2·5 mm. long, glabrous outside. Pods dehiscent, 2–4 × 0·5–0·7 cm., ± falcate, not or only slightly constricted between the seeds, with numerous dark sessile pustular glands scattered over the surface. Seeds olive-brown, 5–7 × 4–4·5 mm., elliptic, compressed; areole 3 × 2 mm.

Botswana. SE: Mahalapye road 11 km. from Shushong, fl. 14.i.1958, *de Beer* 560 (K; SRGH).
Also in the Transvaal. Forming extensive low thickets on black cotton soil; 850–970 m.

31. **Acacia exuvialis** Verdoorn in Bothalia, **6**: 154, fig. 2 (1951).—Codd, Trees & Shrubs Kruger Nat. Park: 52 (1951). Type from the Transvaal.

Small tree, or sometimes a shrub, 1·5–5 m. high; bark peeling on the stems; young branchlets with some scattered inconspicuous dark sessile glands, otherwise glabrous. Stipules spinescent, up to 6 cm. long, whitish, sometimes ± deflexed; " ant-galls " and other prickles absent. Leaves with a small sessile to shortly stipitate gland at the junction of each of the 1–6 pairs of pinnae; petiole and rhachides otherwise eglandular or only with a few small scattered inconspicuous glands; leaflets 3–6 pairs, 3–10 × 1·5–4·5 mm., glabrous, entire, eglandular or almost so, ± spinulose-mucronate at the apex; lateral nerves invisible beneath. Flowers golden-yellow, in axillary pedunculate heads 7–10 mm. in diam. scattered along shoots of the current or previous season; involucel $\frac{2}{3}$–$\frac{5}{6}$-way up the peduncle, 2·5–4 mm. long. Calyx 1·5–2 mm. long, glabrous or subglabrous. Corolla 2·5 mm. long, glabrous outside. Pods dehiscent, 2–3·5 × 0·6–0·8 cm., falcate, ± constricted (sometimes only slightly) between the seeds, eglandular or almost so. Seeds olive-green to olive-brown, c. 6–7 × 4·5–6 mm., elliptic, compressed; areole 3–4 × 2·5–3 mm.

Rhodesia. S: Nuanetsi Distr., Malvernia to Palfrey's, fr. vi.1957, *Davies* 2430 (K; SRGH).
Also in the Transvaal. Roadside scrub in dwarf *Colophospermum mopane* on gravelly ridge; 460 m.

Until more and better material is available, there must remain some doubt about the identification of *Davies* 2430 with *A. exuvialis*. This doubt is not made less by *Patel* 3 (SRGH), Nuanetsi Distr., near Buffalo Bend, fr. 23.iv.1961, which seems very similar to the Davies specimen, but has slightly more glandular pods to 10 mm. wide.

32. **Acacia borleae** Burtt Davy in Kew Bull. **1922**: 325 (1922).—Codd, Trees & Shrubs Kruger Nat. Park: 41, fig. 34 a (1951).—Verdoorn in Bothalia, **6**: 154, fig. 1 (1951).—Boughey in Journ. S. Afr. Bot. **30**: 157 (1964). TAB. **15** fig. 8. Type: Mozambique, Lourenço Marques, *Borle* 271 (PRE).

Shrub 1–5 m. high; young branchlets with numerous pale to reddish sessile pustular glands, sometimes viscid, in addition puberulous with hairs less than 0·25 mm. long; hairs sometimes absent. Stipules spinescent, up to 5 mm. long, slender, whitish, sometimes slightly deflexed; " ant-galls " and other prickles absent. Leaves with a small sessile gland at the junction of each of the 1(2) top pairs of pinnae; a rather large sessile gland at or below the junction of the lowest pair, or often absent; smaller scattered sessile glands present and sometimes numerous on the petiole and rhachides; pinnae 2–10 pairs; leaflets (3) 5–15 pairs, 1·5–6·5 × 1–2 mm., margins clearly crenulate-glandular and usually also minutely ciliolate, shortly spinulose-mucronate or not at the apex; surface also with ± numerous pale or dark sessile glands; lateral nerves invisible beneath. Flowers yellow, in axillary pedunculate heads 8–10 mm. in diam. scattered along the leafy shoots of the current season; involucel $\frac{2}{5}$–$\frac{4}{5}$-way up the peduncle, 1–2 mm. long. Calyx 1·5–2 mm. long, subglabrous. Corolla 2·5 mm. long, glabrous outside. Pods dehiscent, 3·5–7 × 0·6–1 cm., falcate, ± moniliform and constricted between the seeds, with numerous rather small pale to dark sessile pustular glands on the surface, in addition puberulous. Seeds olive, 5–6 × 4·5–5 mm., elliptic to subcircular, compressed; areole 3·5–4 × 2·5–3·5 mm.

Rhodesia. E: Chipinga Distr., Chisumbanje, fl. 20.i.1956, *Mowbray* 99 A (K; SRGH). S: Beitbridge Distr., 13 km. ENE. of Tuli police camp, fl. & fr. 23.iii.1959, *Drummond* 5967 (K; LISC; LM; SRGH). **Mozambique.** SS: Guijá, Lagoa Chinangué, fr. 6.vii.1947, *Pedro & Pedrógão* 1326 (COI; K; PRE; SRGH). LM: Sábiè, Moamba, fr. 5.vi.1948, *Torre* 7948 (BM; K; LISC).

Also in S. Africa (the Transvaal and Natal) and Swaziland. Forming thickets, often in *Acacia* or *Colophospermum* woodland; 430–610 m.

A. borleae is outstanding on account of the numerous sessile glands on the surface and margin of the leaflets.

To the south of our area, the short spreading hairs on the young shoots and leaves appear to be always absent, while in our area puberulence is generally present. Both conditions occur in southern Mozambique. This seems to be an example of minor geographical variation.

33. **Acacia torrei** Brenan in Kew Bull. **21**: 480 (1968). Type: Mozambique, between Inhaminga and Rio Urema, *Torre* 4068 (K; LISC, holotype; PRE).

Shrub 1–2 m. high (? taller and up to 5 m.); young branchlets with numerous conspicuous dark sessile pustular glands, sometimes also with a very few inconspicuous hairs up to 0·5 mm. long. Stipules spinescent, up to 5 cm. long, sometimes rather thicker than in related species, brownish to whitish, often somewhat deflexed; " ant-galls " and other prickles absent. Leaves with a small sessile gland at the junction of each of the top 1–2 pairs of pinnae, often also with a similar gland on the upper side of the petiole, and in addition pubescent and with numerous smaller dark scattered glands; pinnae 7–13(20) pairs; leaflets 6–15 pairs, (2)3–6·5 × 1–2(3) mm., ciliate on the margins, otherwise glabrous, eglandular except often for a few inconspicuous glands on the margin near the apex, spinulose-mucronate at the apex; lateral nerves invisible beneath. Flower-colour uncertain, probably yellow. Calyx c. 2 mm. long. Corolla only known withered and torn. Pods dehiscent, 3–6 × 0·8–1 cm., falcate, constricted (often irregularly) or not between the seeds, with numerous conspicuous dark sessile pustular glands over the surface, as well as numerous spreading setiform hairs up to c. 1 mm. long. Seeds olive-brown, c. 6 × 6 mm., subcircular, compressed; areole 3–3·5 × 2·5–3 mm.

Mozambique. MS: between R. Urema and Inhaminga, fr. 16.v.1942, *Torre* 4068 (BM; K; LISC; PRE).

Apparently endemic to Mozambique. Gregarious in savanna.

34. **Acacia permixta** Burtt Davy in Kew Bull. **1922**: 330 (1922) pro parte excl. var. *glabram*; F.P.F.T. **2**: 340 (1932).—Verdoorn in Bothalia, **6**: 155, fig. 3 (1951).—Boughey in Journ. S. Afr. Bot. **30**: 158 (1964). Type from the Transvaal.

Shrub or small tree 1·2–3(5) m. high; bark reddish-brown; young branches ± densely hairy or tomentose with spreading grey to whitish hairs 0·75–2 mm. long, among which some conspicuous reddish sessile glands are present. Stipules spinescent, up to 6·5 cm. long, whitish, often slightly deflexed; " ant-galls " and other prickles absent. Leaves with a small raised columnar gland at the junction of each pinna-pair, or lacking at a median pair; no special gland on the upper side of the petiole, but conspicuous reddish sessile glands irregularly scattered there and on the rhachides; pinnae 1–4 pairs; leaflets 5–10 pairs, 2–6 × 1–2(3) mm., ciliate, sometimes glandular on the margins towards the apex, some at least spinulose-mucronate at the apex; lateral nerves invisible beneath. Flowers yellow, in axillary pedunculate heads 8–12 mm. in diam. scattered along the leafy shoots of the current season; involucel $\frac{2}{3}$–$\frac{5}{6}$-way up the peduncle, 2·5–3·5 mm. long. Calyx 1·75–2 mm. long, subglabrous. Corolla 3 mm. long, glabrous outside. Pods dehiscent, 2–4(6·5) × 0·6–1 cm., with conspicuous reddish sessile raised glands, otherwise glabrous. Seeds olive or olive-brown, 5–6 × 4–5 mm., elliptic to subcircular, compressed; areole 2–3 × 2–2·5 mm.

Rhodesia. S: Fort Tuli, fr. 23.iii.1959, *Drummond* 5956 (K; LISC; LM; SRGH).
Also in the Transvaal. Dry rocky hills and ridges on shallow gravelly basaltic soil, sometimes with *Commiphora* and *Colophospermum mopane*.

35. **Acacia nebrownii** Burtt Davy in Kew Bull. **1921**: 50 (1921).—Bak. f., Legum. Trop. Afr. **3**: 851 (1930).—O. B. Mill., B.C.L.: 20 (1948); in Journ. S. Afr. Bot. **18**: 23 (1952).—Verdoorn in Bothalia, **6**: 156, fig. 4 (1951).—Boughey in Journ. S. Afr. Bot. **30**: 158 (1964). TAB. **15** fig. 9. Syntypes; Botswana, Kwebe Hills, *Mrs. E. J. Lugard* 16 (K); also from the Transvaal, Swaziland and SW. Africa.
Acacia rogersii Burtt Davy in Kew Bull. **1922**: 331 (1922); F.P.F.T. **2**: 342 (1932).—Bak. f., loc. cit.—O. B. Mill., B.C.L.: 21 (1948). Type from the Transvaal.

Shrub 1–5 m. high (? higher or even a tree to 7 m.); young branchlets with numerous dark sessile pustular glands, otherwise glabrous. Stipules spinescent, up to 6 cm. long, slender, whitish, sometimes slightly arcuate and deflexed; " ant-galls " and other prickles absent. Leaves with a shortly columnar or narrowly cup-shaped gland at the junction of the pinna-pair, or at the upper pair of two; pinnae 1(2) pairs; leaflets 3–5 pairs, 2–8 × 1–5 mm., eglandular or with some small pale inconspicuous glands on the margin and sometimes the surface, mostly very shortly mucronate at the apex; margin otherwise glabrous, entire; lateral nerves invisible beneath. Flowers golden-yellow, in axillary pedunculate heads 8–12 mm. in diam. produced along last season's shoots; involucel basal or up to $\frac{1}{4}$-way up the peduncle, 1–2 mm. long. Calyx 1·25–2 mm. long, glabrous. Corolla 2·5 mm. long, glabrous outside. Pods dehiscent, 2–4·7 × 0·9–1·1 cm., falcate to slightly falcate, not or scarcely constricted between the seeds, with numerous conspicuous dark sessile pustular glands scattered over the surface, otherwise glabrous. Seeds olive to olive-brown, 8–10 × 6–7 mm., elliptic, compressed; areole 4–6 × 3–3·5 mm.

Botswana. N: Kwebe Hills, fl. & fr. viii.1897–ii.1898, *Mrs. E. J. Lugard* 16 (K). SW: Takatshwane Pan, fl. 20.ii.1960, *Wild* 5086 (K; SRGH). **Rhodesia.** W: Matobo Distr., near Tuli R. close to junction with Shashi R., fl. 31.viii.1958, *Darbyshire* 2888 (SRGH). S: Gwanda Distr., Beitbridge, Shashi Camp, fl. 28.viii.1958, *West* 3710 (K).
Also in SW. Africa and the Transvaal N. of the Soutpansberg. Forming thickets in grassland in dry areas, or with *Colophospermum mopane*, *Terminalia prunioides*, *Acacia mellifera* subsp. *detinens*, etc.; 550–910 m.

36. **Acacia swazica** Burtt Davy in Kew Bull. **1922**: 332 (1922); F.P.F.T. **2**: 342 (1932). —Hutch., Botanist in S. Afr.: 365, 370 (1946).—Codd, Trees & Shrubs Kruger Nat. Park: 51 (1951).—Verdoorn in Bothalia, **6**: 156, fig. 6 (1951). Syntypes from Swaziland and the Transvaal.

Shrub or small tree 1–3 m. high; young branchlets glabrous except for ± numerous conspicuous reddish sessile pustular glands. Stipules not spinescent, up to 5 cm. long, whitish, slender, sometimes slightly deflexed; " ant-galls " and other prickles absent. Leaves with a small sessile gland at the junction of a single pair of pinnae, or if more than one pair then gland at the lowest pair sometimes absent; no special gland on the upper side of the petiole, but small inconspicuous pale to reddish sessile glands thinly scattered there and on the rhachides; pinnae 1–2(3) pairs; leaflets (3)4–6(7) pairs, (2)4–9(12) × 1·5–5·5 mm., glabrous, spinulose-mucronate at the apex; margin entire, eglandular or with a few very inconspicuous glands towards the apex; lateral nerves ± prominent and visible beneath. Flowers orange-yellow, in axillary pedunculate heads 8–12 mm. in diam. scattered along the leafy shoots of the current, or sometimes previous, season; involucel $\frac{1}{2}$–$\frac{3}{4}$-way up the peduncle (sometimes slightly above or below this range), large, 2–5 mm. long. Calyx 1·5–2 mm. long, glabrous. Corolla 3 mm. long, glabrous outside. Pods dehiscent, 2–5·5 × 0·8–1·2 cm., falcate, glabrous except for numerous conspicuous dark sessile pustular glands outside. Seeds olive to olive-brown, 6·5–7 × 5–5·5 mm., elliptic, compressed; areole 3·5–4 × 2·5–4 mm.

Mozambique. LM: Sábiè, 75 km. NW. of Lourenço Marques, between Moamba and Movene, fl. & fr. 18.ii.1955, *E.M. & W.* 478 (LISC; SRGH).

Also in the Transvaal and Swaziland. In *Combretum apiculatum–Acacia–Ozoroa insignis* bush.

A. swazica is separated from all its close relatives by the prominent venation on the lower surface of the leaflets, which are often larger than is usual in the other species. *A. giraffae Willd.* has strictly apical involucels and non-spinulose-mucronate leaflets.

37. **Acacia giraffae** Willd., Enum. Hort. Berol.: 1054 (1809).—Burch., Trav. Int. S. Afr. **2**: 240–241 (1824).—Eyles, Trans. Roy. Soc. S. Afr. **5**: 362 (1916).—Bak. f., Legum. Trop. Afr. **3**: 835 (1930).—Burtt Davy, F.P.F.T. **2**: 340, fig. 59 (1932).—Hutch., Botanist in S. Afr.: 178, 341, 386, 412, 418, 424, 425, 481, 543, 547, cum photogr. (1946).—O. B. Mill., B.C.L.: 18 (1948); in Journ. S. Afr. Bot. **18**: 21 (1952).—Fanshawe, Fifty Common Trees N. Rhod.: 8 cum fig. (1952).—Pardy in Rhod. Agric. Journ. **50**: 4 cum photogr. (1953).—Wild, S. Rhod. Bot. Dict.: 47 (1953).—Torre, C.F.A. **2**: 281 (1956).—Story in Mem. Bot. Surv. S. Afr. **30**: 23 (1958).—Leistner in Koedoe, **4**: 101–104 (1961).—Palmer & Pitman, Trees S. Afr.: 153 cum fig. et photogr. (1961).—F. White, F.F.N.R.: 84, fig. 17 (1962).—Mitchell in Puku, **1**: 104 (1963).—de Winter, de Winter & Killick, Sixty-six Transvaal Trees: 46 cum photogr. (1966).—Boughey in Journ. S. Afr. Bot. **30**: 157 (1964). TAB. **15** fig. 10. Type from S. Africa.

Tree (4·5)6–16(22) m. high, often rather large, rounded or flat-topped with a spread up to 18 m.; bark deeply furrowed, grey-black to blackish-brown; young branchlets glabrous to rather densely spreading-pubescent, eglandular. Stipules spinescent, up to 5(8·5) cm. long, often rather stout, often thickened below and fused together at the base into an enlarged " ant-gall " up to c. 1·5–2 × 2–2·5 cm., furrowed or not down the middle; other prickles absent. Leaves with a small sessile gland at the junction of each of the (1)2–5 pairs of pinnae; no petiolar gland; leaflets (6)8–15 pairs, 4·5–13 × 1·5–4·5 mm., glabrous to pubescent especially on the margins, entire, eglandular, rounded to subacute but not spinulose-mucronate at the apex; lateral nerves ± prominent and visible beneath. Flowers bright golden-yellow, in axillary often long-pedunculate heads 8–12 mm. in diam. scattered along shoots usually of the previous season, rarely of the current one; involucel strictly apical. Calyx 2·5 mm. long, glabrous. Corolla 3–3·5 mm. long, glabrous or nearly so. Pods indehiscent, (4)6–11 × 2·5–4·7 cm., straight to falcate, very densely grey-velutinous all over. Seeds deep-chestnut-brown, 8–13 × 7–10 mm., irregular in size and shape, sometimes scarcely compressed; areole variable in size, 3–6·5 × 2–4 mm.

Caprivi Strip. About 80 km. from Katima Mulilo on road to Linyanti, fr. 27.xii.1951, *Killick & Leistner* 3141 (PRE). **Botswana.** N: Chobe R., fl. 28.vii.1950, *Robertson & Elffers* 70 (K; PRE). SW: Damara Pan, fr. 20.iv.1930, *Van Son* 28869 (BM; PRE). SE: Mahalapye, st. ix.1961, *Yalala* 136 (K; SRGH). **Zambia.** B: Sesheke, fr. immat. 31.i.1952, *White* 1983 (K). S: Namwala, fl. 23.vi.1952, *White* 2977 (K). **Rhodesia.** W: Wankie Game Reserve, fl. 13.ix.1958, *Guy* 55/58 (SRGH). S: Nuanetsi Distr., 16 km. W. of. Marhumbini Mission, fl. 12.ix.1967, *Müller* 659 (K; SRGH).

Also in Angola, SW. Africa and S. Africa. Generally on Kalahari Sand; in dry woodland, wooded grassland, and sometimes in *Baikiaea-Acacia* forest; 900–1050 m.

On account of the combination of leaflets with prominent venation beneath, apical involucels and bright-yellow flowers, this species is easily separated from all but *A. farnesiana*, which differs greatly in habit, non-inflated spines, different arrangement of glands on the leaf-rhachis, usually narrower leaflets, and smaller glabrous pods.

A. giraffae is the Camelthorn or Kameeldoorn of S. Africa. Its characteristic thick-walled pods which have transverse partitions between the seeds, are eaten by stock.

Roots, probably of *A. giraffae*, have been reported at a depth of 150 feet in a borehole in SW. Africa (see Mem. Bot. Surv. S. Afr. **27**: 117 (1952)).

The pollen-grains of *Acacia* are multiple-celled. According to Coetzee in S. Afr. Journ. Sci. **52**: 23 (1955), those of *A. giraffae* are anomalous in consisting of 32 cells and not 16 as in the other species studied.

38. **Acacia haematoxylon** Willd., Enum. Hort. Berol.: 1056 (1809).—Bak. f., Legum. Trop. Afr. **3**: 835 (1930).—Hutch., Botanist in S. Afr.: 413 (1946).—O. B. Mill. in

Journ. S. Afr. Bot. **18**: 22 (1952).—Palmer & Pitman, Trees S. Afr.: 15 cum fig. et photogr. (1961).—Volk in Journ. SW. Afr. Wiss. Gesell. Windh. **20**: 43, fig. 7 (1966). TAB. **15** fig. 11. Type from S. Africa.

Shrub or small tree 1–6 m. high; young branchlets ± densely grey-puberulous or tomentellous; branchlets slender, ± flexuous or zig-zag, freely branched. Stipules spinescent, 0·5–5·7 cm. long, slender, divergent, straight, brownish to whitish, never inflated; other prickles absent. Leaves ± (usually very) densely grey-tomentellous, 2-pinnate but the leaflets so small close and imbricate that the pinnae resemble single linear crenulate leaflets; petiole 1–3(5) mm. long; rhachis 0·8–5 cm. along; pinnae of best-developed leaves of mature shoots 15–27 pairs, though ones with fewer pairs often also present; pinnae 5–12 mm. long; leaflets minute, 0·25–0·75 × c. 0·5 mm., closely imbricate, ± tomentellous. Flowers yellow, in globose heads; involucel at or above the middle or at the apex of the densely grey-tomentellous (and somewhat glandular) peduncle. Corolla c. twice as long as the calyx, tomentellous on the lobes outside. Pods indehiscent, 9–14 × 0·9–1·3 cm., arcuate, densely grey-tomentellous outside and with minute red-brown glands, turgid, sometimes almost round in section, ± constricted between the seeds. Seeds deep purplish-brown, 9–11 × 6–8 × 3–4 mm., ellipsoid; areole 6–7 × 4·5 mm., continuous.

Botswana. SW: 16 km. S. of Tsabong, st. 25.ii.1960, *Wild* 5160 (K; SRGH).

Also in SW. Africa and S. Africa (dry northerly parts of Cape Prov. towards the Kalahari). In dry *Acacia-Terminalia* low-bush-grassland.

A very distinctive species only just reaching our area, very easily recognized by its compact grey leaves with minute leaflets. It forms hybrids with *A. giraffae* Willd.

39. **Acacia kirkii** Oliv., F.T.A. **2**: 350 (1871).—Burkill in Johnston, Brit. Centr. Afr.: 245 (1897).—Harms in Warb., Kunene-Samb.-Exped. Baum: 243 (1903).—Bak. f., Legum. Trop. Afr. **3**: 848 (1930).—O. B. Mill., B.C.L.: 19 (1948); in Journ. S. Afr. Bot. **18**: 22 (1952).—Brenan, T.T.C.L.: 333 (1949); in Kew Bull. **12**: 361–3 (1958); F.T.E.A. Legum.-Mimos.: 106 (1959).—Gilbert & Boutique, F.C.B. **3**: 163 (1952).—Wild, Guide Fl. Vict. Falls: 130, 148 (1953).—Torre, C.F.A. **2**: 285 (1956).—F. White, F.F.N.R.: 86, fig. 18 H (1962). TAB. **15** fig. 12. Type: Zambia, Southern Prov., Batoka country, *Kirk* (K, holotype).

Acacia kirkii subsp. *kirkii*.—Brenan in Kew Bull. **12**: 363 (1958); F.T.E.A. Legum.-Mimos.: 106 (1959).

Acacia kirkii subsp. *kirkii* var. *kirkii*.—Brenan in Kew Bull. **12**: 363 (1958).

Acacia kirkii var. *intermedia* Brenan in Kew Bull. **12**: 363 (1958); F.T.E.A. Legum.-Mimos.: 107 (1959).—Boughey in Journ. S. Afr. Bot. **30**: 157 (1964). Type from Kenya.

Tree 2·5–15 m. high, flat-crowned; bark green, peeling or scaling; young branchlets pubescent to sometimes subglabrous, with numerous reddish sessile glands; twigs grey, brown or plum-coloured, not showing yellow bark. Stipules spinescent, varying in length, up to 8 cm. long, straight or almost so; " ant-galls " and other prickles absent. Leaves: rhachis 3–8 cm. long, normally rather densely pubescent above; pinnae 6–14 pairs (some leaves always with 8–9 pairs or more); leaflets numerous, small, 2–5 × 0·5–1(1·25) mm., narrowly oblong or oblong-linear. Flowers with red corolla and white stamen-filaments, in axillary heads whose involucels are conspicuous, 2–3 mm. long and near the base of or $\frac{1}{5}$–$\frac{1}{2}$-way up the peduncle; peduncles rather densely pubescent and with sessile glands throughout, rarely sparingly pubescent. Pods indehiscent, 3·5–9 × 0·8–2·1(2·5) cm., narrowly oblong, straight (or bent only in a plane at right-angles to the flattened plane of the pod), often ± moniliform with the segments mostly as wide as or wider than long; in our area each segment with a medium or small wart-like projection up to 4·5 mm. high in the centre of each of its flat sides. Seeds blackish-olive, 5–7 × 4·5–5·5 mm., smooth, subcircular to elliptic, compressed; areole 3·5–5 × 2·5–3 mm.

All the material from our area belongs to subsp. *kirkii*. Subsp. *mildbraedii* (Harms) Brenan, whose pods entirely lack any wart-like central projections, occurs in the Congo, Rwanda, Burundi, Uganda and NW. Tanzania.

Caprivi Strip. Mpilila I., fr. 15.i.1959, *Killick & Leistner* 3382 (K; SRGH). **Botswana.** N: 5 km. NE. of Makarikari Pan, fr. 20.iv.1957, *Drummond & Seagrief* 5144 (K; SRGH). **Zambia.** E: Luangwa Valley, Ndefu, fr. 20.iv.1952, *White* 2424 (K).

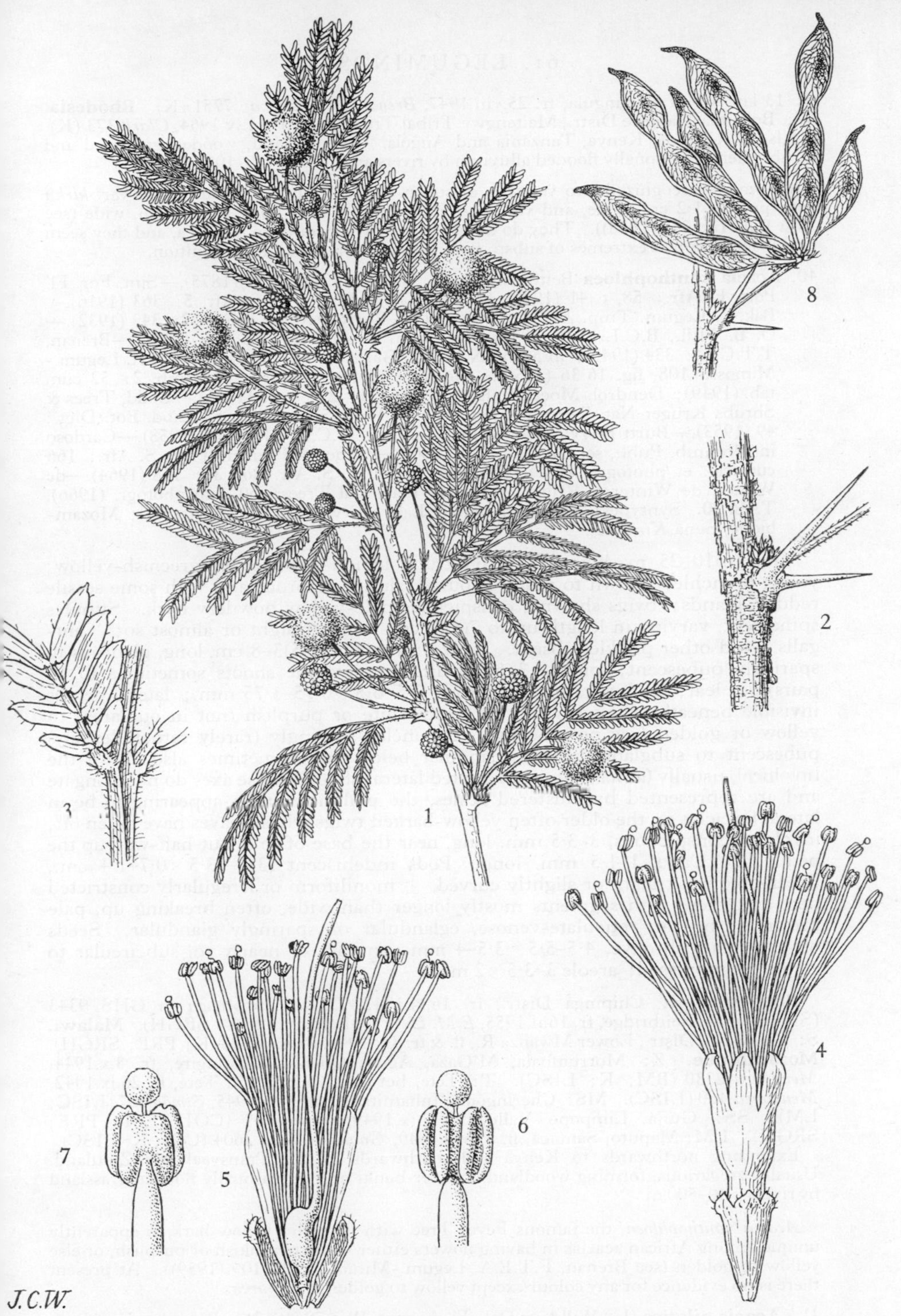

Tab. 20. ACACIA XANTHOPHLOEA. 1, flowering branch (×⅔), *Gomes e Sousa* 3604; 2, part of branch, showing bark and paired spines (×1), *Mendonça* 2330; 3, part of leaf-rhachis and pinnae (×4); 4, flower (×10); 5, flower, opened out to show ovary and stamens (×8); 6, anther, front view (×82); 7, anther, rear view (×82), all from *Gomes e Sousa* 3604; 8, pods (×⅔), *Mendonça* 2330.

S: 13 km. W. of Kazungula, fr. 25.viii.1947, *Brenan & Greenway* 7751 (K). **Rhodesia.** W: Bulalima Mangwe Distr., Maitengwe Tribal Trust land, fr. 11.iv.1964, *Clark* 373 (K).
Also in Uganda, Kenya, Tanzania and Angola. In woodland, wooded grassland and scrub, often in seasonally flooded alluvium by rivers and lakes; 900–1070 m.

I formerly recognized two varieties of subsp. *kirkii* occurring in our area: var. *kirkii* with pods 1–1·2 cm. wide, and var. *intermedia* Brenan with pods 1·2–2·5 cm. wide (see Kew Bull. **12**: 363 (1958)). They do not seem clearly separable in our area, and they seem best looked upon as extremes of subsp. *kirkii* not meriting separate recognition.

40. **Acacia xanthophloea** Benth. in Trans. Linn. Soc. **30**: 511 (1875).—Sim, For. Fl. Port. E. Afr.: 58, t. 41 (1909).—Eyles, Trans. Roy. Soc. S. Afr. **5**: 363 (1916).—Bak. f., Legum. Trop. Afr. **3**: 851 (1930).—Burtt Davy, F.P.F.T. **2**: 343 (1932).—O. B. Mill., B.C.L.: 21 (1948); in Journ. S. Afr. Bot. **18**; 26 (1952).—Brenan, T.T.C.L.: 334 (1949); in Mem. N.Y. Bot. Gard. **8**: 430 (1954); F.T.E.A. Legum.-Mimos.: 108, fig. 16/36 (1959).—Gomes e Sousa, Dendrol. Moçamb. **2**: 52 cum tab. (1949); Dendrol. Moçamb. Estudo Geral, **1**: 231, t. 35 (1966).—Codd, Trees & Shrubs Kruger Nat. Park: 52, figs. 44 c, d, 46 (1951).—Wild, S. Rhod. Bot. Dict.: 49 (1953).—Burtt Davy & Hoyle, rev. Topham, N.C.L. ed. 2: 65 (1958).—Cardoso in Moçamb. Publ., ser. A, **3**: 1–52 (1960).—Palmer & Pitman, Trees S. Afr.: 166 cum fig. et photogr. (1961).—Boughey in Journ. S. Afr. Bot. **30**: 158 (1964).—de Winter, de Winter & Killick, Sixty-six Transvaal Trees: 60 cum photogr. (1966). TAB. **20.** Syntypes: Malawi, E. end of Lake Shirwa (Chilwa), *Meller* (K); Mozambique, Sena, *Kirk* (K).

Tree (6)10–25 m. high; bark on trunk lemon-coloured to greenish-yellow; young branchlets brown to plum-coloured, almost glabrous and with some sessile reddish glands; twigs showing conspicuous pale-yellow powdery bark. Stipules spinescent, varying in length up to 7(8·5) cm. long, straight or almost so; " ant-galls " and other prickles absent. Leaves: rhachis (2·5)3–8 cm. long, glabrous to sparingly pubescent; pinnae 3–6(8) pairs (on juvenile shoots sometimes to 10 pairs) per leaf; leaflets rather numerous, 2·5–6·5 × 0·75–1·75 mm.; lateral nerves invisible beneath. Flowers varying from white or purplish (not in our area) to yellow or golden (see note below). Peduncles sparingly (rarely rather densely) pubescent to subglabrous, and glandular below and sometimes also above the involucel, usually (at least) on abbreviated lateral shoots whose axes do not elongate and are represented by clustered scales, the peduncles thus appearing to be in lateral fascicles on the older often yellow-barked twigs whose leaves have fallen off; involucel conspicuous, 3–3·5 mm. long, near the base of to about half-way up the peduncle. Calyx 1–1·5 mm. long. Pods indehiscent (3)4–13·5 × 0·7–1·4 cm., linear-oblong, straight or slightly curved, ± moniliform or irregularly constricted here and there with segments mostly longer than wide, often breaking up, pale brown or brown, reticulate-venose, eglandular or sparingly glandular. Seeds olive to blackish-olive, 4·5–5·5 × 3·5–4 mm., smooth or nearly so, subcircular to elliptic, compressed; areole 3–3·5 × 2 mm.

Rhodesia. E: Chipinga Distr., fr. 16.xi.1942, *Cooke-Yarborough* in GHS 9343 (SRGH). S: Beitbridge, fr. 16.ii.1955, *E.M. & W.* 441 (BM; LISC; SRGH). **Malawi.** S: Chikwawa Distr., Lower Mwanza R., fl. & fr. 3.x.1946, *Brass* 17928 (K; PRE; SRGH). **Mozambique.** Z: Morrumbala, M'Gaza, Águas Quentes-Messingire, fr. 3.x.1944, *Mendonça* 2330 (BM; K; LISC). T: Tete, between Chioco and Tete, fl. 26.ix.1942, *Mendonça* 460 (LISC). MS: Cheringoma, Inhamitanga, fl. 27.x.1945, *Simão* 607 (LISC; LM). SS: Guijá, Limpopo Valley, fl. 3.ix.1949, *Myre* 799 (COI; LM; PRE; SRGH). LM: Maputo, Santaca, fl. 27.viii.1949, *Gomes e Sousa* 3604 (COI; K; LISC).
Extending northwards to Kenya and southwards to the Transvaal and Zululand. Usually gregarious, forming woodland on river-banks and in seasonally flooded grassland by rivers; 40–500 m.

Acacia xanthophloea, the famous Fever Tree with its pallid yellow bark, is apparently unique among African acacias in having flowers either white to pinkish or purplish, or else yellow to golden (see Brenan, F.T.E.A. Legum.-Mimos.: 108–109 (1959)). At present there is no evidence for any colour except yellow to golden in our area.

41. **Acacia nilotica** (L.) Willd. ex Del., Fl. Aegypt. Ill.: 79 (1813).—Brenan in Kew Bull. **12**: 83 (1957); F.T.E.A. Legum.-Mimos.: 109 (1959). Type from Egypt.
Mimosa nilotica L., Sp. Pl. **1**: 521 (1753). Type as above.

An exceedingly variable species. Tree (1·2)2·5–14 m. high; bark on trunk rough, fissured, blackish or grey or brown, neither powdery nor peeling; young branchlets from almost glabrous to subtomentose; glands inconspicuous or absent; bark of twigs not flaking off, grey to brown. Stipules spinescent, up to 8(11) cm. long, straight or almost so, often ± deflexed; " ant-galls " and other prickles absent. Leaves often with 1(2) petiolar glands and other glands between all or only the topmost of the 2–11(17) pairs of pinnae; leaflets 7–25(30) pairs, 1·5–7 × 0·5–1·5 mm., glabrous to pubescent, not spinulose-mucronate at the apex; lateral nerves invisible beneath. Flowers bright-yellow, in axillary pedunculate heads 6–15 mm. in diam.; involucel from near the base to c. half-way up the peduncle, very rarely somewhat higher up. Calyx 1–2 mm. long, subglabrous to pubescent. Corolla 2·5–3·5 mm. long, glabrous to ± pubescent outside. Pods especially variable, indehiscent, (4)8–17(24) × 1·3–2·2 cm., straight or curved, glabrous to grey-velvety, ± turgid. Seeds deep-blackish-brown, 7–9 × 6–7 mm., smooth, subcircular, compressed; areole 6–7 × 4·5–5 mm.

Typical subsp. *nilotica* of the Sudan, with glabrous necklace-like pods regularly and narrowly constricted between the seeds, is not found in our area.

Subsp. **kraussiana** (Benth.) Brenan in Kew Bull. **12**: 84 (1957); F.T.E.A. Legum.-Mimos.: 110 (1959).—Palmer & Pitman, Trees S. Afr.: 161 cum fig. et photogr. (1961).—F. White, F.F.N.R.: 86, fig. 18 G (1962).—Mitchell in Puku, **1**: 104 (1963).—Boughey in Journ. S. Afr. Bot. **30**: 158 (1964). TAB. **16** fig. 13, **21**. Type from Natal.

Acacia arabica var. *kraussiana* Benth. in Hook., Lond. Journ. Bot. **1**: 500 (1842).—Burtt Davy, F.P.F.T. **2**: 343 (1932).—Codd, Trees & Shrubs Kruger Nat. Park: 38, fig. 33 b (1951).—O. B. Mill. in Journ. S. Afr. Bot. **18**: 18 (1952). Type as above.

Acacia benthamii Rochebr., Toxicol. Afr. **2**: 192 (1898) non Meisn. (1844).—Bak. f., Legum. Trop. Afr. **3**: 850 (1930).—Steedman, Trees etc. S. Rhod.: 12 (1953).—Hutch., Botanist in S. Afr.: 268 (" benthamiana "), 664 (1946).—O. B. Mill., B.C.L.: 16 (1948).—Wild, S. Rhod. Bot. Dict.: 46 (1953). Type as above.

Acacia arabica sensu Sim, For. Fl. Port. E. Afr.; 57, t. 36 (1909).

Acacia nilotica var. *kraussiana* (Benth.) A. F. Hill in Bot. Mus. Leafl. Harvard Univ. **8**: 98 (1940). Type as above.

Acacia subalata sensu Brenan in Mem. N.Y. Bot. Gard. **8**: 430 (1954).—Pardy in Rhod. Agric. Journ. **51**: 489 cum photogr. (1954).

Acacia nilotica subsp. *subalata* sensu Boughey in Journ. S. Afr. Bot. **30**: 158 (1964).

Young branchlets ± densely pubescent. Pods not necklace-like, 1–1·8 cm. wide, oblong, ± pubescent all over at first, with the raised parts over the seeds becoming glabrescent, shining and black when dry, margins ± shallowly crenate.

Botswana. N: Ngamiland, Matabakosi, fl. xii.1930, *Curson* 45* (PRE). SE: E. side of Nata R. c. 8 km. from its mouth, fr. 22.iv.1957, *Drummond & Seagrief* 5181 (K; SRGH). **Zambia.** N: Luangwa Valley, between Mutinondo R. and Chifungwe Plain, fr. 7.vi.1957, *Savory* 170 (SRGH). E: Luangwa Valley, NW. from Chief Nsefu's village, fl. & fr. 11.x.1958, *Robson* 49 (BM; K; LISC; SRGH). S: Great North Road 12 km. N. of Livingstone, fr. 18.iii.1952, *White* 2279 (K). **Rhodesia.** N: Mtoko Distr., just off main Tete–Salisbury road, fr. 16.iv.1951, *Lovemore* 5 (SRGH). W: Victoria Falls, S. of Zambezi R., fr. 31.v.1930, *Milne-Redhead* 413 (K). C: Ngesi-Mondoro Reserve, fr. 28.ii.1956, *Cleghorn* 160 (SRGH). E: Umtali Commonage, fr. 22.vi.1946, *Chase* 238 (SRGH). S: 8 km. NW. of Zimbabwe, fr. 7.v.1963, *Leach* 11667 (SRGH). **Malawi.** N: 97 km. N. of Mzimba, fr. 9.vi.1938, *Pole Evans & Erens* 658 (K; PRE; SRGH). S: Mpatamanga Gorge, Shire R., fr. 13.v.1961, *Leach & Rutherford Smith* 10831 (K; LISC; SRGH). **Mozambique.** N: Metuge road, Muagide Farm, fr. 25.x.1960, *Gomes e Sousa* 4582 (COI; K; PRE). Z: Mocuba, Namagoa, fl. & fr. x.1944, *Faulkner* 60 (COI; K; LMJ; PRE). T: Muatize, fr. 8.v.1948, *Mendonça* 4138 (BM; K; LISC). MS: Chemba, Chien, fr. 22.iv.1960, *Lemos & Macuácua* 133 (BM; COI; K; LISC; PRE; SRGH). SS: Caniçado, fr. 18.vi.1947, *Pedrógão* 301 (COI; K). LM: between Namaacha and Lourenço Marques, fl. & fr. 16.x.1940, *Torre* 1773 (LISC).

Also in Tanzania, Angola, the Transvaal and Natal; an Ethiopian specimen, *Mooney* 5576 (K), is apparently referable to subsp. *kraussiana*. Woodland of various sorts, wooded grassland, scrub and thicket; 10–1340 m.

A. nilotica subsp. *kraussiana* has a wide range of altitude and a considerable range of habitat. As might be expected, it is variable. Forms with numerous pinnae (13–17 pairs) seem most frequent in central Mozambique. The pod also shows considerable variation. It is hard to differentiate concisely in words between subsp. *kraussiana* and subsp. *subalata* (Vatke) Brenan, which occurs in East Africa and perhaps also the Sudan, and some specimens of subsp. *kraussiana* show in certain ways an approach to subsp. *subalata*. *Trapnell*

* Without mature pods. Determination probable but uncertain.

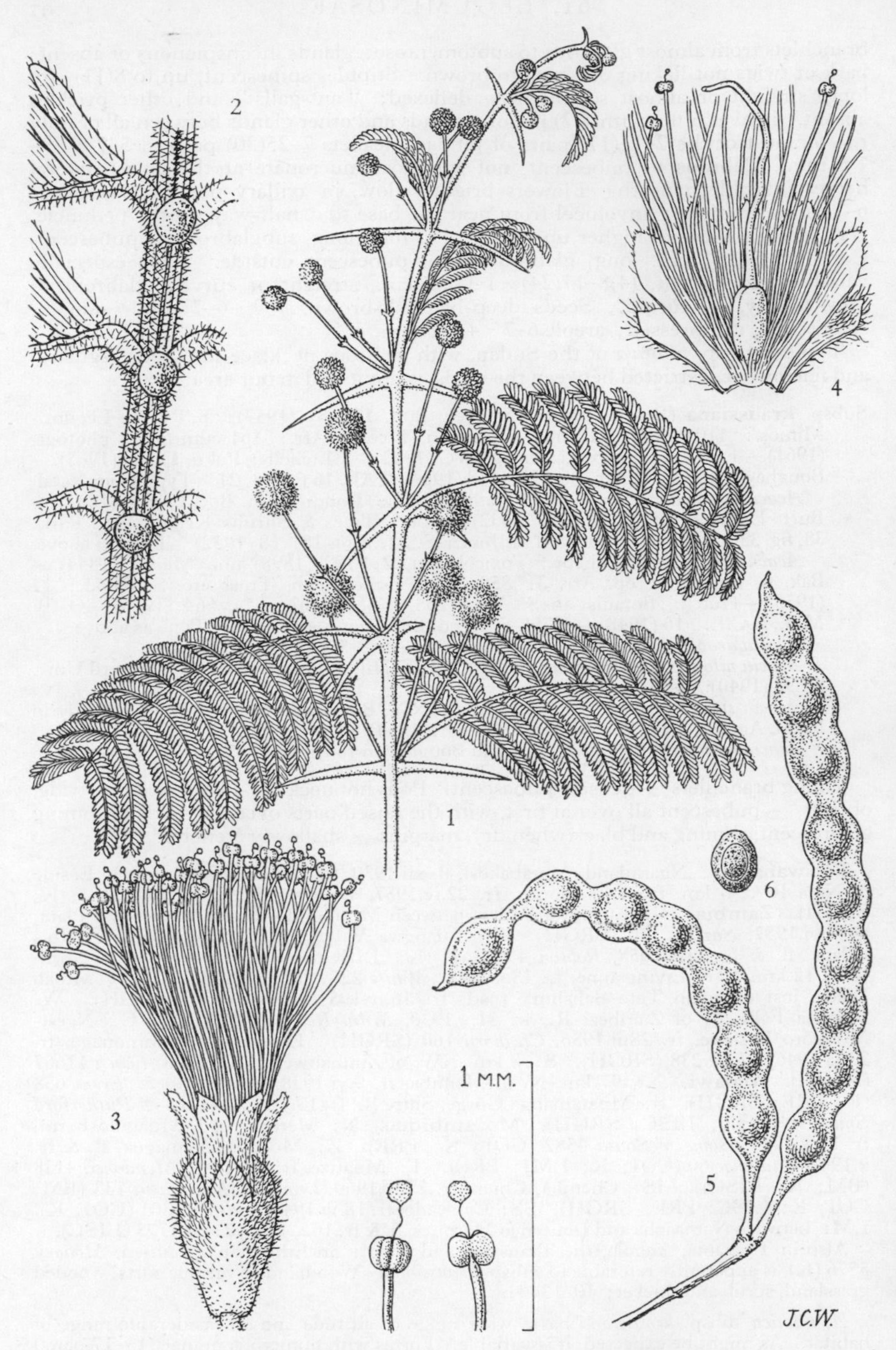

Tab. 21. ACACIA NILOTICA SUBSP. KRAUSSIANA. 1, flowering branch (×⅔); 2, part of leaf-rhachis showing glands (×6); 3, flower (×10) with enlargement of anthers to show glands, all from *Purves* 215; 4, flower, opened out to show ovary (×10) *Robson* 49; 5, pods (×⅔) *White* 2279; 6, seed (×1) *Milne-Redhead* 413.

1438 (K), from Zambia, S: Gwembe, fr. iii.1934, has crenate-margined ± moniliform pods 1·2–1·3 cm. wide as in subsp. *kraussiana*, but they are pubescent all over. *Angus* 2931 (K), from Zambia, 27 km. E. of Sinazeze, fr. 16.iv.1961, has similar pods but up to 1·5–1·8 cm. wide. *Davies* 2074 (SRGH), from Rhodesia, N: Mtoko Distr., fr. viii.1956 and *Pedro & Pedrógão* 6197 (LMJ), from Mozambique, MS: between Munhinga and Dombe, fr. 4.vi.1949, have pods up to 2 cm. wide, but with the characteristic glabrescent bosses over the seeds as in subsp. *kraussiana*. These will serve to indicate subsp. *kraussiana* when, as sometimes happens, the margin of the pod is almost as straight as it is in subsp. *subalata*.

42. **Acacia abyssinica** Hochst. ex Benth. in Hook., Lond. Journ. Bot. **5**: 97 (1846).—Oliv., F.T.A. **2**: 347 (1871).—Bak. f., Legum. Trop. Afr. **3**: 839 (1930).—Gilbert & Boutique, F.C.B. **3**: 167 (1952).—Brenan in Kew Bull. **12**: 81 (1957); F.T.E.A. Legum.-Mimos.: 112 (1959). Type from Ethiopia.

Conspicuously flat-crowned tree 6–15(20) m. high; bark rough and fissured, brown to nearly black; epidermis not peeling on the twigs; bark on young trees papery; indumentum of branchlets variable, pubescent to shortly villous, grey or somewhat yellowish. Stipules spinescent, other prickles absent; spines variable, absent, short or up to 7·2 cm. long, straight, ashen when elongate, never inflated. Leaves: petiole 2–5 mm. long; pinnae of well-developed leaves of mature shoots 15–51 pairs (reduced leaves with fewer pairs also present); leaflets up to 4 × 0·75 mm. Flowers in heads; stamens white; calyx and corolla red (? always). Corolla glabrous or inconspicuously puberulous on the lobes outside. Pods 5–13 × 1·2–2·1(2·8) cm., subcoriaceous, straight or slightly curved, grey or brown, longitudinally veined, ± glandular and sometimes puberulous, narrowed at the base and sometimes at the top. Seeds ± oblique in the pod, olive-brown, 7–10 × 4–6 mm., smooth, elliptic, compressed; areole 6–7 × 2·5–4 mm.

Typical subsp. *abyssinica*, with pinnae 1·5–2·5 cm. long and leaflets 2–4 × 0·5–0·75 mm., is confined to Ethiopia.

Subsp. **calophylla** Brenan in Kew Bull. **12**: 82 (1957); F.T.E.A. Legum.-Mimos.: 112, fig. 16/39 (1959).—Boughey in Journ. S. Afr. Bot. **30**: 157 (1964). Type from Kenya.
Acacia xiphocarpa sensu Brenan in Mem. N.Y. Bot. Gard. **8**: 429 (1954).
Acacia abyssinica sensu Brenan, T.T.C.L.: 335 (1949).

Pinnae very closely set, 0·4–1·5(2) cm. long; leaflets extremely small, up to 2·5(3) × 0·25–0·4(0·5) mm. wide.

Rhodesia. E: Inyanga Distr., 2 km. E. of " London Road ", fl. x.1956, *Miller* 3853 (K; SRGH). **Malawi.** N: Nyika Plateau, fr. 9.v.1952, *White* 2811 (K; SRGH). C: Dedza–Chongoni road, Chitowo stream, fr. 28.vii.1960, *Chapman* 851 (BM; SRGH). S: Zomba Plateau, fr. immat. 5.vi.1946, *Brass* 16234 (K; PRE). **Mozambique.** N: between Vila Cabral and Massangulo mission, fl. 12.x.1942, *Mendonça* 779 (LISC). T: between Vila Coutinho and the north boundary with Malawi, fr. 3.vi.1962, *Gomes e Sousa* 4768 (COI; K; PRE). MS: Macequece, fl. 2.i.1948, *Barbosa* in *Mendonça* 796 (BM; K; LISC).

Also in the Congo, Sudan, Uganda, Kenya and Tanzania. In montane forest and woodland, often gregariously, also sometimes, presumably relict, in high-altitude *Brachystegia* woodland; 900–1980 m.

A. abyssinica subsp. *calophylla* is related to *A. pilispina* P. Sermolli, *A. lasiopetala* Oliv. and *A. rehmanniana* Schinz, but is separated from all three by the fact that the epidermis of the twigs does not fall away to expose yellow or red bark. It has been often confused with *A. rehmanniana*, from which it may be additionally separated by the capitula not being aggregated into terminal " racemes ", and also by the very different ecology.

43. **Acacia rehmanniana** Schinz in Bull. Herb. Boiss. **6**: 525 (1898).—Eyles in Trans. Roy. Soc. S. Afr. **5**: 362 (1916).—Bak. f., Legum. Trop. Afr. **3**: 838 (1930).—Burtt Davy, F.P.F.T. **2**: 343 (1932).—Steedman, Trees etc. S. Rhod.: 14, t. 9 (1933).—O. B. Mill., B.C.L.: 20 (1948); in Journ. S. Afr. Bot. **18**: 24 (1952).—Wild, S. Rhod. Bot. Dict.: 48 (1953).—Pardy in Rhod. Agric. Journ. **51**: 376 cum phot. (1954).—Burtt Davy & Hoyle, rev. Topham, N.C.L. ed. 2: 64 (1958).—F. White, F.F.N.R.: 86, fig. 18 I (1962).—Boughey in Journ. S. Afr. Bot. **30**: 158 (1964). Type from the Transvaal.

Usually a small to medium, flat-crowned tree 3–8(12) m. high; young branchlets densely spreading-hairy, the hairs at first golden then grey; the epidermis later falling off to expose powdery rusty-red bark. Stipules spinescent, up to 5 cm. long, never inflated; other prickles absent. Leaves with petiole 2–4 mm. long which, like the rhachis, is densely clothed with at first golden then grey spreading hairs; pinnae of well-developed leaves of mature shoots 15–44 pairs (reduced leaves with fewer pairs sometimes also present), mostly 1–2·5 cm. long; leaflets numerous, (1)1·5–2·8 × 0·4–0·7 mm. Flowers white to cream, in heads; heads c. 2–20 per axil, aggregated into a sort of terminal " raceme ", subtended by young to scarcely developed leaves; involucel below the middle of the almost or quite eglandular 1–2 cm. long peduncle, rarely basal. Corolla ± densely pubescent on the lobes outside, about 1½ times as long as the calyx. Pods dehiscent, 7·5–14 × 1·2–2·3 cm., straight, glabrous or very slightly pubescent, flattened, not constricted between the seeds, grey-brown to olive, slightly ± longitudinally venose and often somewhat irregularly wrinkled. Seeds brown, 5–9 × 5–6·5 mm., smooth, ellipsoid to suborbicular, somewhat compressed; areole 3·5–5 × 2–2·8 mm.

Botswana. N: 21 km. from Tsessebe to Pan, fl. ii.xii.1929, *Pole Evans* 2578 (K; PRE). **Zambia.** S: Choma Valley, Chibia Stream, fl. 9.vi.1933, *Michelmore* 401 (K). **Rhodesia.** N: Msengezi, fl. i.1949, *Davies* 351 (K; SRGH). W: Bulawayo, fr. 6.ix.1947, *Wild* 1991 (K; SRGH). C: Salisbury, Strathavon, fl. xi.1955, *Drummond* 4939 (K; LM; SRGH). E: Umtali, fl. 26.xii.1955, *Chase* 5923 (BM; K; SRGH). S: 16 km. S. of Chibi, fr. iv.1953, *Vincent* 216 (SRGH).

Also in the Transvaal and Natal. Wooded grassland (*Acacia-Combretum*, *Acacia-Colophospermum mopane*); often near rivers or streams, sometimes on termite mounds; 910–1520 m.

Owing to confusion with *A. sieberana* DC., this species has been often misunderstood. The records in Bak. f., Legum. Trop. Afr. **3**: 838 (1930) from the Sudan and West Africa are due to this cause. In reality *A. rehmanniana* is a very distinctive species, easily separated from *A. sieberana* by the more numerous pinnae, the much thinner texture of the pods and the way in which the capitula are clustered in the axils and aggregated into terminal " racemes ". The latter character will separate *A. rehmanniana* from *A. abyssinica*, which is even more closely related than is *A. sieberana*.

44. **Acacia pilispina** P. Sermolli, Miss. Stud. Lago Tana, Ric. Bot. **1**: 205, t. 43 (1951).—Brenan, F.T.E.A. Legum.-Mimos.: 113, fig. 16/40 (1959).—F. White, F.F.N.R.: 88 (1962). Type from Ethiopia.

Shrub or tree 1–15 m. high (in our area usually a tree of 5 m. or more high); crown flat or spreading; bark on trunk grey to brown and rugose or ± smooth; young branchlets densely clothed with long grey to slightly yellowish spreading hairs mostly 0·5–2 mm. long, red-brown beneath the hair; epidermis falling away to expose a yellow or sometimes greenish powdery bark on the twigs. Stipules spinescent, mostly short, up to 7 mm. long, straight or nearly so, and hairy except towards the tips, or sometimes longer, grey or straw-coloured, up to 5 cm. long; " ant-galls " and other prickles absent. Leaves: petiole 4–8 mm. long; rhachis mostly 2·5–6 cm. long, hairy; pinnae mostly 8–16(26) pairs, 1–2·2 cm. long; leaflets 14–28 pairs, 1·5–3·75(4·5) × 0·5–1 mm., ciliate. Flowers cream or tinged red outside, in heads on axillary peduncles (0·5)1·5–3·5 cm. long and ± hairy but eglandular, whose involucel is basal or in the lower fifth; peduncles solitary or several together in the axils of developed leaves, usually not in apparently ± leafless terminal " racemes ". Corolla glabrous or sparsely puberulous outside. Pods grey to grey-brown or purple-brown, 5–12·5 × 1·2–2·9 cm., narrowed at the base and sometimes at the top, finely and ± longitudinally veined, glabrous or nearly so. Seeds olive-brown, 6–7 × 5–7 mm., smooth, elliptic or subcircular, compressed; areole 4·5–5 × 2·5–4·5 mm.

Zambia. B: Kabompo, fl. & fr. vii.1933, *Trapnell* 1221 (K). N: Luvu R., fr. 26.ix.1938, *Greenway* 5783 (K). C: Lukanga Swamp, fr. immat. 16.viii.1963, *Fanshawe* 7935 (K). E: Fort Jameson, fl. 6.vi.1954, *Robinson* 839 (K). S: Kafue R., Mumbwa, fl. 7.v.1964, *Mitchell* 25/53 (K). **Malawi.** S: Zomba, fl. 1901, *Purves* 153 (K). **Mozambique.** N: Amaramba, Mandimba, fr. 5.x.1958, *Monteiro* 108 (LISC).

Also in the Congo (Katanga), Ethiopia and Tanzania. By rivers, streams and on edges of grassy valleys (dambos), usually on alluvial or colluvial soils—silt, clay-loams, etc.; sometimes on termite mounds in *Brachystegia-Julbernardia* woodland; 1040–1680 m.

Pichi-Sermolli described *A. pilispina* as a shrub 1–3 m. high, with up to 10 pairs of pinnae and 8–16 pairs of leaflets per pinna. These differences seem very possibly due to the higher altitude at which the species was growing in Ethiopia, but this theory needs checking.

The pods are somewhat variable; in Tanzania they are usually rather broad (1·8–2·9 cm.), slightly curved or oblique, and grey when old, while in our area they are generally narrower (1·2–2·2 cm.) and straight.

A. pilispina is striking on account of the yellow powdery bark on its branches, reminiscent of that of *A. xanthophloea* Benth.

45. **Acacia luederitzii** Engl., Bot. Jahrb. **10**: 23, t. 3B (July 1888) pro parte quoad *Marloth* 1328 (lectotype).—Bak. f., Legum. Trop. Afr. **3**: 840 (1930). Type from SW. Africa.

Usually a tree, occasionally only a shrub, 1–12 m. high; bark very rough, coarsely longitudinally fissured and ridged; crown usually ± rounded in outline but with the ultimate branching ± horizontal; young branchlets densely grey- or whitish-spreading-pubescent, older branchlets glabrescent, going purplish to deep-purplish-brown or even blackish. Stipules spinescent, some short and hooked, 3–10 mm. long, some elongate, slender and straight or slightly curved, 1–7 cm. long, or inflated and ± bent towards the apex. Leaves: rhachis 1–4 cm. long; pinnae 2–9(12) pairs; leaflets 10–26 pairs per pinna, 2–5 × 0·5–1·5 mm., linear-oblong or the terminal ones slightly broadened above, rounded or obtuse at the apex, with spreading cilia on the margins, otherwise glabrous or sometimes ± pubescent. Flowers white or cream, in heads on axillary peduncles (1)1·5–3·2 cm. long; hairs on peduncle shorter than its diameter; involucel $\frac{1}{5}$–$\frac{2}{3}$-way up the peduncle. Calyx shortly pubescent or puberulous above. Corolla glabrous outside. Pods dehiscent, grey-brown to purplish-brown, straight or slightly curved, 3·2–13 × 1–1·9 cm., flattened, attenuate at the base, rounded to acuminate at the apex, marked with oblique or longitudinal veins, finely puberulous especially on the margins and near the base, sometimes subglabrous; valves stiffly coriaceous but scarcely woody. Seeds longitudinal in the pod, 6·5–11·5 × 5–10 mm.; areole 3–7 × 2·75–5 mm.

Var. **luederitzii.** TAB. **16** fig. 14.

Acacia goeringii Schinz in Verh. Bot. Verein. Brand. **30**: 239 (1888).—Bak. f., tom. cit.: 841 (1930).—O. B. Mill., B.C.L.: 18 (1948); in Journ. S. Afr. Bot. **18**: 21 (1952). Type from S. Afr.

Acacia retinens sensu O. B. Mill., B.C.L.: 20 (1948); in Journ. S. Afr. Bot. **18**: 24 (1952).

Acacia uncinata sensu O. B. Mill. in Journ. S. Afr. Bot. **18**: 25 (1952).—Story in Mem. Bot. Surv. S. Afr. **30**: 23 (1958).—Boughey in Journ. S. Afr. Bot. **30**: 158 (1964).

Larger spines elongate, often whitish, 1–7 cm. long, 1·5–5 mm. thick, straight or somewhat curved, not at all inflated.

Botswana. N: Motlhatlogo, fr. 11.v.1930, *Van Son* 28870 (PRE). SW: 80 km. N. of Kang, fl. 18.ii.1960, *Wild* 5054 (K; SRGH). SE: Artesia, fr. 12.iv.1931, *Pole Evans* 3170 (20) (PRE). **Zambia.** B: Machili, fr. 14.iv.1962, *Fanshawe* 6761 (K). **Rhodesia.** W: Wankie Game Reserve, fr. vii.1956, *Davison* in GHS 69320 (K; SRGH).

Also in SW. Africa and in parts of S. Africa close to Botswana. Tree savanna and scrub, particularly on Kalahari Sand; 760–1070 m.

Var. **retinens** (Sim) J. Ross & Brenan in Kew Bull. **21**: 72 (1967). Type: Mozambique, " Umbeluzi and Lebombo ", *Sim* 6391 (not found).

Acacia retinens Sim in For. Fl. Port. E. Afr.: 157, t. 40 fig. A (1909).—J. Ross in Journ. S. Afr. Bot. **31**: 219 (1965). Type as above.

Acacia gillettiae Burtt Davy, F.P.F.T. **2**: xvii, 343 (1932).—O. B. Mill., B.C.L.: 18 (1948); in Journ. S. Afr. Bot. **18**: 21 (1952). Syntypes from the Transvaal.

Larger spines 3–5·5 cm. long, 0·8–1·5 cm. wide, thickened, inflated and often whitish, usually ± uncinate-deflexed near the apex; straight elongate non-inflated spines absent.

Mozambique. LM: Maputo, Catuane, fr. 20.iv.1944, *Torre* 6482 (LISC; PRE).

Also in S. Africa (the Transvaal and Natal) and Swaziland. Ecology in our area unknown.

Although the distinction between the two varieties rests mainly on the enlarged spines (which may not always be present), there are other inconstant differential tendencies. Var. *luederitzii* seems often to be rather taller than var. *retinens*, 4·5–12 m. as against often only 1–4·5 m. Var. *luederitzii* has usually 5–8 pairs of pinnae per leaf, while the number in var. *retinens* is more variable both above and below these figures. The pods in var. *luederitzii* are 10–19 mm. wide, while those of var. *retinens* do not seem to exceed 15 mm. The two varieties are sharply separated geographically.

The name *Acacia reficiens* Wawra (1859) has sometimes been used to cover *A. luederitzii*; but true *A. reficiens* appears worth maintaining as a distinct species separated by ist puberulous or pulverulent indumentum and few pairs of leaflets (up to 11(13) pairs per pinna), glabrous or almost so on the margins. It probably also differs in habit. It is not known to occur in our area although it is found from the Sudan to Kenya and in Angola.

46. **Acacia tortilis** (Forsk.) Hayne, Darstell. u. Beschreib. Arzneyk. Gewächse, **10**: t. 31 (1827).—Oliv., F.T.A. **2**: 352 (1871).—Bak. f., Legum. Trop. Afr. **3**: 841 (1930).—Torre, C.F.A. **2**: 284 (1956).—Brenan in Kew Bull. **12**: 86 (1957); F.T.E.A. Legum.-Mimos.: 117 (1959).—F. White, F.F.N.R.: 84, fig. 17 j (1962). Type from Arabia.

Mimosa tortilis Forsk., Fl. Aegypt-Arab.: CXXIII, 176 (1775). Type as above.

Tree 4–21 m. high, occasionally (probably not in our area) a bush 1 m. high; crown flat or spreading;* bark grey to black, fissured; young branchlets glabrous to densely pubescent, going brown to purplish-black. Stipules spinescent, some short ± hooked and up to c. 5 mm. long, mixed with other long straight whitish ones to c. 8(10) cm. long; "ant-galls" and other prickles absent. Leaves; rhachis short, 2 cm. long or less; pinnae 2–10 pairs, 2–17 mm. long; leaflets 6–19 pairs per pinna, usually very small, 0·5–2·5(6) mm. long, ciliate to glabrous. Flowers cream or whitish, in axillary heads 5–10 mm. in diam. on peduncles 0·4–2·4 cm. long; involucel in the lower half of the peduncle. Pods contorted or spirally twisted, longitudinally veined, tomentellous to glabrous. Seeds olive- to red-brown, 7 × 4·5–6 mm., smooth, elliptic, compressed; areole 5–6 × 3–4 mm.

Typical subsp. *tortilis*, with narrow (3–5 mm. wide) pubescent but eglandular pods, occurs in Egypt, the Sudan, Arabia, Aden and perhaps Israel.

Subsp. **heteracantha** (Burch.) Brenan in Kew Bull. **12**: 88 (1957).—Boughey in Journ. S. Afr. Bot. **30**: 158 (1964).—de Winter, de Winter & Killick, Sixty-six Transvaal Trees: 58 cum photogr. (1966). TAB. **16** fig. 15. Type from S. Africa.

Acacia heteracantha Burch., Trav. Int. S. Afr. **1**: 389 (1822).—Bak. f., Legum. Trop. Afr. **3**: 843 (1930).—Burtt Davy, F.P.F.T. **2**: 344 (1932).—Hutch., Botanist in S. Afr.: 398, 428, 664 (1946).—O. B. Mill., B.C.L.: 19 (1948).—Codd, Trees & Shrubs Kruger Nat. Park: 46, figs. 38, h, i, 39 (1951).—Wild, S. Rhod. Bot. Dict.: 47 (1953).—Pardy in Rhod. Agric. Journ. **51**: 375 cum photogr. (1954).—Torre, C.F.A. **2**: 284 (1956).—Story in Mem. Bot. Surv. S. Afr. **30**: 23 (1958). Type as above.

Acacia litakunensis Burch., op. cit. **2**: 452 (1824).—Burtt Davy, tom. cit.: 345 (1932).—O. B. Mill. in Journ. S. Afr. Bot. **18**: 23 (1952). Type from S. Africa.

Tree 2·5–6(10) m. high; young branchlets shortly pubescent with hairs usually less than 0·25 mm. long, a few to 0·5 mm. Petiole and leaf-rhachis similarly shortly pubescent, the latter 0·5–1·4(2) cm. long. Pods c. 5–7 mm. wide, glabrous or nearly so, eglandular.

Botswana. N: Francistown, fl. ii.1952, *Miller* B 1306 (K; PRE); Kwebe Hills, fl. & fr. xii.1897, *Lugard* 49 (K). SE: Mahalapye, fr. 17.ii.1961, *Yalala* 130 (K; SRGH). **Rhodesia.** N: Urungwe Distr., Sanyati R. bank, fl. 22.xii.1955, *Phelps* 93 (K; SRGH). W: Ngamo Forest Reserve, Gwaai R., fr. iii.1960, *Armitage* 118/60 (SRGH). E: Rupisi, Hot Springs, fl. 28.i.1948, *Wild* 2310 (K; SRGH). S: Sabi–Lundi junction, near Sabi R., fr. 7.vi.1950, *Wild* 3438 (K; LISC; SRGH). **Mozambique.** SS: Guijá, between Chamusca and Mejinge, Caniçado, fr. 18.v.1948, *Torre* 7857 (LISC). LM: Sábiè, Moamba, fr. 2.v.1944, *Torre* 6541 (BM; LISC; PRE).

Also in SW. Africa, Swaziland and S. Africa. Woodland, wooded grassland and bushland, often by rivers; 240–1100 m.

Subsp. **spirocarpa** (Hochst. ex A. Rich.) Brenan in Kew Bull. **12**: 88 (1957); F.T.E.A. Legum.-Mimos.: 117, fig. 16/44 (1959).—Boughey in Journ. S. Afr. Bot. **30**: 158 (1964). Syntypes from Ethiopia.

Acacia spirocarpa Hochst. ex A. Rich., Tent. Fl. Abyss. **1**: 239 (1847).—Oliv., F.T.A. **2**: 352 (1871).—Brenan, T.T.C.L.: 334 (1949).—O. B. Mill. in Journ. S.

* *Smith* 1085 (EA), Tanzania, Masai Distr. (*A. tortilis* subsp. *spirocarpa*) is described as having a rounded crown.

Afr. Bot. **18**: 25 (1952).—Burtt Davy & Hoyle, rev. Topham, N.C.L. ed. 2: 64 (1958). Syntypes as above.

Tree 5–15 m. high; young branchlets with longer denser pubescence than in subsp. *heteracantha*; hairs 0·25–0·75 mm. long. Petiole and leaf-rhachis with hairs mostly more than 0·25 mm. long. Pods 6–9(13) mm. wide, tomentellous or pubescent with spreading or curved hairs, among which are numerous dark-red glands clearly visible through a hand-lens.

Botswana. N: Serondela, fr. 29.vii.1950, *Robertson & Elffers* 71 (K; PRE; SRGH). **Zambia.** N: Luangwa Valley, Nabwalia's area, fr. 12.vi.1957, *Savory* 188 (SRGH). C: Feira Boma, fr. 30.v.1952, *White* 2902 (K). S: 10 km. downstream from Chirundu bridge, fr. 12.viii.1957, *Angus* 1649 (K). **Rhodesia.** N: Sebungwe, Kariangwe, Lubu R., fr. 25.vi.1961, *Lovemore* 65 (K; SRGH). E: Chipinga Distr., Sabi R., fr. 6.iv.1959, *Savory* 322 (COI; K; SRGH). **Mozambique.** T: between Caldas Xavier and Tete, fr. 23.vi.1949, *Barbosa & Carvalho* 3245 (LISC; LM; SRGH). MS: between Mandié and Changara, fr. 31.viii.1949, *Pedro & Pedrógão* 8175 (LMJ).

From Eritrea and the Sudan southwards to our area and Angola. Woodland and bushland, often though not always on alluvium; 520–910 m.

Two poor cultivated specimens, both from Rhodesia, C: Salisbury, Greenwood Park, *Pardy* P 6/50, st. 28.iv.1950 (SRGH) and *Pardy* in GHS 31170, fl. 24.i.1951 (SRGH), may well be subsp. *spirocarpa*, but better material is needed.

Acacia petersiana Bolle in Peters, Reise Mossamb. Bot. **1**: 4 (1861), based on material collected by Peters in Mozambique at Sena, Boror, etc. was said by Baker f., Legum. Trop. Afr. **3**: 842 (1930) to seem " nearly allied " to *A. spirocarpa*. No authentic material has survived. It may well be synonymous with *A. tortilis* in a wide sense, though it would be hard to decide which subspecies. See also remarks by Sim, For. Fl. Port. E. Afr.: 58 (1909).

47. **Acacia robusta** Burch., Trav. Int. S. Afr. **2**: 442 (1824).—Oliv., F.T.A. **2**: 349 (1871).—Bak. f., Legum. Trop. Afr. **3**: 841 (1930).—Burtt Davy, F.P.F.T. **2**: 342 (1932).—Verdoorn in Fl. Pl. S. Afr. **22**: t. 851 (1942).—Hutch., Botanist in S. Afr.: 297, 302 (1946).—O. B. Mill., B.C.L.: 20 (1948); in Journ. S. Afr. Bot. **18**: 24 (1952).—Gomes e Sousa, Dendrol. Moçamb. **2**: 51 cum tab. (1949); Dendrol. Moçamb.; Estudo Geral, **1**: 230, t. 34 (1966).—Codd, Trees & Shrubs Kruger Nat. Park: 48, fig. 43 a, b (1951).—Wild, S. Rhod. Bot. Dict.: 49 (1953).—Pardy in Rhod. Agric. Journ. **51**: 109 cum photogr. (1954).—Brenan in Kew Bull. **12**: 365 (1958).—Palmer & Pitman, Trees S. Afr.: 163 cum fig. et photogr. (1961).—Boughey in Journ. S. Afr. Bot. **30**: 158 (1964).—Gordon-Gray in Brittonia, **17**: 202, figs. 1–4 (1965).—de Winter, de Winter & Killick, Sixty-six Transvaal Trees: 54 cum photogr. (1966). Type from S. Africa.

Tree (2)5–25 m. high; crown flat or spreading; bark on trunk grey to dark-brown, fissured or sometimes smooth; young branchlets usually glabrous, eglandular, becoming grey to grey-brown, sometimes grey-purplish; bark of branchlets lenticellate, otherwise rather smooth, neither flaking off nor fissuring to expose red under-bark. Stipules spinescent, mostly short, up to 7 mm., sometimes longer, to 6–11 cm., straight or very slightly curved; " ant-galls " and other prickles absent. Leaves: rhachis (2·5)3–7 cm. long; pinnae (2)3–8(10) pairs; leaflets 9–27 pairs (in our area usually 13 or more pairs), (2)3·5–13(16) × 1–5(7) mm., glabrous or ciliate on the margins, oblong. Flowers white, very sweetly scented, profuse, in heads on axillary, glabrous to shortly pubescent or puberulous, eglandular or very inconspicuously glandular peduncles, whose involucel is basal, shortly above the base or in the lower half of the peduncle. Corolla glabrous outside. Pods dehiscent, 7–19 × 0·7–3 cm., straight to falcate, linear, glabrous; valves rather thin to almost woody, grey- to deep-brown, ± longitudinally veined, otherwise smooth, attenuate to the base. Seeds dark-blackish-olive, 8–15 × 5–9 mm., smooth, quadrate, compressed; areole 6·5–9 × 3·5–6·5 mm.

Leaf-rhachis glabrous:
Pods straight or slightly curved, 1·8–2·5(3) cm. wide; leaflets (2)2·5–5·5(7) mm. wide; peduncles glabrous or nearly so - - - - - - - - - Subsp. *robusta*
Pods ± falcate, 0·7–1·2 cm. wide; leaflets 1–2 mm. wide; peduncles shortly pubescent or puberulous - - - - - - - - - - - Subsp. *usambarensis*
Leaf-rhachis ± densely pubescent; pods usually ± falcate, mostly 1·3–1·7 cm. wide Subsp. *clavigera*

Subsp. **robusta**. TAB. **16** fig. 16.

Leaf-rhachides glabrous or almost so; pinnae mostly 2–4 pairs; leaflets 10–15 pairs, (2)2·5–5(7) mm. wide. Peduncles and calyx-lobes glabrous or almost so. Pods straight or slightly curved, 1·8–2·5(3) cm. wide.

Botswana. N: 4 km. from Francistown, fr. xii.1945, *Miller* B/409 (PRE). SE: Gaberones, fr. 11.xii.1961, *Yalala* 153 (K; SRGH). **Rhodesia.** W: Rhodes Matopos Estate, fl. & fr. 22.ix.1951, *Plowes* 1259 (SRGH).

Also in the Transvaal and Natal. Woodland; c. 1070 m.

Subsp. **clavigera** (E. Mey.) Brenan, comb. nov. Type from Natal.

Acacia clavigera E. Mey., Comm. Pl. Afr. Austr.: 168 (1836).—Brenan in Kew Bull. **12**: 365–367 (1958); F.T.E.A. Legum.-Mimos.: 118 (1959).—F. White, F.F.N.R.: 86, fig. 18 F (1962). Type as above.

Acacia sambesiaca Schinz in Denkschr. Math.-Nat. Kl. Akad. Wiss. Wien, **78**: 50 (1905).—Gomes e Sousa, Pl. Menyharth.: 69 (1936). Type: Mozambique, *Menyharth* 1003 (W, holotype; Z).

Acacia hirtella sensu Sim, For. Fl. Port. E. Afr.: 57, t. 35 (1909).

Acacia clavigera subsp. *clavigera*.—Brenan in Kew Bull. **12**: 367 (1958); F.T.E.A. Legum.-Mimos.: 118 (1959).—Boughey in Journ. S. Afr. Bot. **30**: 157 (1964).—Gordon-Gray in Brittonia, **17**: 202 (1965). Type as for *A. clavigera*.

Leaf-rhachides ± densely pubescent; pinnae often in 5–6 pairs; leaflets mostly 13–25 pairs, (1)1·5–3·5 mm. wide. Peduncles and calyx-lobes ± densely pubescent. Pods usually ± falcate, mostly 1·3–1·7 cm. wide.

Caprivi Strip. Mpilila I., fl. & fr. 15.i.1959, *Killick & Leistner* 3425 (K; SRGH). **Botswana.** N: between Kasane and Kasungula, fl. xii.1950, *Miller* B/1125 (K; PRE). **Zambia.** E: Petauke Distr., Luembe, fl. 13.xii.1958, *Robson* 941 (BM; K; LISC; SRGH). S: 13 km. W. of Kasungula, fr. 25.viii.1947, *Brenan & Greenway* 7750 (K). **Rhodesia.** N: Gokwe Distr., fr. 12.ix.1949, *West* 2996 (SRGH). W: Siatshilaba's Kraal, 24 km. NE. of Sebungwe river confluence, fr. 11–16.v.1956, *Plowes* 1962 (K; SRGH). E: Chipinga Distr., Sabi Valley Experimental Station, fl. ii.1960, *Soane* 288 (K; SRGH). S: Lundi R., Chipinda Pools, fl. & fr. 12.ix.1959, *Farrell* 101 (K; LISC; SRGH). **Mozambique.** N: Montepuez, fr. 28.xii.1963, *Torre & Paiva* 9757 (LISC). T: Tete, fr. 8.vi.1947, *Hornby* 2727 (K; SRGH). MS: Cheringoma, R. Condué near Inhaminga, fl. 25.viii.1962, *Gomes e Sousa* 4788 (COI; K; PRE). SS: Vilanculos, between Mapinhane and Vilanculos, fl. 30.viii.1942, *Mendonça* 36 (BM; K; LISC). LM: between Umbeluzi and Boane, fl. & fr. 9.viii.1959, *Barbosa & Lemos* in *Barbosa* 8642 (COI; K; LISC; PRE).

Also in the Transvaal, Natal and eastern Cape Prov. of S. Africa. Woodland and wooded grassland; 20–900 m.

It appears clear from the work of Gordon-Gray (Brittonia, **17**: 202–212 (1965)) that *A. robusta* and *A. clavigera* are connected by too many intermediates to be considered as separate species, although I adopted the latter view in Kew Bull. **12**: 365–6 (1958). Although it is clear that in Natal intermediate or mixed populations occur, yet *A. robusta* and *A. clavigera* are so distinct in our area that I consider it more helpful to recognize them as two subspecies of *A. robusta*. It is also, I consider, clear that *A. usambarensis*, although representing a geographical race, cannot be maintained as a species separate from *A. clavigera*, as suggested by Gordon-Gray (tom. cit.: 212 (1965)).

Subsp. **usambarensis** (Taub.) Brenan, comb. nov. Type from Tanzania.

Acacia usambarensis Taub. in Engl., Pflanzenw. Ost-Afr. **C**: 195, t. 20H (1895).—Bak. f., Legum. Trop. Afr. **3**: 846 (1930).—Brenan, T.T.C.L.: 338 (1949). Type as above.

Acacia clavigera subsp. *usambarensis* (Taub.) Brenan in Kew Bull. **12**: 369 (1958); F.T.E.A. Legum.-Mimos.: 118, fig. 16/45 (1959).—Boughey in Journ. S. Afr. Bot. **30**: 157 (1964). Type as above.

Similar to subsp. *clavigera* but leaf-rhachides glabrous, leaflets 1–2 mm. wide and pods 0·7–1·2 cm. wide.

Mozambique. N: Porto Amélia, fl. & fr. 19.xii.1963, *Torre & Paiva* 9604 (LISC). Z: Maganja da Costa, fr. 31.vii.1943, *Torre* 5726 (LISC). MS: Dondo, Vila Machado, st. 16.iv.1948, *Mendonça* 3970 (K; LISC). SS: Govuro, fl. i.1912, *Dawe* 522 (K).

Also in Kenya and Tanzania. Woodland; 10–250 m.

Intermediates do occur between subsp. *usambarensis* and subsp. *clavigera*, but most of these, particularly from our area, with pubescent leaf-rhachides discussed in Kew Bull. **12**: 367 (1958) seem best considered as variants of subsp. *clavigera* with narrow leaflets.

48. **Acacia grandicornuta** Gerstner in Journ. S. Afr. Bot. **4**: 55 (1938).—O. B. Mill., B.C.L.: 19 (1948); in Journ. S. Afr. Bot. **18**: 21 (1952).—Codd, Trees & Shrubs

Kruger Nat. Park: 44, fig. 38 c, d, e (1951).—Boughey in Journ. S. Afr. Bot. **30**: 157 (1964). Syntypes from Natal.

Tree 3–10 m. high, sometimes a shrub as low as 1·2 m.; bark greyish to black; young branchlets glabrous, older ones with bark grey to purple, ± lenticellate but otherwise smooth, neither flaking off nor fissuring to expose a red under-bark. Stipules spinescent, up to c. 9 cm. long, straight or very slightly curved, often stout and ± elongate; " ant-galls " and other prickles absent. Leaves: rhachis 0·5–2(2·5) cm. long, usually glabrous; pinnae 1–4 pairs; leaflets 5–15 pairs, (3·5)5–10 × (1)1·5–4 mm., glabrous, obtuse or rounded; lateral nerves invisible or slightly prominent beneath. Flowers white, very sweetly scented, in heads on axillary glabrous or nearly glabrous peduncles 1·5–2·5 cm. long; involucel ⅓–⅔-way up. Corolla glabrous outside. Pods dehiscent, 6–11·5 × 0·6–1·1 cm., falcate, linear, glabrous; valves rather thin, purplish to brown, ± finely and longitudinally veined, otherwise smooth, attenuate to the base. Seeds olive to brown, 9–10 × 5·5–7 mm., ± oblong, compressed; areole 6·5–7 × 3·5–4·5 mm.

Botswana. SE: Mahalapye road 11 km. from Shoshong, fl. & fr. immat. 13.i.1958, *de Beer* 559 (K; SRGH). **Rhodesia.** E: Nyanyadzi, fr. vii.1959, *Davies* 2565 (SRGH). ?S: Birchenough Bridge, fl. & fr. immat. 1938, *Obermeyer* 2398 (PRE; SRGH). **Mozambique.** SS: Limpopo–Nuanetzi rivers, fr. vii.1932, *Smuts* P 326a (K; PRE). LM: Maputo, Catuane, fr. 29.iv.1948, *Torre* 7744a (K; LISC).

Also in Swaziland, the Transvaal and Natal. Woodland; c. 900 m.

A. grandicornuta is very close indeed to *A. robusta* Burch. in its wide sense, and it is probably arguable that it might be better placed under that species. It is nevertheless a very distinctive taxon and for the present I am content to leave it as a separate species.

The normally glabrous leaf-rhachis, the few pinnae and leaflets and the narrow pods will distinguish *A. grandicornuta* from *A. robusta* subsp. *clavigera*. The narrow curved pods and more slender branchlets separate it from *A. robusta* subsp. *robusta*, and the few pinnae and leaflets and glabrous peduncles from *A. robusta* subsp. *usambarensis*. *A. grandicornuta* is furthermore often easily recognized by the characteristic green colour of its foliage when dried.

49. **Acacia gerrardii** Benth. in Trans. Linn. Soc. **30**: 508 (1875).—Bak. f., Legum. Trop. Afr. **3**: 846 (1930).—Burtt Davy, F.P.F.T. **2**: 343 (1932).—O. B. Mill., B.C.L.: 18 (1948); in Journ. S. Afr. Bot. **18**: 21 (1952).—Codd, Trees & Shrubs Kruger Nat. Park: 44, figs. 36, 38 a, b (1951).—Pardy in Rhod. Agric. Journ. **53**: 507 cum photogr. (1956).—Brenan in Kew Bull. **12**: 369 (1958); F.T.E.A. Legum.-Mimos.: 119 (1959).—F. White, F.F.N.R.: 85 (1962).—Mitchell in Puku, **1**: 104 (1963).—Boughey in Journ. S. Afr. Bot. **30**: 157 (1964). TAB. **16** fig. 17. Type from Natal.

Acacia hebecladoides Harms in Engl., Bot. Jahrb. **36**: 208 (1905).—R.E.Fr., Wiss. Ergebn. Schwed. Rhod.-Kongo-Exped. **1**: 63, t. 2 fig. 4 (1914).—Bak. f., Legum. Trop. Afr. **3**: 846 (1930).—Brenan, T.T.C.L.: 337 (1949).—Gilbert & Boutique, F.C.B. **3**: 162 (1952). Type from Tanzania.

Shrub or more usually a tree 3–15 m. high; crown flat, umbrella-shaped or irregular; bark on trunk grey, blackish-brown or black, rough, fissured; young branchlets ± densely grey-pubescent, rarely glabrous or nearly so, epidermis usually splitting or falling away to expose a rusty-red inner layer. Stipules spinescent, usually straight or nearly so, sometimes recurved, rarely hooked, mostly short, to c. 1 cm. long, rarely to c. 6·5 cm. long and then usually grey; " ant-galls " and other prickles absent.* Leaves: rhachis (1·5)2–7 cm. long, ± densely pubescent; pinnae (3)5–10(12) pairs; leaflets (8)12–23(28) pairs, 3–7·5 × 1–2 mm., ± ciliate on the margins, at least near the base, otherwise glabrous or nearly so, sometimes hairy on the surface. Flowers white or cream, scented, in heads on axillary densely grey-pubescent eglandular or inconspicuously glandular, occasionally strongly glandular peduncles; involucel at or shortly above the base or sometimes to ⅓-way up the peduncle. Corolla glabrous or only slightly and inconspicuously pubescent outside. Pods dehiscent, (4·5)7–16(22) × 0·6–1·1(1·7) cm., falcate, linear or linear-oblong; valves rather thin, ± grey-puberulous to -tomentellous,

* A very few specimens have been collected, though not in the area of the Flora, closely resembling *A. gerrardii* except for the presence of a few " ant-galls ". The taxonomic status of these specimens is still uncertain.

rarely subglabrous or glabrous. Seeds olive-brown, 9–12 × 7 mm., smooth, ± irregularly quadrate, compressed; areole 6·5–7 × 3·5–4·5 mm.

Botswana. N: Ngamiland, fl. xii.1930, *Curson* 581 (PRE). SW: Kgalagadi Distr., 37 km. E. of Tsane, fl. iii.1950, *Miller* B/1004 (K; PRE). SE: Ngwato Distr., 10 km. S. of Debeeti Hill, fl. i.1951, *Miller* B/1152 (K). **Zambia.** N: Isoka Distr., fl. & fr. 8.i.1938, *Trapnell* 1816 (K). C: Lusaka–Kafue, mile 9, fl. 12.i.1960, *Angus* 2120 (K; PRE; SRGH). E: 48 km. from Fort Jameson on Nsefu–Luangwa road, fr. 11.v.1963, *Van Rensburg* KBS 2118 (K). S: Mochipapa near Choma, fr. 15.iii.1962, *Astle* 1566 (K). **Rhodesia.** N: Mazoe, fl. & fr. 21.ii.1956, *Guy* 4/56 (SRGH). W: Matopos, fl. xii.1955, *McKay* in GHS 67959 (SRGH). C: Bromley, Gardner Farm, fr. 13.vi.1932, *Eyles* 7135 (K; SRGH); Ewanrigg, fr. 31.vii.1946, *Verdoorn* 2250 (PRE). S: Gutu Reserve, fl. 2.x.1959, *Phipps* 2244 (K; SRGH). **Malawi.** N: Rumpi, fr. vi.1953, *Chapman* 121 (K). C: 3 km. N. of Kasungu fl. 14.i.1959, *Robson & Jackson* 1169 (BM; K; LISC; SRGH). **Mozambique.** N: Malema, Mutuali, fl. 25.iii.1953, *Gomes e Sousa* 4086 (COI; K; LMJ; PRE; SRGH). T: Macanga, Massamba, fr. 6.vii.1949, *Barbosa & Carvalho* in *Barbosa* 3449 (K; LM; SRGH). MS: Cheringoma, Inhamitanga, fl. 20.ii.1948, *Andrada* 1077 (LISC). SS: 25 km. from Manjacaze, fl. & fr. 7.iii.1948, *Torre* 7505 (BM; LISC). LM: Maputo, fl. 27.ii.1947, *Hornby* 2521 (K; LMJ; PRE; SRGH).

Widespread in tropical Africa from the Sudan southwards to Natal and westwards to Nigeria. Woodland and wooded grassland; from near sea-level to 1500 m.

A. gerrardii is a variable species but there do not appear to be the local variants worthy of taxonomic recognition that there are in East Africa: all our material seems best considered as var. *gerrardii*. Two remarkable extremes have, however, been collected:

(1) *Goodier* 1072 from Rhodesia, Chipinga Distr., near Marya School, fr. 28.vi.1961 (SRGH), with pods 5·5–8 cm. long but only 4·5–5·5 mm. wide. The seeds appear to be ill-formed.

(2) *Duff* in *Eyles* 5270, from Rhodesia, Salisbury, fr. 24.i.1928 (K; SRGH) with very grey-tomentellous pods 4–7 × 1–1·3 cm.

In addition to the above, two other specimens deserve mention: *Sim* 19214, from Rhodesia, Bulawayo, fr. vii.1920 (PRE) and *Keay* FHI 21220, from Rhodesia, Bubi, Inyati Mission, fr. 21.iv.1947 (SRGH). In both of these specimens the pods are nearly straight or only slightly curved, 3–5 × 0·9–1 cm. The indumentum on the pods of Sim's specimen is rather dense and coarser than in Keay's. They may represent a distinct race of *A. gerrardii* or, as the appearance of the plants suggests, crosses between that species and *A. luederitzii* Engl.

50. **Acacia lasiopetala** Oliv., F.T.A. **2**: 346 (1871).—Burkill in Johnston, Brit. Centr. Afr.: 245 (1897).—Sim, For. Fl. Port. E. Afr.: 56 (1909).—Bak. f., Legum. Trop. Afr. **3**: 847 (1930).—Brenan, T.T.C.L.: 337 (1949); F.T.E.A. Legum.-Mimos.: 120, fig. 16/47 (1959).—Burtt Davy & Hoyle, rev. Topham, N.C.L. ed. 2: 64 (1958). Type: Malawi, Mpemba Mt., *Kirk* (K, holotype).

Small tree 2–6 m. high; bark rusty-red; branchlets persistently grey-tomentellous or densely villous (indumentum often yellowish when young); then epidermis flaking away to expose powdery rusty-red bark. Stipules spinescent, to 2·3 cm. long, never inflated; other prickles absent. Leaves with gleaming silky pale-golden indumentum when young, grey-pubescent when older; petiole 3–8 mm. long (to 1·7 cm. in juvenile leaves); pinnae of well-developed leaves of mature shoots 15–40 pairs (reduced leaves with fewer pairs usually also present), mostly 2–3·5 cm. long; leaflets very numerous, 2·5–5 × 0·6–1 mm. Flowers white, in heads; involucel at the base of the nearly or quite eglandular 2·5 cm. long peduncle; peduncles mostly solitary in leaf-axils. Corolla densely pubescent on the lobes outside, c. 1½ times as long as the calyx. Pods dehiscent, c. 9–15 × 0·8–1·1 cm., mostly arcuate, grey-tomentellous, ± turgid over the seeds, with constrictions between them c. 1–2 cm. apart. Seeds olive-brown, usually 8–9 × 5–6 mm., smooth or nearly so, irregularly quadrate or elliptic, compressed; areole 5·5–6 × 2·5–3 mm.

Malawi. C: between Kota Kota and Dowa, on the scarp, fr. 21.vii.1936, *Burtt* 6049 (BM; K; PRE). S: Mpemba Mt., fl. xi.1935, *Clements* H 2997/35 (K; PRE). **Mozambique.** N: Maniamba, Metangula, 22.v.1948, *Pedro & Pedrógão* 3835 (EA; LMJ).

Also in the Congo and Tanzania. On hillsides in *Brachystegia* woodland: c. 1370 m.

51. **Acacia arenaria** Schinz in Mém. Herb. Boiss. **1**: 108 (1900).—Bak. f., Legum. Trop. Afr. **3**: 839 (1930).—Hutch., Botanist in S. Afr.: 523 (1946).—O. B. Mill., B.C.L.: 16 (1948); in Journ. S. Afr. Bot. **18**: 18 (1952).—Torre, C.F.A. **2**: 282 (1956).—Brenan, F.T.E.A. Legum.-Mimos.: 126, fig. 17/55 (1959).—Boughey in Journ. S. Afr. Bot. **30**: 157 (1964). Syntypes from SW. Africa.

Acacia hermannii Bak. f. in Journ. of Bot. **67**: 198 (1929); Legum. Trop. Afr. **3**: 847 (1930).—Brenan, T.T.C.L.: 337 (1949).—Wild, S. Rhod. Bot. Dict.: 47 (1953). Type from Tanzania.

Shrub or small tree 1·2–9 m. high, with a very short bole, branching near the ground; bark on bole dark and rough; branchlets with short inconspicuous puberulence or pubescence, purplish to brownish, soon going grey or sometimes brownish, zig-zag, their epidermis not peeling or flaking away. Stipules spinescent, to 6 cm. long, divergent, slender, straight, never inflated; other prickles absent. Leaves with inconspicuous dull pubescence; petiole 3–14 mm. long; rhachis (5)10–21 cm. long; pinnae of well-developed leaves of mature shoots 15–35 pairs (reduced leaves with fewer pairs sometimes also present), 0·7–2·2(3) cm. long; leaflets 1–5 × (0·5)0·7–1 mm., glabrous or ciliolate. Flowers white or pale pink, in heads; involucel at or above the middle or at the apex of the pubescent and glandular peduncle. Corolla 2–4 times as long as the calyx, glabrous outside. Pods dehiscent, 8–18(22) × 0·5–0·8 cm., arcuate, glabrous to slightly pubescent and glandular, deep-red-brown outside, flat or slightly constricted between the seeds. Seeds olive-grey, 7–9 × 3–5 mm., smooth, quadrate or oblong, compressed; areole 3·5–4·5 × 1·5–2·25 mm.

Caprivi Strip. Linyanti, fl. 27.xii.1958, *Killick & Leistner* 3127 (K; PRE; SRGH). **Botswana.** N: NE. corner of Makarikari Pan, fl. 15.i.1959, *West* 3834 (K; SRGH). SE: 3 km. N. of Makoro Siding, fl. 20.i.1960, *Leach & Noel* 260 (K; SRGH). **Rhodesia.** W: Nyamandhlovu, fl. & fr. immat. 29.i.1954, *West* 3228 (K; SRGH).

Also in Tanzania, Angola and SW. Africa. In the drier types of woodland, grassland and scrub, sometimes with *Colophospermum mopane*; 900–1460 m.

The heads of flowers are characteristically borne only on shoots of the current year, in the axils towards the end of the shoots, so that there appears to be a terminal " raceme " of capitula.

52. **Acacia sieberana** DC., Prodr. **2**: 463 (1825).—Oliv., F.T.A. **2**: 347 (1871).—Bak. f., Legum. Trop. Afr. **3**: 836 (1930).—Brenan, T.T.C.L.: 335 (1949); F.T.E.A. Legum.-Mimos.: 127, fig. 17/57 (1959).—Gilbert & Boutique, F.C.B. **3**: 166 (1952). —Burtt Davy & Hoyle, rev. Topham, N.C.L. ed. 2: 64 (1958).—White, F.F.N.R.: 84, fig. 17 K (1962).—Mitchell in Puku, **1**: 105 (1963).—Type from Senegal.

Tree 5–18(25) m. high; bark usually grey and rough on the trunk, sometimes light-brown, or yellowish and flaking, especially on the branches; young branchlets glabrous to tomentose, eglandular, green to grey or yellowish, later grey; outer bark then usually flaking away to expose an olive or yellow inner layer. Stipules spinescent, whitish, up to 9(12·5) cm. or more long, straight; " ant-galls " and other prickles absent. Leaves: rhachis 2·5–10 cm. long; pinnae mostly 6–23(35) pairs; leaflets 14–45(52) pairs, 2–6·5 × (0·5)0·6–1·5 mm., glabrous to ciliate, narrowly oblong, rounded to obtuse at the apex; midrib, and sometimes small lateral nerves also, somewhat prominent on both surfaces. Flowers white or very pale yellow, in heads on axillary peduncles 1·5–5 cm. long which are variable in indumentum but eglandular, and whose involucel is normally apical or in the upper half of the peduncle. Pods very slow in dehiscing, (8)9–21 × (1·5)1·7–3·5 cm., straight or sometimes ± falcate, flattened but thick and almost woody in texture when dry, ± smooth and glossy, without raised veins, glabrous or somewhat hairy. Seeds olive-grey, 9–12 × 7–8 mm., smooth, elliptic to subcircular, compressed: areole 7–9·5 × 5–6 mm.

Young branchlets glabrous or nearly so; branches of crown usually ascending — var. *sieberana*

Young branchlets ± hairy, usually densely so; branches of crown usually widely spreading:

- Indumentum on branchlets usually neither markedly golden nor villous — var. *vermoesenii*
- Indumentum on branchlets normally villous and markedly golden, especially when young - - - - - - - - - var. *woodii*

Var. **sieberana**.—Keay, F.W.T.A. ed. 2, **1**: 499 (1958).—Brenan, F.T.E.A. Legum.-Mimos.: 127 (1954).

Crown with ascending branches, less spreading than in the following. Young branchlets glabrous or almost so.

Malawi. N: Vipya, along the main Vipya road between Mtongatonga and Chilangawa, fl. xii., *Chapman* 264 (K). C: Dowa Distr., Chitala to Domera Bay, fl. 29.x.1941, *Greenway* 6379 (K; PRE). **Mozambique.** N: Porto Amélia, between Metuge and Mahate, fl. 3.x.1948, *Barbosa* in *Mendonça* 2330 (BM; K; LISC). Z: Pebane, R. Muligudié, 41·8 km. from Naburi, fr. 6.x.1949, *Barbosa & Carvalho* in *Barbosa* 4326 (K; LM).

Also in Senegal, Nigeria, Cameroon, the Congo, the Sudan, Ethiopia, Uganda, Kenya and Tanzania. In woodland of various types; 580–800 m.

See note under var. *woodii*.

Var. **vermoesenii** (De Wild.) Keay & Brenan in Kew Bull. **5**: 364 (1951).—Pardy in Rhod. Agric. Journ. **48**: 406 cum photogr. (1951).—Wild, S. Rhod. Bot. Dict.: 49 (1953).—Brenan, F.T.E.A. Legum.-Mimos.: 128 (1959).—Boughey in Journ. S. Afr. Bot. **30**: 158 (1964). Type from the Congo.

Acacia vermoesenii De Wild., Pl. Bequaert. **3**: 69 (1925).—Brenan, T.T.C.L.: 335 (1949). Type as above.

Crown usually with spreading branches, broad, flat or mushroom-shaped. Young branches ± hairy, usually densely so; indumentum usually neither markedly golden nor villous. Pods glabrous or nearly so, even when young.

Zambia. B: Nangweshi, fr. 19.vii.1952, *Codd* 7125 (BM; K; PRE; SRGH). N: edge of Chambeshi flats 120 km. E. of Kasama, fl. 11.x.1960, *Robinson* 3960 (K; SRGH). W: Kasempa, fl. x.1934, *Trapnell* 1627 (K). C: Chilanga Distr., Quien Sabe, fl. 12.x.1929, *Sandwith* 58 (K; SRGH). S: Mazabuka, fl. 7.x.1930, *Milne-Redhead* 1213 (K; PRE). **Rhodesia.** N: Urungwe, fl. 24.x.1952, *Lovemore* 280 (SRGH). W: 8 km. above Victoria Falls, fl. 22.x.1959, *Wild* 4849 (SRGH). S: Zimbabwe, fr. 20.vi.1924, *Galpin* 9202 (PRE). **Malawi.** N: Vipya Plateau, fl. xii.1959, *Willan* 67 (K; SRGH). S: Mlanje, fl. x.1905, *Purves* 217 (K). **Mozambique.** N: Moma, Chalaua, fr. 6.viii.1948, *Barbosa & Lemos* 1765 (LISC; LMJ). Z: between Nicuadala and Régulo Simogo, fr. 29.viii.1949, *Barbosa & Carvalho* in *Barbosa* 3895 (K; LISC). T: Angónia, fr. 26.vii.1941, *Torre* 3339 (BM; K; LISC). MS: Gorongosa, Parque Nacional de Caça, Chitengo, fl. 13.x.1963, *Torre & Paiva* 9221 (LISC). SS: Morrumbene, fr. 3.x.1947, *Pedro & Pedrógão* 2306 (LMJ).

From the Sudan and Ethiopia southwards to our area. Woodland and wooded grassland, often in river flood-plains; 40–1830 m.

See note under the following variety.

Var. **woodii** (Burtt Davy) Keay & Brenan in Kew Bull. **5**: 364 (1951).—Pardy in Rhod. Agric. Journ. **48**: 406 (1951).—Wild, S. Rhod. Bot. Dict.: 49 (1953).—Palgrave, Trees of Central Afr.: 254 (1956).—Torre, C.F.A. **2**: 281 (1956).—Brenan, F.T.E.A. Legum.-Mimos.: 128 (1959).—Palmer & Pitman, Trees S. Afr.: 163 cum fig. et photogr. (1961).—Boughey in Journ. S. Afr. Bot. **30**: 158 (1964).—de Winter, de Winter & Killick, Sixty-six Transvaal Trees: 56 cum photogr. (1966). TAB. **16** fig. 18. Type from Natal.

Acacia amboensis Schinz in Mém. Herb. Boiss. **1**: 105 (1900).—Bak. f., Legum. Trop. Afr. **3**: 838 (1930).—O. B. Mill., B.C.L.: 16 (1948); in Journ. S. Afr. Bot. **18**: 18 (1952). Type from SW. Africa.

Acacia woodii Burtt Davy in Kew Bull. **1922**: 332 (1922); F.P.F.T. **2**: 344 (1932).—Steedman, Trees etc. S. Rhod.: 15 (1933).—Hutch., Botanist in S. Afr.: 394 (1946).—O. B. Mill., B.C.L.: 21 (1948); in Journ. S. Afr. Bot. **18**: 26 (1952).—Brenan, T.T.C.L.: 335 (1949).—Codd, Trees & Shrubs Kruger Nat. Park: 51, figs. 44 a, b, 45 (1951).—Burtt Davy & Hoyle, rev. Topham, N.C.L. ed. 2: 65 (1958). Type as for *A. sieberana* var. *woodii*.

Acacia lasiopetala sensu Suesseng. & Merxm. in Proc. & Trans. Rhod. Sci. Ass. **43**: 16 (1951).

Crown as in var. *vermoesenii*. Young branchlets ± densely hairy; indumentum normally villous and markedly golden especially when young. Pods densely pubescent when young and usually slightly so even when old.

Caprivi Strip. E. of the Cuando R., fr. x.1945, *Curson* 1236 (PRE). **Botswana.** N: 123 km. NE. of Maun, fr. 14.vi.1930, *Van Son* 28873 (PRE). **Zambia.** W: Kitwe, fl. 13.x.1960, *Linley* 13 (K; LISC; SRGH). C: Chilanga, fl. 24.x.1957, *Angus* 1772 (K; SRGH). E: Petauke Distr., Kapoche stream, fl. 26.x.1938, *Trapnell* 1948 (K). S: Simasunda, Mapanza, fl. 21.ix.1957, *Robinson* 2440 (PRE; SRGH). **Rhodesia.** N: Urungwe, fl. 12.x.1957, *Phipps* 775 (COI; LISC; LM; SRGH). W: Wankie Game Reserve, fl. 20.x.1958, *West* 3741 (K; SRGH). C: Salisbury, fl. xi.1919, *Eyles* 1927 (K; PRE; SRGH). E: Umtali, fl. 8.xii.1957, *Chase* 6780 (K; SRGH). S: Fort Victoria, fl., *Rogers* 2149 (K). **Malawi.** N: Songwe R., fl. x.1896, *Nicholson* (K). S: Blantyre, fl. xi, *Buchanan* 93 (K). **Mozambique.** N: Malema, st. *Hornby* 2302 (LM).

Z: 2 km. W. of Gúruè fork, fl. ix–x.1941, *Hornby* 2302 (PRE). T: Macanga, Chiuta, fl. 15.x.1943, *Torre* 6021 (BM; K; LISC). MS: R. Curumadzi, Jihu, fl. 18.xi.1906, *Swynnerton* 60 (BM; K).

Also in Tanzania, Angola, the Transvaal and Natal; probably occurring in the Congo. Habitat similar to that of var. *vermoesenii*; 700–1620 m.

As pointed out in F.T.E.A. Legum.-Mimos.: 128–9 (1959), var. *sieberana*, comparatively uncommon in our area and only in the north-eastern part, is more distinct from var. *vermoesenii* than that is from var. *woodii*. Indeed the two last-named varieties are linked by puzzling intermediates. It is uncertain how far the differences in habit and pod shown by var. *sieberana* in East Africa will apply in our area, and field observations are needed to test this.

The records of var. *woodii* from the Caprivi Strip and Botswana are rather doubtful—indeed the former is based on pods alone and is the only evidence for the species in the Caprivi Strip. *Miller* 5376 (SRGH), from Rhodesia, Fort Rixon, fr. viii.1958, is apparently referable to var. *woodii* but is very unusual in having pods with a dense short velvety indumentum outside.

West 2421 (SRGH), from Rhodesia, referred to *A. davyi* N.E.Br. by Boughey in Journ. S. Afr. Bot. **30**: 157 (1964), is in fact *A. sieberana*. The specimen is sterile and it would be unwise to determine it more precisely.

53. **Acacia hebeclada** DC., Cat. Hort. Monspel.: 73 (1813).—Oliv., F.T.A. **2**: 348 (1871).—Harms in Warb., Kunene-Samb.-Exped. Baum: 243 (1903).—Gomes e Sousa, Pl. Menyharth.: 69 (1936).—Hutch., Botanist in S. Afr.: 398, 399 (photogr.) (1946).—O. B. Mill., B.C.L.: 19 (1948).—Wild, Guide Fl. Vict. Falls: 148 (1953).—F. White, F.F.N.R.: 85, fig. 18 a (1962).—Boughey in Journ. S. Afr. Bot. **30**: 157 (1964). TAB. 16 fig. 19. Type from S. Africa.

Acacia stolonifera Burch., Trav. Int. S. Afr. **2**: 241 (1824).—Burtt Davy, F.P.F.T. **2**: 340, fig. 58 (1932).—Hutch., Botanist in S. Afr.: 398, 418, 433, 632 (1946).—O. B. Mill., B.C.L.: 21 (1948); in Journ. S. Afr. Bot. **18**: 25 (1952). Type from S. Africa.

Acacia stolonifera var. *chobiensis* O. B. Mill. in Journ. S. Afr. Bot. **18**: 25 (1952). Type: Botswana, Serondela, *Miller* B/1069 (K, holotype).

Shrub or small tree, 0·4–7 m. high, branched near ground-level, forming thickets; bark dark-grey, longitudinally fissured, flaking; young branchlets densely pubescent to tomentose with grey spreading hairs; epidermis on twigs not splitting or if (rarely) so then inner layer not markedly rusty-red as in *A. gerrardii*. Stipules spinescent, straight to arcuate or hooked, either short and up to c. 1·5 cm. long or up to c. 6 cm. and then straight; " ant-galls " and other prickles absent. Leaves: rhachis 0·5–5 cm. long, like the petiole ± densely spreading-pubescent; pinnae 2–9(13) pairs; leaflets 7–16 pairs per pinna, (1·5)2–7 × (0·75)1–2 mm., linear-oblong, often slightly broadened above, rounded or obtuse at the apex, ± ciliate on the margins at least near the base, otherwise glabrous or nearly so. Flowers white to cream, in heads on axillary eglandular peduncles; hairs on peduncle equalling or longer than its diameter (in our area); involucel at or shortly above the base, occasionally and sporadically up to ¼-way up the peduncle. Calyx densely rather long-pubescent above. Corolla glabrous or rarely slightly hairy outside. Pods finally dehiscent, 4–15 × (1·3)1·6–4 cm., straight or nearly so, turgid, oblong-ellipsoid, cylindric or fusiform; valves thick, hard, densely grey-tomentellous outside, faintly but often closely longitudinally nerved. Seeds ± transverse in the pod, brown, 10–13 × 8–11 mm., irregular in size and shape; areole 7–9 × 5–6·5 mm.

Caprivi Strip. Linyanti, fr. 28.xii.1958, *Killick & Leistner* 3164 (K; PRE; SRGH). **Botswana.** N: 24 km. S. of Sigara Pan overlooking Makarikari Pan, fr. 25.iv.1957, *Drummond & Seagrief* 5234 (K; SRGH). SW: Tsabong camp, fr. 3.iii.1958, *de Beer* 711 (SRGH). SE: Springfield Farm, 3 km. S. of Lobatsi, fr. 17.i.1960, *Leach & Noel* 141 (SRGH). **Zambia.** B: 11 km. SW. of Senanga, fl. & fr. 5.viii.1952, *Codd* 7407 (BM; K; PRE). S: Bombwe, fr. 1937, *Martin* 711 (K). **Rhodesia.** W: Victoria Falls, Cataract I., fr. 23.xi.1949, *Wild* 3159 (K; SRGH).

Also in SW. Africa and S. Africa. Usually on sand or alluvium in arid areas, often by rivers; 760–1190 m.

Within *A. hebeclada*, as interpreted here, there are two extremes of variation which both occur in our area:

(1) A low spreading shrub up to c. 1·5 m. high with short internodes mostly up to c. 1–1·5 cm. long, small foliage, small shortly pedunculate heads, and small pods up to c. 9 × 1·5 cm.

(2) A taller shrub or small tree more than 1·5 m. high with large internodes c. 1·5–4 cm. long, larger laxer foliage, large heads on longer peduncles and larger pods up to 15 × 4 cm.

A. stolonifera var. *chobiensis* O. B. Mill. corresponds with No. 2, and this is probably the prevalent variant in Zambia and Rhodesia and the northern part of Botswana. In southern Botswana No. 1 is more general.

Nos. 1 and 2 are connected by intermediates, and the basis of the variation is not clear. It may be that No. 1 is merely a state growing in a more extreme climate, or (and this appears more probable) that the two variants are imperfectly differentiated ecotypes. Further study and observation are needed, and it therefore seems preferable at present to record this variation but not to give it taxonomic recognition.

A. hebeclada when in pod cannot be mistaken. Flowering specimens, however, particularly with inadequate notes on the habit, can be confused with *A. gerrardii* Benth. and and *A. luederitzii* Engl. *A. gerrardii* can usually be recognized by the habit, the epidermis of the twigs splitting to expose a rusty-red inner layer, and by the usually more numerous and more parallel-sided leaflets, while *A. luederitzii* has less grey-green often narrower and more parallel-sided leaflets, and (at least in our area) the indumentum is shorter, the hairs on the peduncles being shorter than its diameter.

54. **Acacia stuhlmannii** Taub. in Engl., Pflanzenw. Ost-Afr. **C**: 194, t. 21, E, F (1895). —Bak. f., Legum. Trop. Afr. **3**: 836 (1930).—Brenan, T.T.C.L.: 334 (1949); F.T.E.A. Legum.-Mimos.: 131, fig. 19 (1959).—Boughey in Journ. S. Afr. Bot. **30**: 158 (1964). TAB. **16** fig. 20. Type from Tanzania.

1–6(7·5) m. high, varying from a low spreading shrub to a small ± obconical-crowned tree (in our area always a shrub c. 2 m. high); young shoots with spreading golden villous hairs up to 1·5–3 mm. long, hairs later going grey; branchlets becoming glabrescent, olive- to grey-brown, marked with pale dot-like lenticels, longitudinally wrinkled, but epidermis neither cracking nor peeling; the old stems in the tree-form, however, may have papery-peeling golden-brown bark over a green layer. Stipules spinescent, 0·7–4·5(6·5) cm. long, straight; " ant-galls " and other prickles absent. Leaves: rhachis usually 2–5 cm. long, spreading-hairy; pinnae 4–8(12) pairs*; leaflets 7–25 pairs, 2–5·5 × (0·6)1–1·5(2) mm., ciliate. Flowers white with reddish-buff or mauve anthers, in heads on axillary densely hairy or tomentose eglandular peduncles 0·4–3 cm. long, often produced when the plant is without leaves; involucel basal or in the lower half of the peduncle†; bracteoles conspicuously ciliate or pubescent. Calyx ± pubescent outside. Corolla-lobes conspicuously pubescent outside. Pods indehiscent, (2)4–9(10·5) × (1·1)1·2–2·5(3) cm., somewhat curved or sometimes straight, thick, hard and woody, usually much attenuate at the base, densely clothed with long spreading hairs. Seeds olive, 6–9 mm. in diam., minutely punctate, ellipsoid to subglobose; central areole 6–7 × 4·5–5 mm.

Botswana. SE: Sowa Flats, Serowe, fr. 26.iii.1957, *de Beer* 55 (K; SRGH). **Rhodesia.** S: 24 km. NW. of Beitbridge on main Bulawayo road, fl. 18.ix.1967, *Müller* 675 (K).

Also in Somalia, Kenya, Tanzania and the Transvaal. " Vlei area in sandveld "; 930 m.

A. stuhlmannii, although a rather variable species in East Africa, is relatively uniform in the Botswana–Transvaal part of its range, being always shrubby and with short peduncles 1–1·5 cm. long and small pods up to c. 6·5 cm. long (including the stipe) and 2 cm. wide. The Botswana–Transvaal area is also widely disjunct geographically from the rest of the range of the species.

Cultivated species

Various Australian species (" wattles ") in particular are cultivated for ornament or their use in, for example, tanning. It is highly probable that species other than those mentioned here occur, or will occur in the future, in cultivation in our area.

* Occasional leaves on apparently juvenile non-flowering shoots may have up to 17 pairs of pinnae and a rhachis up to 8 cm. long.

† One specimen from Somalia, *Peck* 72, apparently *A. stuhlmannii*, has the involucel in the upper half of the peduncle or even apical. I have not seen such an abnormality from our area and have not allowed for it in the key to the species.

55. **Acacia farnesiana** (L.) Willd. in L., Sp. Pl. ed. 4, **4**: 1083 (1806).—Oliv., F.T.A. **2**: 346 (1871).—Bak. f., Legum. Trop. Afr. **3**: 835 (1930).—Brenan, T.T.C.L.: 334 (1949); F.T.E.A. Legum.-Mimos.: 111, fig. 16/38 (1959).—Gilbert & Boutique, F.C.B. **3**: 164 (1952).—Torre, C.F.A. **2**: 278 (1956). Type uncertain, probably a plant from Domingo cultivated at Uppsala.
Mimosa farnesiana L., Sp. Pl. **1**: 521 (1753). Type as above.

Shrub 1·5–4 m. high; young branchlets glabrous or nearly so, purplish to grey; epidermis not obviously peeling off; glands (as on peduncles) few and inconspicuous. Stipules spinescent, usually short, up to 1·8(3) cm. long, never inflated; other prickles absent. Leaves with a small gland on the petiole and sometimes one on the rhachis near the top pair of pinnae; pinnae 2–7 pairs; leaflets 10–21 pairs, 2–7 × 0·75–1·75 mm., very rarely larger, with both midrib and lateral nerves visible and somewhat raised beneath. Flowers bright golden-yellow, sweetly scented, in axillary pedunculate heads; involucel at the apex of the peduncle. Calyx and corolla glabrous outside except for extreme tips of lobes. Pods indehiscent, 4–7·5 × 0·9–1·5(2) cm., straight or curved, subterete and turgid, dark-brown to blackish, glabrous, finely longitudinally striate. Seeds chestnut-brown, 7–8 × 5·5 mm., smooth, elliptic, thick, only slightly compressed; areole 6·5–7 × 4 mm.

Rhodesia. C: Salisbury, fl. i.1920, *Eyles* 5755 (SRGH). **Mozambique.** LM: Lourenço Marques, fl. & fr. 1946, *Pimenta* 15311 (LISC; LM). Division uncertain: fl. & fr., *Kirk* (K).
Probably native of tropical America, doubtfully so in Africa (not in our area) and Australia. Widely introduced in the tropics and often becoming wild. With us probably only planted or an escape from cultivation.

Grown for ornament and for its fragrant flowers which are used to make perfume. The pods of *A. farnesiana* are most distinctive and make the species easy to recognize. If they are absent, then it may be helpful to recall that no African acacia but this has the following combination of features: absence of " ant-galls ", leaflets with the lateral nerves raised and somewhat prominent beneath, apical involucels, and bright-yellow flowers in non-paniculate heads.
A further outstanding but less easily seen difference is that the anthers of *A. farnesiana* lack, even in bud, the small often caducous apical gland which is present in all the other capitate flowered acacias occurring in our area.

56. **Acacia elata** A. Cunn. ex Benth. in Hook., Lond. Journ. Bot. **1**: 383 (1842).—Brenan, F.T.E.A. Legum.-Mimos.: 50 (1949). Type from Australia.

Tree 6–18 m. high, unarmed; young branchlets puberulous or pubescent. Leaves 2-pinnate, large; pinnae 3–5 pairs; leaflets mostly 2·5–5 × 0·4–1 cm., lanceolate to linear-lanceolate, appressed-puberulous. Flowers pale-yellow, in heads arranged in axillary racemes or panicles.

Rhodesia. E: Melsetter Distr., Mawenji, fl. 15.viii.1956, *McGregor* 48/56 (K; SRGH).
A native of Australia cultivated for ornament and shade.

57. **Acacia schinoides** Benth. in Hook., Lond. Journ. Bot. **1**: 383 (1842).—Brenan, F.T.E.A. Legum.-Mimos.: 50 (1949). Type from Australia.
Acacia pruinosa sensu auct. pro parte.

Tree up to 15 m. high, unarmed; young branchlets glabrous. Leaves 2-pinnate; petiole 2–7·5 cm. long; pinnae mostly 5–14 cm. long, not crowded; leaflets mostly 1–2 cm. long, glabrous. Flowers in racemose or paniculate heads.

Rhodesia. E: Umtali, fl. 27.x.1953, *Eley* GHS 44237 (SRGH).
A native of Australia occasionally cultivated.

58. **Acacia mearnsii** De Wild., Pl. Bequaert. **3**: 62 (1925).—Brenan, F.T.E.A. Legum.-Mimos.: 95, fig. 15/21 (1959).—Tindale in Beadle, Evans & Carolin, Handb. Vasc. Pl. Sydney Distr. & Blue Mts.: 231 (1962). Type from Kenya.
Acacia decurrens var. *mollis* Lindl. in Bot. Reg.: t. 371 (1819).—Gilbert & Boutique, F.C.B. **3**: 168 (1952). Type a cultivated plant, apparently not preserved.
Acacia mollissima sensu Burtt Davy, F.P.F.T. **2**: 345 (1932).—Brenan, T.T.C.L.: 333 (1949).—F. White, F.F.N.R.: 82 (1962).

Tree 2–15 m. high, unarmed; crown conical or rounded; all parts (except flowers) ± densely pubescent or puberulous. Leaves: petiole 1·5–2·5 cm. long, often with a gland above; rhachis usually 4–12 cm. long, with numerous raised glands all along its upper side both at and between insertions of pinnae-pairs; pinnae (8)12–21 pairs; leaflets usually in 16–70 pairs, 1·5–4 × 0·5–0·75 mm., linear-oblong. Flowers pale-yellow, fragrant, in heads 5–8 mm. in diam. on peduncles 2–6 mm. long, panicled. Pods dehiscing along one margin only, usually c. 3–10 × 0·5–0·8 cm., with 3–12 joints, ± grey-puberulous, jointed, almost moniliform (in Australia forms with less moniliform, almost glabrous pods occur). Seeds black, 5 × 3·5 mm., smooth, elliptic, compressed; caruncle conspicuous; areole 3·5 × 2 mm.

Rhodesia. C: Marandellas, Forshaw's road, fl. 10.ix.1949, *Corby* 457 (SRGH). E: Inyanga Distr., near Pungwe R., fl. viii.1956, *Pardy* 25/56 (K; SRGH). **Mozambique.** MS: Manica, Penhalonga, near Guido's Farm, fl. & fr. viii.1911, *Dawe* 396 (K).

A native of Australia, this is the well-known Black Wattle, economically important on account of its tan bark, and also used for building and firewood. In East Africa it has become locally naturalized in Kenya and Tanzania, as it is in the Eastern Division of Rhodesia in our area.

59. **Acacia dealbata** Link, Enum. Hort. Berol. **2**: 445 (1822).—Burtt Davy, F.P.F.T. **2**: 346 (1932).—Brenan, T.T.C.L.: 332 (1949); F.T.E.A. Legum.-Mimos.: 50 (1959).—Tindale in Beadle, Evans & Carolin, Handb. Vasc. Pl. Sydney Distr. & Blue Mts.: 231 (1962). Type a plant cultivated at Berlin.

Shrub or small tree 2·5–10 m. high, rarely more; young branchlets usually densely short-pubescent, rarely subglabrous, and also ± grey-pruinose; pubescence grey, or yellowish at first then grey. Leaves: petiole (0·5)1–2 cm. long, eglandular; rhachis 2·5–9·5 cm. long, with a raised gland above at the junction of each pair of pinnae, but without other glands in between as in *A. mearnsii*; pinnae (5)12–26 pairs; leaflets in 17–50 pairs, 2·5–5·5 × 0·4–0·7 mm., linear-oblong. Flowers bright-yellow, in heads 4–7 mm. in diam. on peduncles up to 6 mm. long, panicled or racemose. Pods dehiscent along one margin, 3–8 × 0·7–1·3 cm., not or only slightly moniliform. Seeds brown to blackish-brown, 5–6 × 3–3·5 mm.; caruncle conspicuous; areole 3·5–4 × 0·75–1·5 mm.

Rhodesia. E: Rhodes Inyanga Hotel, fl. vii.1956, *Pardy* 25/56 (K; LISC; SRGH). **Mozambique.** MS: Manica, Macequece, fl. immat. & fr. 25.iii.1948, *Garcia* in *Mendonça* 726 (K; LISC).

A native of Australia, the Silver Wattle, occasionally planted in our area.

A. dealbata is close to and sometimes confused with *A. mearnsii*. It is easily separated by being more pruinose, lacking the intercalary glands along the leaf-rhachis and having wider usually less moniliform pods.

60. **Acacia baileyana** F. Muell. in Trans. & Proc. Roy. Soc. Victoria, **24**: 168 (1887).—Brenan, F.T.E.A. Legum.-Mimos.: 50 (1959). Type from Australia.

Shrub or tree 5–10 m. high, unarmed; young branchlets ± inconspicuously pubescent. Leaves 2-pinnate, glaucous; petiole very short, c. 2 mm.; pinnae in 2–5 pairs, crowded, c. 0·8–3 cm. long; leaflets 3–8 × 0·8–1·5 mm., glabrous or subciliate on the margins only. Flowers yellow, in heads arranged in axillary racemes longer than the leaves.

Rhodesia. E: Inyanga Distr., Nyangani, fl. buds 21.iii.1949, *Chase* 1664 (SRGH).

A native of Australia cultivated for ornament.

61. **Acacia cyanophylla** Lindl., Bot. Reg. **25**: Misc. 45 (1839).—F. White, F.F.N.R.: 82 (1962). Type from Australia.

Shrub or small tree up to 6 m. high, unarmed; young branchlets angular, glabrous. Leaves phyllodic, apparently simple, mostly c. 6–20 × 0·5–1·5 (lower to 3·5 or more) cm., linear-lanceolate to linear, with a single midrib and finely but distinctly penninerved, glabrous, straight or slightly falcate. Flowers yellow, in heads 6–12 mm. in diam. on peduncles 7–22 mm. long, the heads arranged in short axillary racemes.

Mozambique. LM: Lourenço Marques, fl. & fr. immat. xii.1945, *Pimenta* 15304 LISC).

A native of Australia cultivated for ornament.

Turner in *Eyles* 1470, from Rhodesia, E: Inyanga, st. i.1919 (SRGH), may be this species. According to F.F.N.R.: 82 (1962), *A. cyanophylla* is planted at Lundazi, Zambia.

62. **Acacia podalyriifolia** A. Cunn. ex G. Don, Gen. Syst. **2**: 405 (1832).—Brenan, T.T.C.L.: 332 (1949); F.T.E.A. Legum.-Mimos.: 51 (1959). Type from Australia.

Shrub or small tree 3–6 m. high, unarmed; young branchlets densely grey-pubescent. Leaves glaucous, phyllodic, apparently simple, mostly 1·5–5·5 × 1–2·5 cm., ovate to elliptic or elliptic-oblong, softly ± pubescent. Flowers yellow, in heads arranged in axillary racemes which are ± aggregated terminally.

Rhodesia. W: Bulawayo, fl. viii.1958, *Miller* 5378 (SRGH). C: Marandellas, fl. 20.ii.1962, *Corby* 1035 (LISC; SRGH). E: Umtali, fl. 1.vii.1949, *Chase* 1663 (LISC; SRGH). **Malawi.** S: Zomba, fl. buds 13.iii.1961, *Sichali* 21 (SRGH).
A native of Australia cultivated for ornament.

63. **Acacia cultriformis** A. Cunn. ex G. Don, Gen. Syst. **2**: 406 (1832). Type from Australia.

Shrub, unarmed; young branchlets glabrous, angular. Leaves glaucous, phyllodic, apparently simple, 0·8–3 × 0·5–1·1 cm., obliquely obovate-lanceolate to ovate-triangular with the apex of the triangle on the upper margin, glabrous. Flowers yellow, in heads arranged in axillary racemes which are larger than the leaves and often ± aggregated terminally.

Rhodesia. C: Salisbury, st. 19.iii.1951, *Pardy* in GHS 32492 (SRGH).
A native of Australia cultivated for ornament.

SPECIES NOT SUFFICIENTLY KNOWN

64. **Acacia purpurea** Bolle in Peters, Reise Mossamb. Bot. **1**: 6 (1861).—Oliv., F.T.A. **2**: 343 (1871).—Sim, For. Fl. Port. E. Afr.: 55 (1909).—Bak. f., Legum. Trop. Afr. **3**: 831 (1930). Types: Mozambique, Chupanga, Sena, Tete, etc., *Peters* (B, ?syntypes †).

Tree up to 7·5 m. high; young branchlets tomentose, older ones grey-puberulous. Prickles paired, somewhat bent downwards, dark-glossy-brown. Leaves with 8 pairs of pinnae with a gland on the rhachis between each pair; leaflets 12 pairs, c. 6 × 3 mm., oblong-ovate, obtuse at the apex, somewhat hairy. Inflorescence a dense spike. Calyx with 5 pointed teeth, yellow-hairy. Corolla small, enclosed by the calyx, of 5 lobes free almost to the base, glabrous, purple. Stamen-filaments twice as long as the calyx, purple.

Mozambique. T & MS: S. side of the Zambeze at Chupanga, Sena, Tete, etc., *Peters* (B†).
By water and on plains.
I cannot place this species with certainty in the absence of authentic material. The flower-colour suggests *A. galpinii*, but the small corolla in relation to the calyx would be wrong for this species. Except for the purple corolla and stamens, the description might suggest *A. burkei*. Bolle compared his *A. purpurea* with *A. caffra*.

13. ALBIZIA Durazz.

(By J. P. M. Brenan)

Albizia Durazz., Magazz. Tosc. **3** (4) (vol. 12): 10, 13, illustr. (1772).*

Trees, sometimes shrubs, very rarely climbing (not so in Africa); prickles or spines absent in the African spp. (except for a very small prickle beneath the node in *A. harveyi* and except that in *A. anthelmintica* some branchlets may be sharp and spinescent at the ends); sharp hooks apparently representing petiole-bases present in a very few extra-African spp. Leaves 2-pinnate; pinnae each with one to many pairs of leaflets. Inflorescences of round heads, or (not in native African spp.)

* I have not seen this very rare work. Reference from Little in Amer. Midl. Nat. **33**: 510 (1945).

spikes or spiciform racemes, pedunculate, axillary and solitary or much more often fascicled, often aggregated near the ends of branchlets which may be lateral and much shortened, sometimes paniculately arranged. Flowers ☿ or occasionally ♂ and ☿; 1–2 central flowers in each head frequently larger, different in form from the others and apparently ♂. Calyx gamosepalous, with normally 5 teeth or lobes (rarely 4, 6 or 7). Corolla gamopetalous, infundibuliform or campanulate, with normally 5 lobes (rarely 4 or 6, or in *A. coriaria* and *A. tanganyicensis* the lobes may be irregularly connate among themselves). Stamens numerous (19–50), fertile, their filaments united in their lower part into a slender tube sometimes projecting beyond, sometimes shorter than the corolla. Pods oblong, straight, flat, dehiscent or not, not septate inside, the valves papery to rigidly coriaceous but not thickened or fleshy. Seeds usually ± flattened.

A genus of c. 100–150 spp. throughout the tropics, a few in the subtropics.

The generic name is often misspelt *Albizzia*; for the reasons for rejecting this spelling see Little in Amer. Midl. Nat. **33**: 510 (1945).

Root-nodules recorded in spp. 2, 3, 5, 6, 7, 9, 10, 14, 17 and 18.

Insufficient attention has been hitherto paid to the dehiscence or indehiscence of the pods. This varies from species to species and provides a useful basis for grouping our native and naturalized species:

I. *Staminal tube not or scarcely projecting beyond the corolla*

(*a*) Species 1–7. Pod dehiscent, though in *A. lebbeck* this may take place only after it has fallen from the tree. Stamen-filaments 1·5–5 cm. long.

(*b*) Species 8–14. Pod apparently always indehiscent. Stamen-filaments 0·5–1·5 cm. long.

II. *Staminal tube projecting about 0·7–2·8 cm. beyond the corolla. Pod dehiscent*

(*c*) Species 15–17.

In the following key the descriptions of the floral parts must not be taken to apply to the 1–2 larger modified flowers commonly present in the centre of the heads; in these the staminal tube is not or scarcely exserted even when it is long-exserted in the others.

Inflorescence spicate; spikes paniculate; flowers sessile; staminal tube not exserted — 20. *falcataria*

Inflorescence capitate:

- Staminal tube not or scarcely projecting beyond the corolla; pods indehiscent or dehiscent:
  - Leaflets small or very small, 0·5–3(4) mm. wide, usually in numerous pairs (pairs (5)12–48; pinnae 2–56 pairs:
    - Midrib of leaflet running along or almost along the upper margin of the leaflet from base to apex; leaflets acute at the apex, ± pubescent or puberulous beneath; stipules large and conspicuous but quickly falling - - - - 18. *chinensis*
    - Midrib of leaflet not marginal, though often placed very asymmetrically; leaflets mostly rounded to subacute at the apex (acute in *A. harveyi*); stipules small and inconspicuous:
      - The leaflets mostly very small, 0·5–1·5 mm. wide and 2–6(7) mm. long:
        - Apex of leaflets acute, asymmetric, the point turned towards the pinna-apex; stamen-filaments c. 1·5–2 cm. long; bracteoles persistent until the flowers open; lateral nerves of leaflets ± raised and visible beneath; pods dehiscent, glabrous or nearly so except for a little pubescence near the base and along the margins - - - - - - - - - - 7. *harveyi*
        - Apex of leaflets obtuse or subacute, symmetric; stamen-filaments c. 0·5–1·2 cm. long; bracteoles already fallen when the flowers open; lateral nerves of leaflets not distinct beneath; pods indehiscent, glabrous to minutely puberulous over the surface:
          - Young branchlets clothed with rather short dense spreading grey to golden pubescence; calyx puberulous or pubescent outside; stipe of pod 0·5–0·8 cm. long:
            - Pinnae 4–12(16) pairs; leaflets 10–29 pairs — 9. *amara* subsp. *amara*
            - Pinnae (7)14–46(56) pairs; leaflets (12)21–48 pairs — 9. *amara* subsp. *sericocephala*

Young branchlets sparsely to rather densely appressed-grey-puberulous; pinnae (3)6–10(17) pairs; leaflets 15–30 pairs; calyx glabrous to ± puberulous; stipe of pod (0·5)1–2·5 cm. long - - 10. *brevifolia*

The leaflets mostly 1·25–4 mm. wide and (4)6·5–12 mm. long, rounded to subacute at the apex; bracteoles normally already fallen when the flowers open; pods normally puberulous over the surface; lateral nerves of leaflets usually ± raised and visible beneath:

Calyx 3–5 mm. long; pods closely transversely venose, the veins ± parallel and c. 2–4 mm. apart; seeds 4·5–6·5 mm. wide, nearly twice as long as broad; flowers sessile or almost so; leaflets asymmetric at the apex 8. *forbesii*

Calyx 1–2·5 mm. long; seeds 7 mm. wide or more:

Pedicels of flowers 1·5–6 mm. long:

Leaflets 1·25–4·5 mm. wide; pedicels 1·5–2 mm. long; indumentum on the outside of the calyx and corolla grey - - 11. *isenbergiana*

Leaflets mostly 4 mm. wide or more; pedicels 2–6 mm. long; indumentum on the outside of the calyx and corolla grey or, more commonly, brown 13. *schimperana*

Pedicels of flowers 0–1·5 mm. long; indumentum on the outside of the calyx and corolla ± rusty (when dry) - - - 12. *zimmermannii*

Leaflets medium to large, (3·5)4–45 mm. wide, in 1–20(23) pairs; pinnae 1-8(11) pairs:

Rhachides of leaves and pinnae of all or most leaves projecting at the ends as a short rigid persistent deflexed or downwards-bent hook or claw; often a single stipel similarly bent near the base of the pinnae; flowers usually precocious on almost or quite leafless shoots, with calyces and corollas glabrous or sparsely puberulous outside; pinnae 1–4 pairs; leaflets usually c. 3 pairs (range 1–5 pairs); if indumentum on flowers rusty, compare with *A. versicolor* and *A. tanganyicensis*, below - - - - - - - - 6. *anthelmintica*

Rhachides of leaves and pinnae not projecting at the ends, or else projections straight, not hooked or deflexed (except rarely and casually in *A. versicolor*) and usually caducous; calyces and corollas usually ± densely puberulous to tomentose outside, or if glabrous then flowers not precocious (except sometimes in *A. tanganyicensis*):

Stamen-filaments 1·5–5 cm. long; calyx normally 3–7 mm. long (in *A. lebbeck* sometimes only 2·5 mm.); corolla frequently more than 8 mm. long; pods dehiscent (sometimes only after falling in *A. lebbeck*), glabrous and often glossy on the surface, or with a few hairs along the margins and at the base only:

Filaments of stamens red above and white below; leaflets mostly in 6–11 pairs, subglabrous or thinly puberulous, oblong to elliptic- or ovate-oblong. 1. *coriaria*

Filaments of stamens white to green or greenish-yellow, not red:

Leaflets beneath grey to whitish and very glaucous, glabrous on both sides, ovate to rhombic-ovate or elliptic-oblong; flowers sessile or up to 2 mm. pedicellate - - - - 5. *antunesiana*

Leaflets not very glaucous beneath, or if so then ± pubescent to subtomentose:

Indumentum on outside of calyx and corolla conspicuously rusty (when dry), at least on the lobes; pods brown to crimson:

Leaflets in 3–5 (occasionally 6) pairs, densely pubescent to tomentose beneath, mostly broadly obovate to subcircular; young branchlets densely rusty-tomentose; pods rounded or abruptly narrowed at the base; bark usually rough, not peeling - - - 3. *versicolor*

Leaflets (at least of the distal pinnae) always in 7 or more pairs, usually glabrous to thinly pubescent beneath, ovate- to obovate-elliptic or ovate-oblong; young branchlets glabrous to pubescent, pods gradually narrowed at the base; bark smooth, peeling off in brown papery pieces - - - - - - - 2. *tanganyicensis*

Indumentum on outside of calyx grey to whitish, not rusty; pods pale-straw-coloured, marked with bumps; leaflets subglabrous beneath or rarely pubescent - - - - - - 4. *lebbeck*

Stamen-filaments 0·5–1·3 cm. long ; calyx 1–2·5(3) mm. long; corolla 3–7·5 mm. long; pods indehiscent, usually puberulous over their surface and not glossy:

Pedicels 1·5–7 mm. long:

Indumentum on outside of calyx and corolla wholly ashen-grey; leaflets of 2 distal pairs of pinnae 3–5(6) pairs (in our area) 14. *glaberrima* var. *glabrescens*

Indumentum on outside of calyx and corolla wholly or partially brown or brownish; leaflets of 2 distal pairs of pinnae 6–23 pairs
13. *schimperana*

Pedicels 0–1 mm. long:

Indumentum on outside of calyx and corolla brown or brownish; leaflets 7–15(17) mm. long, with slightly asymmetric midrib; peduncles 2–5 cm. long - - - - - - - - - 12. *zimmermannii*

Indumentum on outside of calyx and corolla wholly grey; leaflets 15–37(44) mm. long, with very asymmetric midrib; peduncles 1·4–2·5(3·5) cm. long - - - - - - - 19. *odoratissima*

Staminal tube projecting beyond the corolla for a length of c. 0·7–2·5 cm. (usually more than 1 cm.), usually red, pink or greenish, at least partly; pods dehiscent:

Calyx and corolla ± puberulous to pubescent outside, the former (1·5)2·5–5 mm. long:

Leaflets of the 2 distal pairs of pinnae 8–17 pairs:

The leaflets not auriculate at the base on the proximal side, though the proximal margin may be ± rounded to the insertion on the petiole; young branchlets and rhachides of leaves and pinnae densely fulvous-pubescent; leaflets ± pubescent all over the lower surface; stipules ovate, c. 5–12 × 3–8 mm.; pods ± densely and persistently pubescent - - - 16. *adianthifolia*

The leaflets markedly auriculate at the base on the proximal side; young branchlets and rhachides of leaves finely and shortly brownish-pubescent; leaflets pubescent beneath only on the midrib and margins, glabrous between or rarely, especially when young, some occasional hairs on the primary lateral nerves; stipules lanceolate, c. 6–7 × 2–2·5 mm.; pods glabrescent
15. *gummifera*

Leaflets of the 2 distal pairs of pinnae 2–6 pairs 17. *petersiana* subsp. *evansii*

Calyx (except for margins) and corolla glabrous outside, the former 1–2 mm. long
17. *petersiana* subsp. *petersiana*

1. **Albizia coriaria** Welw. ex Oliv., F.T.A. **2**: 360 (1871).—Bak. f., Legum. Trop. Afr. **3**: 861 (1930).—Brenan, T.T.C.L.: 342 (1949); F.T.E.A. Legum.-Mimos.: 143 (1959).—Gilbert & Boutique, F.C.B. **3**: 187 (1952).—Torre, C.F.A. **2**: 291 (1956).—F. White, F.F.N.R.: 432 (1962). Types from Angola (Cuanza Norte).

Tree 6–36 m. high; crown spreading, flat; bark rough, flaking off; young branchlets puberulous or shortly pubescent, later glabrescent. Leaves: rhachis ± thinly crisped-puberulous or shortly pubescent; pinnae (2)3–6(8) pairs; leaflets (4)6–11(12) pairs, 13–33 × 5–14(17) mm., oblong to elliptic- or ovate-oblong, rounded at the apex, subglabrous except for a few hairs on the midrib beneath, or sometimes ± thinly puberulous beneath especially towards the base. Flowers subsessile or on pedicels 0·5–2 mm. long; bracteoles already fallen by flowering time, minute, mostly 1·5–2 mm. long. Calyx 3·5–6·5 mm. long, not slit unilaterally, puberulous outside, with a few shortly stipitate glands (× 20 lens necessary) principally on the outside of the lobes. Corolla white or whitish, 8–13·5 mm. long, puberulous outside. Staminal tube not or scarcely exserted beyond the corolla; filaments 1·7–4 cm. long, red above, white below. Pod dehiscent, (10)14–21 × (2·3)3·2–3·7(4·8) cm., oblong, glabrous or nearly so, ± glossy, obscurely venose, brown or purplish-brown, usually ± tapered and acute at the base and sometimes at the apex. Seeds c. 9–12 × 8–9 mm., flattened.

Zambia. N: Mweru Wantipa, fl. 4.ix.1960, *Bands* in *Mutimushi* DB 624 (FHO).

Also from the Ivory Coast eastwards to the Sudan, Uganda and Tanzania; in the W. extending southwards to Angola. In tall-grass woodland.

2. **Albizia tanganyicensis** Bak. f. in Journ. of Bot. **67**: 199 (1929); Legum. Trop. Afr. **3**: 862 (1930).—Brenan, T.T.C.L.: 342 (1949); F.T.E.A. Legum.-Mimos.: 144 (1959).—Torre, C.F.A. **2**: 293 (1956).—Codd in Bothalia, **7**: 75 (1958).—F. White, F.F.N.R.: 90 (1962).—Mitchell in Puku, **1**: 108 (1963).—Boughey in Journ. S. Afr. Bot. **30**: 158 (1964). Type from Tanzania.

Albizia rhodesica Burtt Davy, F.P.F.T. **2**: xviii, 348 (1932).—Codd, Trees & Shrubs Kruger Nat. Park: 56 (1951).—O. B. Mill. in Journ. S. Afr. Bot. **18**: 27 (1952).—Wild, Guide Fl. Vict. Falls: 148 (1953).—Pardy in Rhod. Agric. Journ. **51**: 4, cum photogr. (1954).—Palgrave, Trees of Central Afr.: 269, cum tab. et photogr. (1956).—Burtt Davy & Hoyle, rev. Topham, N.C.L. ed. 2: 66 (1958). Syntypes: Rhodesia, Matopos, *Galpin* 7082 (PRE); Victoria Falls, *Allen* 174 (K); *Rogers* 5319 (K).

Albizia lebbeck var. *australis* Burtt Davy in Burtt Davy & Hoyle, N.C.L.: 53 (1936) *nom. nud.*—Burtt Davy & Hoyle, rev. Topham, N.C.L., ed. 2: 65 (1958) *nom. nud.*

Tree (3)9–20 m. high, deciduous and usually flowering when quite leafless; trunk smooth except at the base where burned, with old bark peeling off in brown papery pieces, the young bark creamy-white to ochre-yellow or yellow-green; crown flat or rounded; young branchlets glabrous to pubescent. Leaves: rhachis clothed like the young branchlets, not hooked or clawed at the ends; pinnae 3–7 pairs; leaflets (4)7–13(17) pairs, somewhat asymmetric, 11–55 × 6–29(32) mm., ovate- to obovate-elliptic or ovate-oblong, rounded to subacute at the apex, glabrous to ± crisped-pubescent on both surfaces. Flowers white, usually produced before the young leaves, sessile or up to 1 mm. pedicellate; bracteoles spathulate, c. 2 mm. long, already fallen when the flowers open. Calyx 4–6 mm. long, sometimes slit unilaterally, densely brown-tomentellous on the lobes; tube glabrous to ± pubescent, occasionally with sessile glands. Corolla pale-green, 7–11 mm. long, ± brown-tomentellous on the lobes; tube glabrous to ± pubescent, occasionally with sessile glands. Staminal tube not or scarcely exserted beyond the corolla; filaments 1·5–4 cm. long, white below, greenish-white near the apex. Pod dehiscent, 10–35 × 2·5–5·2 cm., oblong, glabrous, brown, ± glossy, not or only obscurely venose. Seeds c. 10–17 × 8–13 mm., flattened.

Botswana. N: Kazane, fr. v.1936, *Miller* B. 135 (K). **Zambia.** C: Kafue R. Gorge, fr. 6.x.1957, *Angus* 1740 (K). S: Mumbwa Distr., between Nambala Mission and Matala, fl. 15.ix.1947, *Brenan & Greenway* 7858 (K). **Rhodesia.** N: Darwin Distr., Hill of Doves near Winda Pools, fl. & fr., 6.ix.1958, *Phipps* 1317 (K; SRGH). W: Insiza Distr., Filabusi, fl. & fr., x.1954, *Davies* in GHS 48309 (K; SRGH). E: Umtali Distr., Zimunya's Reserve, fl. 7.x.1956, *Chase* 6209 (K; SRGH). S: Gwanda Distr., Ntolole, fl., *Howden* 36/58 (SRGH). **Malawi.** S: Ncheu Distr., Benni's Village, st. 26.xi.1957, *Jackson* 2120 (K; SRGH). **Mozambique.** N: Amaramba, Cuamba, fl. & fr. 15.x.1942, *Mendonça* 866 (LISC). T: Moatize, Mt. Zóbuè, 11.iii.1964, *Torre & Paiva* 11149 (LISC).

Also in Tanzania, Angola and the Transvaal. Characteristic of rocky hills and rock-outcrops and also, sometimes gregariously, on Kalahari Sand in Matabeleland (Pardy, loc. cit.); 450–1520 m.

A very distinct species with remarkable smooth papery-peeling bark and seeds usually larger than those of any other of our *Albizia* species. The flowers are usually produced when the tree is leafless, and specimens collected in this state, without notes about the bark, are liable to be confused with *A. antunesiana* Harms. The latter species may be distinguished (at least when dry) by the more prominent and raised nerves on the calyx-tube and by its smaller flowers.

There is much variation in the indumentum of this species. Glabrous to sparingly pubescent leaflets appear predominant in our area, but in *Jackson* 2120 from Malawi, cited above, they are rather densely pubescent and smaller than usual.

Six-month old seedlings collected at the forest-nursery at Kitwe, Zambia and said to belong to *A. tanganyicensis*, show the rootstock with a remarkable abrupt tuberous thickening c. 3–3·5 × 1·5–2 cm. It would be interesting to know if this is normal at this stage of development of the species.

3. **Albizia versicolor** Welw. ex Oliv., F.T.A. **2**: 359 (1871).—Eyles in Trans. Roy. Soc. S. Afr. **5**: 361 (1916).—Bak. f., Legum. Trop. Afr. **3**: 863 (1930).—Burtt Davy, F.P.F.T. **2**: 348 (1932).—Steedman, Trees etc. S. Rhod.: 16 (1933).—Gomes e Sousa, Pl. Menyharth.: 69 (1936); Dendrol. Moçamb. **1**: 94, cum tab. (1948); Dendrol. Moçamb. Estudo Geral, **1**: 237, t. 41 (1966).—Brenan, T.T.C.L.: 343 (1949); F.T.E.A. Legum-Mimos.: 146 (1959).—Codd, Trees & Shrubs Kruger Nat. Park: 57, figs. 52–53 (1951); in Bothalia, **7**: 73 (1958).—Gilbert & Boutique, F.C.B. **3**: 182, fig. 7 (1952).—O. B. Mill. in Journ. S. Afr. Bot. **18**: 27 (1952).—Wild, Guide Fl. Vict. Falls: 148 (1953).—Torre, C.F.A. **2**: 293 (1956).—Burtt Davy & Hoyle, rev. Topham, N.C.L., ed. 2: 66 (1958).—Mogg in Macnae & Kalk, Nat. Hist. Inhaca I., Moçamb.: 145 (1958).—Fanshawe, Fifty Common Trees N. Rhod.: 20 cum tab. (1962).—F. White, F.F.N.R.: 89, fig. 16 J (1962).—Mitchell in Puku, **1**: 108 (1963). —Boughey in Journ. S. Afr. Bot. **30**: 158 (1964).—Needham & Lawrence in Rhod. Agric. Journ. **63**, 6: 137–140 (1966). TAB. **23** fig. D. Type from Angola (Cuanza Norte).

Albizia versicolor var. *mossambicensis* Schinz in Bull. Herb. Boiss., Sér. 2, **2**: 946 (1902). Type: Mozambique, Boroma, Nhaondue, *Menyharth* 77 b (Z, holotype).

Albizia mossambicensis Sim, For. Fl. Port. E. Afr.: 59, tab. 60 (1909).—Eyles in Trans. Roy. Soc. S. Afr. **5**: 361 (1916). Type: Mozambique, *Sim* 6392 (NU, holotype).

Tree (3)5–18(20) m. high, deciduous; crown spreading, ± flat or rounded; bark usually rough, greyish-brown; young branchlets densely rusty-tomentose.

Leaves: rhachis clothed like the young branchlets; pinnae 1–4(5) pairs; leaflets 3–6 pairs, 14–63(70) × 12–49(55) mm., broadly and obliquely obovate to subcircular, sometimes broadly oblong, rounded and mucronate to emarginate at the apex, rarely subacute, becoming coriaceous, pubescent above, densely tomentose or pubescent beneath. Flowers white to greenish-yellow; pedicels 0–2(2·5) mm. long; bracteoles at flowering time present or already fallen. Calyx 4·5–7 mm. long, densely rusty-pubescent or -tomentose outside, not slit unilaterally. Corolla clothed like the calyx, 8–12 mm. long. Staminal tube not or scarcely exserted beyond the corolla; filaments 2·5–4·5 cm. long (to 5·5 cm. *fide* F.C.B.). Pod dehiscent, 10–27(30 *fide* F.C.B.) × 3·2–6·5 cm., oblong, glabrous (or with some hairs on the stipe and margins only), ± glossy, obscurely venose, chestnut-brown or crimson. Seeds c. 9–13 × 8–11 mm., flattened.

Botswana. N: 4 km. NE. of Serondela, Chobe, fl. x.1944, *Miller* B/335 (PRE). **Zambia.** B: Balovale, fl. & fr. 11.x.1952, *Angus* 622 (BM; K; PRE). N: Kasama Distr., Mbesuma Ranch, fl. 26.x.1961, *Astle* 1000 (K; SRGH). W: Solwezi Distr., W. of Mumbezhi R., fr. 27.vii.1930, *Milne-Redhead* 788 (K; PRE). C: Chilanga Distr., Old Argosy Mine near Quien Sabe, fl. 12.x.1929, *Sandwith* 56 (K; SRGH). E: Luangwa Valley, 16 km. W. of Jumbe, fl. & fr. 13.x.1958, *Robson & Angus* 90 (BM; K; SRGH). S: Mapanza W., fr. 31.v.1953, fl. 17.x.1954, *Robinson* 929 (K). **Rhodesia.** N: Urungwe, fl. 15.x.1952, *Phelps* 27 (SRGH). W: Victoria Falls, fr. xii.1913, *Rogers* 13048 (K). E: Chipinga Distr., Sabi Valley Experimental Station, fl. xi.1959, *Soane* 157 (K; SRGH). S: Nuanetsi Distr., Malipata, fr. xi.1955, *Davies* 1644 (K; SRGH). **Malawi.** N: Rumpi Boma, fl. xi.1953, *Chapman* 179 (K). C: Chia Lagoon road below Nchisi, fr. 16.vi.1961, *Chapman* 1375 (K; SRGH). S: Zomba, fl. xi.1915, *Purves* 254 (K). **Mozambique.** N: Imala, between Mocuburi and Namina, fl. 26.x.1948, *Barbosa* 2580 (LISC; LM). Z: between Gúruè and Namarroi, fr. 16.ix.1949, *Barbosa & Carvalho* 4109 (K; LISC; LM). T: between Tete and Casula, fr. 27.viii.1941, *Torre* 3357 (BM; K; LISC). MS: Manica, between Mavita and R. Moçambize, fl. 25.x.1944, *Mendonça* 2619 (BM; LISC). SS: Guijá, Massingir, Rio dos Elefantes, fl. & fr. 2.xii.1944, *Mendonça* 3229 (LISC). LM: Lourenço Marques, fl. & fr. i.1946, *Pimenta* 15309 (LISC; LM; SRGH).

Also in the Congo, Uganda, Kenya, Tanzania, Angola, the Transvaal and Natal. In mixed woodland of various types; near sea-level to 1520 m.

A very distinct and easily recognized species, marked by the combination of tomentum or coarse pubescence, usually ± rust-coloured, over the vegetative parts, and the comparatively few and broad leaflets. The ± glossy glabrous or subglabrous pods are also characteristic. There is not much variation: the leaflets vary somewhat in size, and the indumentum beneath may be comparatively short and coarsely pubescent, or longer and tomentose. The variation does not seem to be at all correlated with geography.

The bark is usually rough, but said occasionally to be smooth, for example in *Davies* 1644 and 2156 from Rhodesia, Nuanetsi Distr. In our area the pods are usually c. 10–20 cm. long.

The unripe pods and seeds of *A. versicolor* have been proved to be toxic to stock (Needham & Lawrence in Rhod. Agric. Journ. **63**, 6: 137–140 (1966)).

There is a resemblance between *A. versicolor* and *Samanea saman* (Jacq.) Merr., which is sometimes planted in our area. The latter may be separated by its indumentum being yellowish (when dry), not rusty; by the gland on the petiole being smaller and at or near the insertion of the lowest pair of pinnae, not well below them; by the leaflets being more glabrescent above, and by the different pods. In habit the two are very different, *A. versicolor* being deciduous with a lightly branched crown, while *Samanea saman* is evergreen with a dense heavy crown.

4. **Albizia lebbeck** (L.) Benth. in Hook., Lond. Journ. Bot. **3**: 87 (1844).—Sim, For. Fl. Port. E. Afr.: 60 (1909).—Bak. f., Legum. Trop. Afr. **3**: 862 (1930).—Brenan, T.T.C.L.: 342 (1949); F.T.E.A. Legum.-Mimos.: 147 (1959).—Gilbert & Boutique, F.C.B. **3**: 187 (1952).—Torre, C.F.A. **2**: 292 (1956).—Burtt Davy & Hoyle, rev. Topham, N.C.L., ed. 2: 65 (1958).—Codd in Bothalia, **7**: 81 (1958).—F. White, F.F.N.R.: 90 (1962). Type from Egypt.
Mimosa lebbeck L., Sp. Pl. **1**: 516 (1753). Type as above.

Tree 2·5–15 m. high; bark grey, rough; young branchlets puberulous, sometimes pubescent. Leaves: rhachis subglabrous, puberulous or sometimes pubescent; pinnae (1)2–4(5) pairs; leaflets 3–11 pairs, 15–48 × (6)8–24(33) mm., oblong or elliptic-oblong (terminal leaflets ± obovate), somewhat asymmetric with midrib nearer the upper margin, rounded at the apex, glabrous or rarely thinly pubescent above, beneath subglabrous or puberulous, rarely pubescent. Flowers pedicellate; pedicels 1·5–4·5(7·5) mm. long, puberulous; bracteoles minute, c. 2–3 mm. long, falling in early bud. Calyx (2·5)3·5–5 mm. long, not slit unilaterally, ± puberulous outside. Corolla 5·5–9 mm. long, glabrous outside except for puberulence on the outside of the lobes. Staminal tube not or scarcely exserted beyond the corolla; filaments 1·5–3 cm. long, pale-green or greenish-yellow in the upper part, white below. Pod dehiscent, though perhaps not until after falling from the tree, (9)15–33 × (2·4)2·9–5·5(6) cm., oblong, glabrous or almost so except near the base, coriaceous, glossy, ± venose, pale straw-coloured, not (or rarely to 5 mm.) stipitate at the base. Seeds 7–11·5 × 7–9 mm., flattened, marked by bumps on the outside of the valves, the alternate ones more projecting on each valve.

Zambia. S: Livingstone, fl. & fr. 22.ix.1955, *Gilges* 436 (PRE; SRGH). **Rhodesia.** C: Salisbury, fr. x.1924, *Eyles* 4039 (K; SRGH). E: Umtali Park, fr. 13.v.1947, *Chase* 322 (BM; K; SRGH). **Malawi.** S: Fort Johnston, fr. 27.v.1961, *Leach & Rutherford-Smith* 11029 (K; SRGH). **Mozambique.** N: Mossuril, Lumbo, on road to Nampula, fr. 5.v.1948, *Pedro & Pedrógão* 3141 (PRE). Z: Mocuba, Namagoa, fl. & fr. x.1946, *Faulkner* PRE 32A (COI; K). T: Tete, by R. Zambeze, fl. & fr. 20.x.1965, *Neves Rosa* 105 (LM). LM: Lourenço Marques, fl. 4.ii.1947, *Pedro & Pedrógão* 580 (PRE).

Pantropical, but probably nowhere native in Africa and originating from tropical Asia. In some of the above localities it is doubtless only a planted tree; in others it appears to be naturalized; but in some its status is uncertain. It is probably always more or less closely associated with human habitations; 440–1460 m.

The very characteristic stiff pale yellow-brown pods are marked with bumps usually projecting alternately on one side and the other of the pod. The pods, with their included seeds, are said when agitated by the breeze to make an incessant rattle that has been compared with women's chatter and the sound of fish being fried.

The epithet of *A. lebbeck* has often been misspelt " *lebbek* ".

5. **Albizia antunesiana** Harms in Engl., Bot. Jahrb. **30**: 317 (1901).—R.E.Fr. in Wiss. Ergebn. Schwed. Rhod.-Kongo Exped. **1**: 63 (1914).—Eyles in Trans. Roy. Soc. S. Afr. **5**: 361 (1916).—Bak. f., Legum. Trop. Afr. **3**: 861 (1930).—Steedman, Trees etc. S. Rhod.: 15 (1933).—Brenan, T.T.C.L.: 342 (1949); F.T.E.A. Legum.-Mimos.: 148 (1959).—Pardy in Rhod. Agric. Journ. **48**: 398, cum photogr. (1951).—Gilbert & Boutique, F.C.B. **3**: 189, fig. 10 C–D (1952).—O. B. Mill., in Journ. S. Afr. Bot. **18**: 27 (1952).—Wild, Guide Fl. Vict. Falls: 148 (1953).—Palgrave, Trees of Central Afr.: 261, cum tab. et photogr. (1956).—Torre, C.F.A. **2**: 291 (1956).—Burtt Davy & Hoyle, rev. Topham, N.C.L., ed. 2: 65 (1958).—Codd in Bothalia, **7**: 74 (1958).—Fanshawe, Fifty Common Trees N. Rhod.: 18 cum tab. (1962).—F. White, F.F.N.R.: 90, fig. 16 M (1962).—Mitchell in Puku, **1**: 108 (1963).—Boughey in Journ. S. Afr. Bot. **30**: 158 (1964). Syntypes from Tanzania and Angola.

Tree (1·5)6–18 m. high; bark rough, or sometimes smooth, reticulate; branches spreading; young branchlets glabrous or nearly so, or very shortly pubescent. Leaves: rhachis glabrous or subglabrous; pinnae 1–3(4) pairs; leaflets 4–8(9) pairs (15)23–50(68) × (7)11–25(41) mm., oblique, ovate to rhombic-ovate or elliptic-oblong, usually rounded or slightly emarginate at the apex, glabrous, papery to subcoriaceous, venose, very glaucous and often pale-grey beneath. Flowers greenish-yellow, ochre when over, with whitish filaments, sessile or up to 2 mm. pedicellate; bracteoles already fallen by flowering time, minute, up to 1·7 mm. long. Calyx (3)3·5–5·5 mm. long, rather densely puberulous or minutely pubescent outside, not slit unilaterally. Corolla (5)5·5–11 mm. long, densely minutely appressed-pubescent outside. Staminal tube not or scarcely exserted beyond the corolla; filaments c. 1·5–3 cm. long. Pod dehiscent, 12–23 × 2·7–4·6 cm., oblong, glabrous except for some hairs near the base and margins, slightly venose, ± transversely plicate, thin, usually pale-brown. Seeds c. 7–9 mm. in diam., flattened.

Botswana. N: Chobe, *Miller* B/188 (*fide* Miller, loc. cit., specimen not seen). **Zambia.** B: Sesheke Distr., Lusu, fl. 14.ix.1947, *Rea* 107 (K). N: Abercorn, fl. & fr. 22.ix.1949, *Bullock* 1030 (K; SRGH). W: Solwezi Distr., Mutanda Bridge, fr. 21.vi.1931, *Milne-Redhead* 558 (K; PRE). C: Mt. Makulu Research Station, fr. 10.vi.1957, *Angus* 1618 (K). E: Fort Jameson, fl. & fr., *Bococks* in FD 6661 (PRE). S: Mapanza W., fl. 12.ix.1954, *Robinson* 885 (K). **Rhodesia.** N: Mazoe, fl. x.1924, *Eyles* 4036 (K; SRGH). W: Matopos, fl. ix.1909, *Rogers* 5343 (K; SRGH). C: Salisbury, fl. x.1917, *Eyles* 850 (BM; K; SRGH). E: Umtali, fl. 26.ix.1948, *Chase* 1504 (BM; COI; K; LISC; SRGH). S: Buhera, st. iv.1954, *Masterson* 92 (SRGH). **Malawi.** N: SE. slope of Mafinga Mts., fl. 10.xi.1958, *Robson & Fanshawe* 539 (BM; K; LISC; SRGH).

C: Lilongwe, fl. 28.ix.1951, *Jackson* 597 (K). S: Shire Highlands, fl. & fr., *Buchanan* (K). **Mozambique.** T: Marávia, between Chicoa and Fíngoè, fr. 26.vi.1949, *Barbosa & Carvalho* 3306 (K; LM; PRE; SRGH).

Also in the Congo, Tanzania, Angola and SW. Africa. In woodland, particularly *Brachystegia-Julbernardia*, but also in other types of mixed woodland with *Combretum* and *Terminalia*; (270)900–1680 m.

When in foliage, easily known by its glabrous, very discolorous leaflets, but the flowers are often produced when the tree is leafless.

6. **Albizia anthelmintica** Brongn. in Bull. Soc. Bot. Fr. **7**: 902 (1860).—Sim, For. Fl Port. E. Afr.: 60 (1909).—Eyles in Trans. Roy. Soc. S. Afr. **5**: 361 (1916) pro parte. —Bak. f., Legum. Trop. Afr. **3**: 859 (1930).—Gomes e Sousa, Pl. Menyharth.: 68 (1936).—Brenan, T.T.C.L.: 341 (1949); F.T.E.A. Legum.-Mimos.: 148 (1959). —Codd, Trees & Shrubs Kruger Nat. Park: 53 (1951); in Bothalia, **7**: 77 (1958).—Pardy in Rhod. Agric. Journ. **53**: 952, cum phot. (1956).—Torre, C.F.A. **2**: 289 (1956).—Burtt Davy & Hoyle, rev. Topham, N.C.L., ed. 2: 65 (1958).—F. White, F.F.N.R.: 90, fig. 16 L (1962).—Mitchell in Puku, **1**: 108 (1963).—Boughey in Journ. S. Afr. Bot. **30**: 158 (1964). TAB. **23** fig. C. Type from Ethiopia.

Besenna anthelmintica A. Rich., Tent. Fl. Abyss. **1**: 253 (1847) *nom. provis.*—Walp., Ann. **2**: 461 (1851–2).

Albizia umbalusiana Sim, For. Fl. Port. E. Afr.: 59, t. 55 (1909). Type: Mozambique, Lourenço Marques and Maputo Districts, up to the Libombos, *Sim* 6200. Type-specimen not traced (see Codd in Bothalia, **7**: 78 (1958)).

Albizia anthelmintica var. *australis* Bak. f., Legum. Trop. Afr. **3**: 859 (1930).—Torre, C.F.A. **2**: 290 (1956). Syntypes from Angola and SW. Africa.

Albizia anthelmintica var. *pubescens* Burtt Davy, F.P.F.T. **2**: xvii, 348 (1932).—O. B. Mill. in Journ. S. Afr. Bot. **18**: 27 (1952). Syntypes from the Transvaal.

Bush or tree 2–9(12) m. high, deciduous; bark smooth, grey to brown; young branchlets glabrous or sometimes shortly pubescent; twigs often with short divaricate almost spinescent-tipped lateral branches. Leaves: rhachides of leaves and pinnae glabrous to shortly pubescent, in all or most leaves projecting at the ends in a short rigid persistent deflexed or downwards-bent hook or claw; often a single stipel similarly bent near the base of the pinnae; pinnae 1–2(4) pairs; leaflets 1–4(5) pairs (7)10–36(42) × (4)6–31 mm., obliquely obovate or elliptic to subcircular, mucronate at the apex, venose, glabrous to ± sparsely shortly pubescent. Flowers usually on leafless twigs, on pedicels 0·5–5·5 mm. long. Calyx pale-greenish, (very rarely 2)3–5 mm. long, glabrous to sparsely finely pubescent outside, irregularly denticulate at the apex and usually slit unilaterally to c. 1–2·5 mm. Corolla pale-greenish, 6–12 mm. long, glabrous, or puberulous on or near the lobe-margins. Staminal tube not or scarcely exserted beyond the corolla; filaments c. 1·5–2·5 cm. long, white. Pod dehiscent, (6)7–18 × 1·5–2·9 cm., oblong, quite glabrous or occasionally puberulous all over, straw-coloured when mature. Seeds 9–13 mm. in diam., flattened, round.

Botswana. N: Kwebe Hills, fl. viii, fr. ix.1897, *Lugard* 15 (K). SE: Kgatla Distr., Ramonaka, 13 km. N. of Sikwane, fl. 13.ix.1955, *Reynolds* 407 (K; PRE). **Zambia.** B: Nangweshi, fl. 21.vii.1952, *Codd* 7139 (BM; K; PRE). N: Luangwa Valley, Mwunyamazi R., fr. 30.ix.1933, *Michelmore* 612 (K). C: Chingombe, fr. 26.ix.1957, *Fanshawe* 3740 (K). E: Foothills E. of Machinje Hills, fr. 12.x.1958, *Robson* 67 (BM; K; LISC; SRGH). S: Zambezi Sawmills Railway, near Ngwesi R. crossing, fl. 7.viii.1947, *Brenan & Keay* 7637 (K). **Rhodesia.** N: Kavira Hotsprings, Mlibizi R., Binga, Sebungwe, fl. 31.ix.1958, *Lovemore* 550 (K; LISC; SRGH). W: Wankie, fl. iv.1932, *Levy* 16 (PRE; SRGH). E: Tanganda R., fl. 15.viii.1961, *Lord Methuen* 65 (K). S: West Nicholson Distr., Liebig's Ranch, fr. 26.x.1952, *Plowes* 1520 (K; SRGH). **Malawi.** C: Dedza, Mua, fl. 31.vii.1960, *Chapman* 858 (SRGH). S: 26 km. S. of Fort Johnston, fl. 5.viii.1960, *Leach* 10407 (SRGH). **Mozambique.** N: Malema, Mutuali, fl. & fr. 2.ix.1953, *Gomes e Sousa* 4103 (COI; K; LMJ; PRE). Z: Morrumbala, between Aguas Quentes and Megaza, fl. 12.viii.1942, *Torre* 4557 (LISC). T: Mutarara, Sinjal, fl. 16.viii.1947, *Simão* 1482 (LISC; LM; SRGH). MS: Gorongosa, between Zangorga and Kanga-N'Thole, fr. 23.x.1956, *Gomes e Sousa* 4320 (COI; K; LISC; SRGH). SS: Govuro, between Mabote and Zimane, fl. 2.x.1944, *Mendonça* 1953 (LISC). LM: Namaacha, Goba, fl. 23.viii.1944, *Mendonça* 1846 (LISC).

From the Sudan and Ethiopia (Eritrea) southwards to the Transvaal and Zululand. In the drier types of woodland, also in wooded grassland, thickets and bushland; c. 30–1070 m.

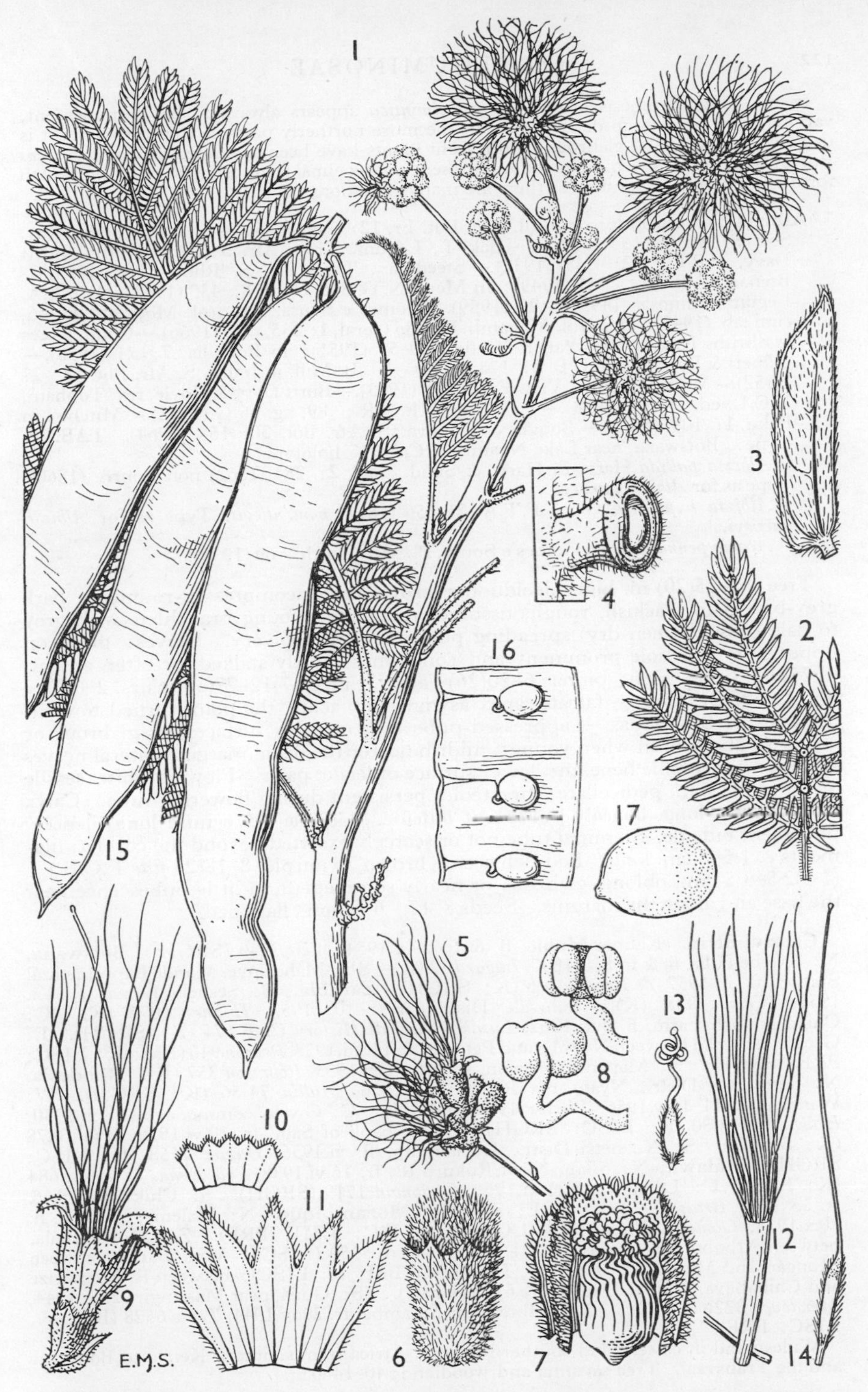

Tab. 22. ALBIZIA HARVEYI. 1, flowering branch (×1); 2, part of leaf showing glands on rhachis (×2); 3, leaflet (×8), all from *Burtt* 3809; 4, gland on rhachis (×24); 5, young flower head (×2); 6, flower bud (×8); 7, flower bud opened, showing arrangement of stamens (×8); 8, anthers from bud, front & back views (×40); 9, open flower (×4); 10, calyx opened out (×4); 11, corolla opened out (×4); 12, stamen filaments and tube (×4); 13, ovary from bud (×8); 14, ovary from mature flower (×4), all from *Burtt* 5037; 15, pods (×⅔) *Burtt* 1661; 16, part of valve of pod, seen from inside (×⅔); 17, seed (×2), all from *Legat* 65. From F.T.E.A.

In the Flora Zambesiaca area *A. anthelmintica* appears always to be ± pubescent, varying from sparsely to densely so. In the more northerly parts of tropical Africa it is usually (and typically) glabrous. Pubescent plants have been separated as var. *pubescens* Burtt Davy, but this seems at most to represent a weak clinal trend from N. to S., not worth formal taxonomic recognition as there are many gradations.

7. **Albizia harveyi** Fourn. in Bull. Soc. Bot. Fr. **12**: 399 (?1866).—Eyles in Trans. Roy. Soc. S. Afr. **5**: 361 (1916).—Bak. f., Legum. Trop. Afr. **3**: 865 (1930).—Burtt Davy, F.P.F.T. **2**: 348 (1932).—Steedman, Trees etc. S. Rhod.: 16 (1933).—Brenan, T.T.C.L.: 341 (1949); in Mem. N.Y. Bot. Gard. **8**: 430 (1954); F.T.E.A. Legum.-Mimos.: 149, fig. 20 (1959).—Gomes e Sousa, Dendrol. Moçamb. **4**: 46, cum tab. (1949); Dendrol. Moçamb. Estudo Geral, **1**: 235, t. 39 (1966).—Codd, Trees & Shrubs Kruger Nat. Park: 56, figs. 49–51 (1951); in Bothalia, **7**: 71 (1958).—Gilbert & Boutique, F.C.B. **3**: 173 (1952).—O. B. Mill. in Journ. S. Afr. Bot. **18**: 27 (1952).—Wild, Guide Fl. Vict. Falls: 148 (1953).—Burtt Davy & Hoyle, rev. Topham, N.C.L. ed. 2: 65 (1958).—F. White, F.F.N.R.: 89, fig. 16 G (1962).—Mitchell in Puku, **1**: 108 (1963).—Boughey in Journ. S. Afr. Bot. **30**: 158 (1964). **TAB. 22.** Type: Botswana, near Lake Ngami, *McCabe* (K, holotype).

Albizia pallida Harv. in Harv. & Sond., F.C. **2**: 284 (1862) non Fourn. (1860). Type as for *Albizia harveyi.*

Albizia hypoleuca Oliv., F.T.A. **2**: 356 (1871) *nom. illegit.* Type as for *Albizia harveyi.*

Acacia pennata sensu Gomes e Sousa, Pl. Menyharth.: 69 (1936).

Tree 1·5–15(20) m. high, deciduous; crown flat or compressed-rounded; bark grey-brown to blackish, rough, fissured, reticulate; young branchlets with grey to pale-brown (when dry) spreading pubescence, not silvery. Leaves: gland on upper side of petiole prominent and sometimes shortly stalked, or often absent, (0·25)0·5–1 mm. high; pinnae 6–20(26) pairs; leaflets (7)12–27(30) pairs, 2–6(7) × (0·6)1–1·25(2) mm., ± falcate, apex asymmetric, acute, the point turned towards the apex of the pinna, ± appressed-pubescent on both surfaces, or glabrous or nearly so above even when young; midrib nearer the distal margin; lateral nerves ± raised and visible beneath; lower surface of leaflet paler. Flowers white, sessile or up to 0·5 mm. pedicellate; bracteoles persistent during flowering time. Calyx 1·5–2·5 mm. long, densely pubescent outside. Corolla 3·5–6 mm. long, densely pubescent outside. Staminal tube not or scarcely exserted beyond the corolla; filaments c. 1·5–2 cm. long. Pod dehiscent, brown to purple, 8–18(25 *fide* F.C.B.) × (1·5)2·5–3·5 cm., oblong, glabrous or nearly so except for a little pubescence near the base and along the margins. Seeds 8–12 × 6–9 mm., flattened.

Caprivi Strip. Katima Mulilo, fl. & fr. 23.x.1954, *West* 3249 (SRGH). **Botswana.** N: Kwebe Hills, fl. & fr. 8.xi.1897, *Lugard* 32 (K). SE: Mahalapye, Morale Experimental Station, fr. iii.1957, *de Beer* Bob 6 (K; SRGH). **Zambia.** B: Sesheke Boma, fr. 31.i. 1952, *White* 1988 (K).; Balovale Distr., fr. 8–10.v.1953, *Holmes* 1076 (K). C: Chilanga Fish Farm, fl. 4.xi.1962, *Lusaka Natural History Club* 174 (K; SRGH). E: Great East Road between Nyimba and Petauke, fl. 14.xii.1958, *Robson* 946 (BM; K; LISC; PRE; SRGH). S: Mapanza Mission, fl. & fr. 1.xi.1953, *Robinson* 357 (K). **Rhodesia.** N: Urungwe Distr., Nyanyanja R., fl. 27.x.1956, *Mullin* 74/56 (K; SRGH). W: Wankie, fr. 11–16.v.1956, *Plowes* 1961 (SRGH). C: Gwelo commonage, fl. xi.1950, *Hodgson* H 9/50 (K; LISC; SRGH). E: E. bank of Sabi, fr. 10.vi.1950, *Wild* 3478 (K; SRGH). S: Nuanetsi Distr., Devata dip, fl. xi.1956, *Davies* 2158 (K; LISC; SRGH). **Malawi.** N: 5 km. N. of Rukuru R., fr. 16.vi.1938, *Pole Evans & Erens* 684 (K; PRE). C: Lilongwe, fl. 1.xi.1962, *Chapman* 1719 (SRGH). S: Chikwawa, fl. & fr. 3.x.1946, *Brass* 17912 (K; PRE; SRGH). **Mozambique.** N: Malema, Mutuáli, fl. 30.x.1954, *Gomes e Sousa* 4272 (COI; K; LISC; LMJ; PRE). Z: Morrumbala, between Mbobo and Morire, fr. 21.v.1943, *Torre* 5350 (LISC). MS: Báruè, between Mungári and Mandiè, fl. 26.x.1943, *Torre* 6083 (BM; K; LISC); between R. Mossurize and Chibabava, fl. 10.xi.1943, *Torre* 6125 (LISC). SS: Guijá, near Massingir, 2.xii.1944, *Mendonça* 3226 (LISC). LM: Sábiè, near Moamba, fr. 29.iv.1944, *Torre* 6528 (BM; K; LISC; PRE).

Widespread in eastern and southern tropical Africa from southern Kenya to Botswana and the Transvaal. Tree savanna and woodland; 40–1460 m.

8. **Albizia forbesii** Benth. in Hook., Lond. Journ. Bot. **3**: 92 (1844).—Sim, For. Fl. Port. E. Afr.: 58, t. 39 A (1909).—Burtt Davy, F.P.F.T. **2**: 348 (1932).—Codd, Trees and Shrubs Kruger Nat. Park: 54, fig. 48 a (1951); in Bothalia, **7**: 72 (1958).—Mogg in Macnae & Kalk, Nat. Hist. Inhaca I., Moçamb.: 145 (1958).—Brenan, F.T.E.A. Legum.-Mimos.: 151 (1959).—Boughey in Journ. S. Afr. Bot. **30**: 158

(1964).—Gomes e Sousa, Dendrol. Moçamb. Estudo Geral, **1**: 236, t. 40 (1966). Type: Mozambique, Delagoa Bay, *Forbes* (K, holotype).

Tree 2–21 m. high; bark grey to blackish, thick, rough (said to be smooth in *Davies* 2220); young branchlets densely grey-pubescent. Leaves: rhachis ± densely grey-pubescent; pinnae (2)3–7(8) pairs; leaflets 5–16(20) pairs 4–9 × 1·5–4 mm., obliquely oblong to oblong-elliptic (the terminal pair obovate), with the midrib nearer the distal margin, rounded and mucronate to subacute at the apex, which is turned towards the apex of the pinna, glabrous or sometimes pubescent above, beneath glabrous except for pubescence on the midrib and recurved margins, sometimes pubescent all over. Flowers creamy-white, sessile or almost so; bracteoles 1·5–2 mm. long, linear or oblanceolate, falling before the flowers open. Calyx 3–5 mm. long, densely and shortly pubescent outside, not or slightly slit unilaterally. Corolla 5–8 mm. long, densely and shortly ± appressed-pubescent outside. Staminal tube not or scarcely exserted beyond the corolla; filaments 1–1·5 cm. long. Pod apparently indehiscent, 9–20 cm. (including 1–2 cm. long stipe) × 3·2–5 cm., oblong, ± puberulous over the surface (use × 10 lens), sometimes nearly glabrous except on the margins and stipe, dark-brown, closely and prominently transversely venose, the veins ± parallel and c. 2–4 mm. apart. Seeds 11–12 × 4·5–6·5 mm., slightly flattened, oblong-ellipsoid or ellipsoid.

Rhodesia. E: Chipinga Distr., Sabi-Dott's Drift to Rupisi, fr. 7.iv.1959, *Savory* 342 (COI; K; SRGH). S: Nuanetsi R. below gorge, fr. xi.1956, *Davies* 2220 (COI; K; SRGH). **Mozambique.** N: Porto Amélia, Nangororo, fl. 1.xi.1959, *Gomes e Sousa* 4498 (COI; K; LISC; PRE). Z: Pebane, fl. & fr. 24.x.1942, *Torre* 4671 (BM; LISC). MS: Dondo, Vila Machado, fl. & fr. 31.x.1963, *Gomes e Sousa* 4812 (COI; K; LMU; PRE). SS: Massinga, between Rio das Pedras and Mapinhane, fl. & fr. 17.xi.1941, *Torre* 3831 (LISC). LM: Lourenço Marques, Ponta Vermelha, fl. & fr. 18.xi.1959, *Lemos & Balsinhas* 1 a (BM; COI; K; LISC; SRGH).
Also in Tanzania, the Transvaal and N. Natal. In woodland of various sorts, bushland and thicket, often coastal.

As far as our area is concerned this species is almost confined to Mozambique, where it appears to be locally common. It just reaches the eastern part of Rhodesia along the Sabi and Nuanetsi river-valleys. The pods are most distinctive, and the seeds usually narrow in proportion to their length.

9. **Albizia amara** (Roxb.) Boiv. in Encycl. XIX-me Siècle, **2**: 34 (1834?)*.—Brenan, F.T.E.A. Legum.-Mimos.: 151 (1959). Type from India.
Mimosa amara Roxb., Pl. Corom. **2**: 13, t. 122 (1799). Type as above.

Tree, rarely shrubby, 1·5–18 m. high, deciduous; crown rounded or flat; bark fissured, rough; young branchlets with rather short dense spreading grey to golden pubescence. Leaves: gland on upper side of petiole low, sessile, up to c. 0·25 mm. high; pinnae 4–46(56) pairs; leaflets 12–48 pairs, 2–5(7·5) × 0·5–1·5(2·5) mm., oblong-linear to linear, symmetric and obtuse or sometimes subacute at the apex, ± appressed-pubescent on one or both surfaces or on the margins only, glabrescent or not later; midrib nearly central (except at the base); lateral nerves not distinct beneath, rarely slightly raised. Flowers white or flushed-pink, subsessile or up to 1·5 mm. pedicellate; bracteoles very caducous, fallen by flowering time. Calyx 1–2 mm. long, puberulous or pubescent outside. Corolla 3·5–7 mm. long, puberulous or pubescent outside. Staminal tube not or scarcely exserted beyond the corolla; filaments c. 0·5–1·2 cm. long. Pod apparently indehiscent, 10–28 × 2–5 cm., linear-oblong, puberulous over the surface, brown. Seeds 8–13 × 7–8 mm., flattened.

Albizia amara sensu Steedman, Trees etc. S. Rhod.: 15, t. 10 (1933) is certainly not *A. amara*. I am uncertain of its identity. It might be *A. tanganyicensis*, but that has the bark smooth, pale and peeling, not grey and rough as described by Steedman.

Subsp. **amara**.—Brenan in Kew Bull. **10**: 189 (1955); F.T.E.A. Legum.-Mimos.: 151 (1959).
Albizia gracilifolia Harms in Notizbl. Bot. Gart. Berl. **8**: 146 (1922).—Bak. f., Legum. Trop. Afr. **3**: 866 (1930).—Brenan, T.T.C.L.: 341 (1949). Types from Tanzania.

* I have not personally seen this work.

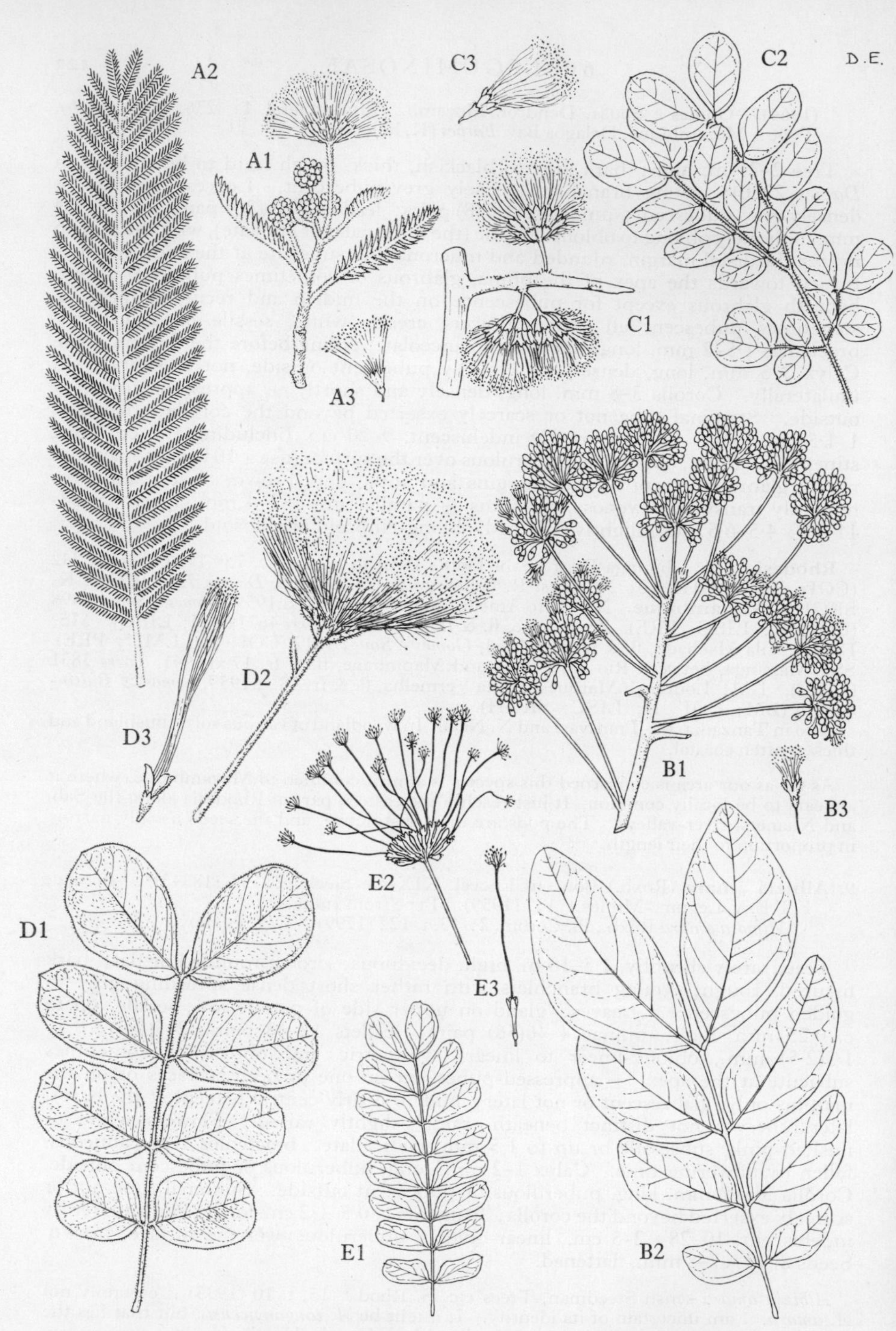

Tab. 23. A.—ALBIZIA AMARA SUBSP. SERICOCEPHALA. A1, flowering branch ($\times \frac{1}{2}$) *Wild* 3017; A2, leaf ($\times \frac{1}{2}$) *White* 2299; A3, flower ($\times \frac{2}{3}$) *Wild* 3017. B.—ALBIZIA GLABERRIMA VAR. GLABRESCENS. B1, part of flowering branch ($\times \frac{1}{2}$); B2, pinna ($\times \frac{1}{2}$); B3, flower ($\times \frac{2}{3}$), all from *Gomes e Sousa* 4488. C.—ALBIZIA ANTHELMINTICA. C1, part of flowering branch ($\times \frac{2}{3}$) *Brenan & Keay* 7637; C2, leaf ($\times \frac{1}{2}$) *Barbosa & Carvalho* 3188; C3, flower ($\times \frac{2}{3}$) *Brenan & Keay* 7637. D.—ALBIZIA VERSICOLOR. D1, pinna ($\times \frac{1}{2}$); D2, flower head ($\times \frac{1}{2}$); D3, flower ($\times \frac{2}{3}$), all from *Van Rensburg* 2599. E.—ALBIZIA PETERSIANA. E1, pinna ($\times \frac{1}{2}$); E2, flower head ($\times \frac{1}{2}$); E3, flower, ($\times \frac{2}{3}$), all from *Drummond & Hemsley* 3547.

Pinnae and leaflets comparatively few, the former 4–12(16) pairs, the latter 10–29 pairs; leaflets 3–7·5(10) × 0·75–1·5(2) mm., sometimes ± glaucous beneath when mature; lateral nerves, especially the basal, rarely slightly raised beneath.

Mozambique. N: Memba, fl. 8 fr. 9.xii.1963, *Torre & Paiva* 9484 (LISC). MS: Cheringoma, Condué, fl., *Simão* 1232 (LM).

Also in Kenya, Tanzania, India and Ceylon. Woodland; c. 10–550 m.

Subsp. **sericocephala** (Benth.) Brenan in Kew Bull. **10**: 190 (1955); F.T.E.A., Legum.-Mimos.: 152 (1959).—Palgrave, Trees of Central Afr.: 258, cum tab. et photogr. (1956).—Codd in Bothalia, **7**: 70 (1958).—F. White, F.F.N.R.: 89, fig. 16 H (1962).—Boughey in Journ. S. Afr. Bot. **30**: 158 (1964). TAB. **23** fig. A. Syntypes from the Sudan and Ethiopia.

Albizia sericocephala Benth. in Hook., Lond. Journ. Bot. **3**: 91 (1844).—Milne-Redh. in Kew Bull. **1934**: 301 (1934).—Brenan, T.T.C.L.: 341 (1949).—Pardy in Rhod. Agric. Journ. **49**: 14, cum photogr. (1952). Types as above.

Albizia amara sensu Oliv., F.T.A. **2**: 356 (1871).—Eyles in Trans. Roy. Soc. S. Afr. **5**: 361 (1916).—Bak. f., Legum. Trop. Afr. **3**: 865 (1930).—Gilbert & Boutique, F.C.B. **3**: 172 (1952).—Mitchell in Puku, **1**: 107 (1963).

Albizia struthiophylla Milne-Redh. in Kew Bull. **1933**: 144 (1933).—O. B. Mill. in Journ. S. Afr. Bot. **18**: 27 (1952) (" struthiofolia "). Type: Zambia, Mazabuka, *Milne-Redhead* 1207 (K, holotype; PRE).

Pinnae and leaflets comparatively many, the former (7)14–46(56) pairs, the latter (12)21–48 pairs; leaflets 2–4(4·5) × 0·5–1(1·25) mm., usually green beneath; lateral nerves not distinct.

Botswana. N: Francistown to Bosoti, at Wilmots, fr. 17.iv.1931, *Pole Evans* 3232 (30) (K; PRE). SE: 72 km. W. of Francistown, *Miller* B/412 (PRE). **Zambia.** W: Ndola, young fr. 11.v.1953, *Ferreira* 11 (PRE). C: Mt. Makulu near Chilanga, fr. 15.vi.1957, *Angus* 1629 (COI; K; PRE). S: Mapanza, fl. & fr. 11.x.1953, *Robinson* 347 (K). **Rhodesia.** N: Urungwe Distr., Magunge, fr. 4.v.1955, *Shiff* 11 (K; PRE; SRGH). W: Bulawayo, Hillside School, fr. v.1962, *Furness* 8/62 (SRGH). C: Salisbury, fl. 2.xi.1945, *Wild* 335 (K; SRGH). E: Umtali Distr., Eastern Commonage, fl. 8 fr. 5.xi.1949, *Chase* 1825 (BM; COI; K; LISC; SRGH). S: 16 km. E. of Fort Victoria, fl. & fr. 9.x.1949, *Wild* 3017 (K; SRGH). **Malawi.** C: Lilongwe, fl. 29.x.1962, *Chapman* 1716 (SRGH). S: Mchese Mt., fr. 2.vii.1958, *Chapman* W/599 (K; SRGH). **Mozambique.** N: Amaramba, Mecanhelas, fl. 20.x.1948, *Andrada* 1431 (BM; COI; K; LISC). T: Macanga, between Massamba and Casula, fr. 7.vii.1949, *Barbosa & Carvalho* in *Barbosa* 3481 (K; LISC). MS: Manica, Matarara do Lucite, fr. 13.x.1953, *Pedro* 4284 (PRE).

From the Sudan and Eritrea southwards to Botswana and the Transvaal. In woodland, wooded grassland and thickets; 600–1500 m. Said by Pardy (loc. cit.) to occur characteristically in small clumps, though it may also be found growing singly.

10. **Albizia brevifolia** Schinz in Bull. Herb. Boiss., Sér. 2, **2**: 945 (1902).—Codd in Bothalia, **7**: 69 (1958).—Boughey in Journ. S. Afr. Bot. **30**: 158 (1964). Type: Mozambique, Boroma, on the Nhasinde, *Menyharth* 994 (K; Z, holotype).

Albizia rogersii Burtt Davy, F.P.F.T. **2**: xviii, 348 (1932).—O. B. Mill. in Journ. S. Afr. Bot. **18**: 27 (1952).—F. White, F.F.N.R.: 89, fig. 16 I (1962). Type from the Transvaal.

Albizia parvifolia Burtt Davy, F.P.F.T. **2**: xvii, 348 (1932). Type from the Transvaal.

A rounded bush or small tree c. 3–16 m. high, deciduous; bark grey to black, smooth or very shallowly fissured; trunk often forking near the base into several to many ascending branches; young branchlets sparsely to rather densely appressed-grey-puberulous. Leaves: gland on upper side of petiole squat, sessile, c. 0·25 mm. high; pinnae (3)6–10(17) pairs; leaflets 15–30 pairs, 3–9 × 0·75–1·8 mm., narrowly oblong to linear-oblong, symmetric and obtuse to subacute at the apex, glabrous or with margins ± appressed-ciliate; midrib nearly central (except at the base); lateral nerves not distinct beneath. Flowers white to creamy-yellow, 1–1·5 mm. pedicellate; bracteoles very caducous, fallen by flowering time. Calyx 1–1·5 mm. long, glabrous to ± puberulous outside. Staminal tube not or scarcely exserted beyond the corolla; filaments c. 1–1·2 cm. long. Pod apparently indehiscent, 9–27 × 1·8–3·6 cm., linear-oblong, glabrous to finely puberulous, brown. Seeds 8–10 × c. 6·5 mm., flattened.

Botswana. N: Francistown Distr., between Sukwe and Nata, fr. 23.iii.1962, *de Beer & Yalala* 33 (SRGH). **Zambia.** C: near Feira Boma, fr. 30.v.1952, *White* 2904 (BM;

K). S: Sinazongwe, fr. 27.v.1961, *Fanshawe* 6620 (K). **Rhodesia.** N: Urungwe Distr., Kariba, fr. i.1959, *Goldsmith* 2/59 (BM; K; LISC; LM; SRGH). W: Wankie, fl. & fr. 21.x.1951, *Lovemore* 123 (K; LISC; SRGH). E: Melsetter Distr., Hotsprings, fr. 24.ii.1952, *Chase* 4383 (BM; K; LISC; SRGH). S: Malipati, Nuanetsi R., fl. 2.xi.1955, *Wild* 4696 (K; SRGH). **Mozambique.** T: Mutarara, fr. 15.vi.1949, *Andrada* 1584 (COI; LISC). MS: Báruè, between Mandiè and Mungári, fl. & fr. 30.x.1941, *Torre* 3712 (BM; K; LISC).

Also in the Transvaal. Usually recorded from rocky places, often basalt but also sandstone; also in mopane woodland; 130–1000 m.

A. brevifolia is unquestionably closely allied to *A. amara*, particularly the subsp. *amara*. The two species have a number of significant characters in common, including the unusual apparently indehiscent pods. *A. brevifolia* differs from *A. amara* subsp. *amara* in having much shorter sparse indumentum on the branchlets, and more glabrous leaflets (usually) and calyx, and in the more stipitate pods (stipe (0·5)1–2·5 cm. as against 0·5–0·8 cm.).

11. **Albizia isenbergiana** (A. Rich.) Fourn. in Ann. Sci. Nat., Sér. 4, **14**: 373 (1850).—Brenan, F.T.E.A. Legum.-Mimos.: 152 (1959).—F. White, F.F.N.R.: 432 (1962). Type from Ethiopia.
Inga isenbergiana A. Rich., Tent. Fl. Abyss. **1**: 236 (1847). Type as above.

Tree up to 15 m. high; crown flat or umbrella-shaped; young branchlets densely and rather shortly pubescent. Leaves: gland on upper side of petiole squat and sessile, or somewhat raised, 0·25–0·75 mm. high; pinnae (3)6–12 pairs; leaflets (12)16–33 pairs, (5·5)6·5 × 1·25–4·5 mm., slightly falcate to oblong, with apex usually slightly asymmetric, obtuse to subacute, ± appressed-pubescent on both surfaces or glabrescent above; lateral nerves ± raised and visible beneath; lower surface of leaflet paler. Flowers white, on pedicels 1·5–2 mm. long; bracteoles soon caducous, fallen by the time the flowers open. Calyx 1·5–2·25 mm. long, densely shortly pubescent outside. Corolla 4–5 mm. long, densely short-pubescent or puberulous outside. Staminal tube not or scarcely exserted beyond the corolla; filaments c. 1 cm. long or less. Pod apparently indehiscent, 10–27 × 2–4 cm., oblong, puberulous over the surface. Seeds c. 12 × 7 mm., flattened.

Zambia. W: Kitwe, fl. 2.xi.1957, *Fanshawe* 3824 (K; LISC). E: Fort Jameson, fr. 1.vi.1958, *Fanshawe* 4486 (K). S: Mapanza, Choma, fl. 27.x.1957, *Robinson* 2481 (K; PRE; SRGH). **Mozambique.** N: Macondes, between Namaua and Mutamba dos Macondes, fl. 26.ix.1948, *Pedro & Pedrógão* 5362 (LMJ).

Also in Ethiopia, Uganda, Kenya and Tanzania. Ecology imperfectly known: *Fanshawe* 3824 from a termite mound, the others from a stream- and river-bank respectively. The Mozambique specimen has no habitat information.

A poorly-known species, with an oddly discontinuous distribution, known from Kenya, Uganda and Tanzania by single gatherings only.

12. **Albizia zimmermannii** Harms in Engl., Bot. Jahrb. **53**: 455 (1915).—Bak. f., Legum. Trop. Afr. **3**: 864 (1930).—Brenan, T.T.C.L.: 343 (1949); F.T.E.A. Legum.-Mimos.: 153 (1959).—Boughey in Journ. S. Afr. Bot. **30**: 158 (1964). Syntypes from Tanzania.
Albizia nyasica Dunkley in Kew Bull. **1937**: 469 (1937).—Burtt Davy & Hoyle, rev. Topham, N.C.L., ed. 2: 65 (1958). Type: Malawi, Mangoche Mt., *Clements* 574 (K).

Tree 6–15 m. high; crown flat, spreading; bark smooth, finely fissured, grey to grey-brown; young branchlets sparsely to densely rusty-puberulous or -pubescent, sometimes nearly glabrous; indumentum going grey with age. Leaves: rhachis clothed like the branchlets, not hooked or clawed at the end; pinnae 3–6(10) pairs; leaflets 8–17 pairs, 7–15(17) × (3)4–8 mm., rounded at apex, oblong-elliptic, slightly oblique, beneath paler, ± glaucous and appressed-puberulous (occasionally glabrous or nearly so). Heads often aggregated on short leafless branches; pedicels 0–1 mm. long; bracteoles very minute, but sometimes present when the flowers open. Flowers white or pink. Calyx 1–2·5 mm. long, densely rusty-puberulous outside, not slit unilaterally. Corolla 3·5–5 mm. long, densely rusty-puberulous outside. Staminal tube not or scarcely exserted beyond the corolla; filaments 0·7–1·3 cm. long. Pod apparently indehiscent, 15–32 × (3)3·8–7 cm., oblong, ± puberulous, crimson near maturity, turning brown with age, typically (though apparently not in our area) with very prominent transverse veins much raised, particularly in the centre, sometimes almost wing-like, and anastomosing. Seeds c. 10 × 7 mm., flattened.

Zambia. S: c. 6 km. N. of Chaanga on road to Mapangazia and Mazabuka, fr. 9.vi.1963, *Bainbridge* 824 (SRGH). **Rhodesia.** N: Mtoko, fr. v.1953, *Phelps* in GHS 43162 (K; SRGH). **Malawi.** C: Dowa Distr., Lake Nyasa Hotel, fr. 3.viii.1951, *Chase* 3867 (BM; K; SRGH). S: Fort Johnston, Boadzulu I., st. 18.v.1954, *Jackson* 1311 (K). **Mozambique.** Z: Morrumbala (Massingire), fr. 15.v.1943, *Torre* 5312 (BM; K; LISC). T: between Mungari and Changara, fl. 26.x.1943, *Torre* 6091 (BM; K; LISC). MS: Gorongosa. Dumbo, st. 9.vii.1947, *Simão* 1402 (LISC).

Also in Kenya and Tanzania. Ecology imperfectly known: recorded from riverine and dry forest, granite kopjes, and dunes by Lake Nyasa; 430–820 m.

The specimen from Malawi (S) cited above is sterile and rather doubtful*.

The peculiar and characteristic venation of the pod of typical *A. zimmermannii* does not appear to be fully developed in our area. In Zambia, Rhodesia and Malawi there is a tendency towards rather narrow pods c. 3–4 cm. wide with sparse puberulence. These are not known elsewhere, but trees with more densely puberulous pods c. 4·5 cm. wide have also been collected in Malawi.

13. Albizia schimperana Oliv., F.T.A. **2**: 359 (1871).—Bak. f., Legum. Trop. Afr. **3**: 866 (1930).—Gilbert & Boutique, F.C.B. **3**: 183 (1952).—Burtt Davy & Hoyle rev. Topham, N.C.L., ed. 2: 66 (1958).—Brenan, F.T.E.A. Legum.-Mimos.: 154 (1959).—Boughey in Journ. S. Afr. Bot. **30**: 158 (1964). Type from Ethiopia.

Albizia maranguensis Taub. ex Engl. in Abh. Preuss. Akad. Wiss. **1891**: 241 (1892).—Bak. f. tom. cit.: 867 (1930).—Brenan, T.T.C.L.: 341 (1949).—Burtt Davy & Hoyle, tom. cit.: 65 (1958). Syntypes from Tanzania.

Tree 5–23(30) m. high; crown flat or not; bark smooth, grey or sometimes brownish and rough; young branchlets densely, or sometimes sparsely, and shortly brown-pubescent (sometimes pubescence grey but not in our area), later glabrescent. Leaves: rhachis shortly and densely to sparsely pubescent; pinnae (1)2–7 pairs; leaflets of the 2 distal pairs of pinnae 6–21(23) pairs (sometimes as few as 5 pairs on the lower pinnae), 7–21(30) × 3·5–8·5(16) mm., variable in shape and size, obliquely oblong, or rhombic to falcate-oblong, acute to rounded and mucronate at the apex, which is turned towards the pinna-apex, with diagonal midrib, ± appressed-pubescent beneath and often with the lower surface paler than the upper or even whitish when dry, glabrescent above. Flowers white or pale-yellow, pedicellate; pedicels 2–6 mm. long, densely and shortly brown- (in our area) pubescent or sometimes puberulous, as are the calyces and corollas. Calyx 1·5–2·5 mm. long, not slit unilaterally. Corolla 4–7·5 mm. long. Staminal tube not or scarcely exserted beyond the corolla; filaments c. 0·7–1·2 cm. long. Pod apparently indehiscent, 15–34 × (2)2·8–5·9 cm., oblong, puberulous (sometimes sparsely so), not glossy, venose, brown. Seeds 9–11 × 6·5–8 mm., flattened.

Var. **schimperana**.—Brenan in Kew Bull. **10**: 190 (1955); F.T.E.A. Legum.-Mimos.: 154 (1959).

Young shoots brown-pubescent. Each pinna somewhat narrowing towards its apex; leaflets of 2 distal pairs of pinnae 8–21(23) pairs with the terminal pair somewhat smaller than the rest, 7–21 × 3·5–8·5 mm., rarely more, 2–3 times as long as wide. Indumentum on pedicels, calyx and corolla brown.

Rhodesia. E: Umtali, fl. 29.x.1956, *Chase* 6223 (K; SRGH). **Malawi.** C: Dowa Distr., Nchisi, fr. 9.v.1961, *Chapman* 1297 (K; SRGH). S: Mt. Mlanje, Chipalombe Valley, fr. 20.v.1958, *Chapman* 575 (K; PRE; SRGH). **Mozambique.** N: Maniamba, Serra Jéci, fr. 3.iii.1964, *Torre & Paiva* 10987 (LISC). T: Angónia, Monte Dómuè, fr. 9.iii.1964, *Torre & Paiva* 11092 (LISC). MS: Manica, Monte Xiroso, Mavita, fl. 26.x.1944, *Mendonça* 2640 (BM; K; LISC).

Also in the Congo, Sudan, Ethiopia, Somalia, Kenya and Tanzania. Typically in and on margins of montane forest, but apparently occasionally occurring in montane *Brachystegia* woodland; 900–1830 m.

A. schimperana var. *schimperana* shows considerable variation in foliage. The leaflets vary in number from 8–23 pairs per pinna, their length/breadth ratio varies from 2–3, their apex may be acute or rounded, and their lower surface varies from green to whitish. These trends do not appear correlated with each other or with geography. Typical *A. schimperana* has rather numerous narrow acute to obtuse leaflets. It should be emphasized that the whole range of variation does not occur on one tree, and probably not even in one population.

* Professor H. Wild writes that he has seen and collected good fruiting material of this species at Monkey Bay in this division of Malawi.

A. schimperana appears prone to a malformation affecting a part or the whole of single pinnae, whose leaflets are then ± confluent into a lobed lamina resembling a fern pinnule.

A. maranguënsis may in part be referable to var. *amaniënsis* rather than to var. *schimperana*, since the leaflets were described as 15–25 × 7–15 mm., although in up to 14 pairs.

Var. **amaniënsis** (Bak. f.) Brenan in Kew Bull. **10**: 191 (1955); F.T.E.A. Legum.-Mimos.: 155 (1959). Type from Tanzania.

Albizia amaniënsis Bak. f. in Journ. of Bot. **70**: 255 (1932).—Brenan, T.T.C.L.: 341 (1949). Type as above.

Indumentum as in var. *schimperana*. Each pinna ± broadening towards the apex; leaflets in 5–11 pairs, with the terminal pair somewhat larger than the rest, mostly 10–32 × 5–16 mm.; leaflets c. twice as long as wide.

Rhodesia. E: Umtali, Murambi Gardens Suburb, fl. 27.x.1962, *Chase* 7880 (K; LISC; SRGH). **Malawi.** C: Dedza–Mpata Milande road, fr. 10.x.1960, *Chapman* 916 (K; SRGH). **Mozambique.** MS: Mossurize, between Espungabera and Macuiana, fl. 30.x.1944, *Mendonça* 2675 (BM; K; LISC).

Also in Tanzania. Ecology apparently similar to that of var. *schimperana*.

The status of this variety is doubtful. The material is rather heterogeneous, but apparently distinct from the other varieties by the pinnae somewhat broadening upwards, with the terminal leaflets somewhat larger than the others. In some ways var. *amaniënsis* is similar to *A. glaberrima* and the strong possibility of var. *amaniënsis* being the product of crossing between *A. glaberrima* var. *glabrescens* and *A. schimperana* var. *schimperana* should be borne in mind. The apparent scarcity of occurrence and wide discontinuous distribution may also perhaps indicate a rarely but repeatedly formed interspecific hybrid rather than a single taxonomic entity. Field observation of this tree might be very helpful.

14. **Albizia glaberrima** (Schumach. & Thonn.) Benth. in Hook., Lond. Journ. Bot. **3**: 88 (1844).—Keay in Kew Bull. **8**: 489–490 (1954); F.W.T.A. ed. 2, **1**: 502 (1958).—Torre, C.F.A. **2**: 292 (1956).—Brenan, F.T.E.A. Legum.-Mimos.: 156 (1959). Type from Ghana.

Mimosa glaberrima Schumach. & Thonn., Beskr. Guin. Pl.: 321 (1827). Type as above.

Tree 9–25 m. high; crown ± flattened; bark smooth, grey or grey-brown; young branchlets usually ± sparingly puberulous or shortly pubescent. Leaves: rhachis clothed like the young branchlets; pinnae 1–3(4) pairs; leaflets 3–6(8) pairs, 18–70(90) × 9–33(42) mm., obliquely rhombic-ovate to rarely -obovate or slightly falcate-oblong or asymmetrically elliptic, with the midrib often ± diagonal, narrowed to an obtuse (or, but not in the Flora area, subacute to acute) apex, glabrous or nearly so above, the midrib puberulous to almost glabrous, beneath glabrous or nearly so except for sparse puberulence or rarely short pubescence on the midrib, rarely finely appressed-puberulous over the surface, often pale beneath when dry. Flowers white or whitish, pedicellate; pedicels 1·5–7 mm. long, densely covered with grey puberulence or very short fine pubescence as are the outsides of the calyces and corollas. Calyx 1·5–2·5(3) mm. long, occasionally slit unilaterally. Corolla 3–6 mm. long. Staminal tube not or scarcely exserted beyond the corolla; filaments c. 0·6–1 cm. long. Pod apparently indehiscent, 12–29 × 3–4·2 cm., oblong, puberulous over the surface, not or only slightly glossy, somewhat venose, usually brown. Seeds c. 9–11 × 6–7 mm., ± flattened.

Typical var. *glaberrima*, with leaflets with a minute appressed puberulence on the lower surface, occurs from Ghana eastwards to the Sudan and Uganda, and extends southwards in western Africa to Cameroun and Angola.

Var. **glabrescens** (Oliv.) Brenan in Kew Bull. **10**: 192 (1955); F.T.E.A. Legum.-Mimos.: 156 (1959).—F. White, F.F.N.R.: 89, fig. 16 K (1962).—Boughey in Journ. S. Afr. Bot. **30**: 158 (1964). TAB. **23** fig. B. Syntypes from Tanzania and Mozambique: R. Kongone, *Kirk* (K).

Albizia glabrescens Oliv., F.T.A. **2**: 357 (1871).—Bak. f., Legum. Trop. Afr. **3**: 862 (1930).—Brenan, T.T.C.L.: 342 (1949).—Gomes e Sousa, Dendrol. Moçamb. **4**: 40, cum tab. (1949); Dendrol. Moçamb. Estudo Geral, **1**: 240, t. 44 (1966).—Gilbert & Boutique, F.C.B. **3**: 184 (1952) pro parte.—Burtt Davy & Hoyle, rev. Topham, N.C.L. ed. 2: 65 (1958). Syntypes as above.

Albizia glaberrima sensu Mitchell in Puku, **1**: 108 (1963).

Leaflets with puberulence or sometimes short pubescence on the midrib and rarely on the margins, otherwise glabrous or nearly so beneath; apex obtuse; midrib above glabrous or nearly so, with puberulence only near the base, or sometimes puberulous to the apical part of the leaflet.

Zambia. N: Luapula R. at Fort Rosebery road crossing, fl. 28.x.1952, *White* 3518 (BM; COI; K; PRE). C: Kafue R. gorge, fl. 6.x.1957, *Angus* 1731 (K; PRE). S: Kafue R. at Puku Flats, fr. 11.vi.1960, *Angus* 2417 (K; PRE). **Rhodesia.** E: Chipinga Distr., Mangazi R. valley SW. of Chirinda, fl. x.1962, *Goldsmith* 205/62 (COI; K; LISC; PRE). S: Sabi–Lundi junction, Chitsa's Kraal, fr. 4.vi.1950, *Wild* 3430 (K; SRGH). **Malawi.** N: Rumpi Distr., Rukuru R., fr. 30.iv.1952, *White* 2537 (K). S: Mt. Mlanje, Tuchila R., Malere Rock, fr. 25.vii.1957, *Chapman* 390 (BM; K; PRE). **Mozambique.** N: Porto Amélia, R. Ridi, Nangororo. fl. 26.x.1959, *Gomes e Sousa* 4488 (COI; K; LISC; LMU; PRE). Z: Mocuba, Namagoa, fl. ix.1944, *Faulkner* 65 (BM; K; PRE). T: between Tete and Casula, fl. 8, fr. 21.ix.1943, *Torre* 6073 (LISC). MS: Buzi, Grudja, fr. 2.vi.1941, *Torre* 2777 (LISC; LM). SS: Govuro, Nova Mambone, fl. & fr. 5.ix.1944, *Mendonça* 2002 (BM; K; LISC).

The variety also in the Congo, Kenya, Tanzania and Zanzibar, in lowland rain-forest, particularly by rivers; occasionally recorded from woodland, perhaps as a relict; 60–910 m.

15. **Albizia gummifera** (J. F. Gmel.) C. A. Sm. in Kew Bull. **1930**: 218 (1930) pro parte excl. syn. *Mimosa adianthifolia* ("adiantifolia"), *Zygia fastigiata* et *Albizia fastigiata*.—Brenan, T.T.C.L.: 339–40 (1949) pro parte quoad "Species A" tantum; in Kew Bull. **7**: 511 (1953); in Mem. N.Y. Bot. Gard. **8**: 430 (1954); F.T.E.A. Legum.-Mimos.: 157, figs. 21/1–5, 22/1 (1959).—Gilbert & Boutique, F.C.B. **3**: 181 (1952).—Pardy in Rhod. Agric. Journ. **49**: 255, cum photogr. (1952).—Palgrave, Trees of Central Afr.: 265, cum tab. et photogr. (1956).—Burtt Davy & Hoyle, rev. Topham, N.C.L. ed. 2: 65 (1958).—Boughey in Journ. S. Afr. Bot. **30**: 158 (1964). TAB. **24** fig. A. Type from Ethiopia.

Sassa gummifera J. F. Gmel. in L., Syst. Nat. ed. 13, **2**, 2: 1038 (1792). Type as above.

Albizia fastigiata sensu Eyles in Trans. Roy. Soc. S. Afr. **5**: 361 (1916).—Steedman, Trees etc. S. Rhod.: 15, t. 11 (1933). Non *Albizia fastigiata* (E. Mey.) Oliv.

Medium or large tree up to 30 m. high; crown flattened; bark smooth, very rarely rough, grey; young branchlets finely and shortly brownish-pubescent, soon glabrescent and usually deep- or blackish-purple, ultimately grey-barked. Leaves: pinnae 5–7(8) pairs (rarely only 3 on occasional reduced leaves), each pinna ± narrowing upwards; leaflets of 2 distal pairs of pinnae 9–16(17) pairs, mostly c. 10–20(25) × 4–8(13) mm., obliquely rhombic-quadrate to rhombic-subfalcate, auricled (or sometimes not, but not in our area) on the proximal side, obtuse to acute at the apex, subglabrous or somewhat pubescent on the midrib and margins, rarely, especially when young, some occasional hairs on the primary lateral nerves; raised venation beneath lax. Stipules and bracts at base of peduncles lanceolate, up to c. 6–7 × 2–2·5 mm. Peduncles puberulous or finely pubescent; bracteoles mostly caducous, linear, inconspicuous, (1)2–6 mm. long, normally shorter than the flower-buds except when extremely young. Flowers subsessile; pedicels puberulous or sometimes glabrous, 0·25–0·75(1) mm. long. Calyx 2·5–5 (very rarely indeed 1·5–2·5) mm. long, minutely, shortly and rather appressedly brownish-pubescent to subglabrous outside. Corolla 6·5–12 mm. long, minutely pubescent outside, white. Staminal tube exserted c. 1·5–2·8 cm. beyond the corolla, white below, crimson above. Pod dehiscent, (8)10–23·6 × 2–3·4(4) cm., oblong, flat or slightly transversely plicate, glabrescent, glossy, eglandular, less prominently and closely venose than in *A. adianthifolia*, pale-brown to reddish-brown or purplish. Seeds 8–12 × 8–10 mm., flattened.

Rhodesia. E: Umtali Distr., Commonage, fl. 11.x.1956, *Chase* 6211 (K; LISC; SRGH). S: Bikita Distr., Old Bikita, Donga, fl. 16.xii.1953, *Wild* 4403 (K; LISC; SRGH). **Malawi.** N: Misuku Hills, Mugesse Forest Reserve, fl. x.1953, *Chapman* 171 (K). S: Zomba Plateau, fr. 5.vi.1946, *Brass* 16255 (BK; K; PRE; SRGH). **Mozambique.** N: between Nacala and Mossuril, fl. 28.x.1952, *Barbosa & Balsinhas* 5219 (K; LISC; LMJ). Z: between Nhamarroi and Ile, fl. & fr. 23.ix.1941, *Torre* 3511 (BM; K; LISC). T: Moatize, Zóbuè, st. 7.vi.1941, *Torre* 2876 (LISC). MS: Manica, Mavita, valley of R. Rotanda, fl. 23.viii.1957, *Pedro* 4971 (K).

In eastern Africa from the Sudan and Ethiopia southwards to our area and westwards to the eastern Congo; also in Cameroon (Bamenda), S.E. Nigeria and Madagascar. In upland rain-forest and riverine forest; 1130–1580 m.

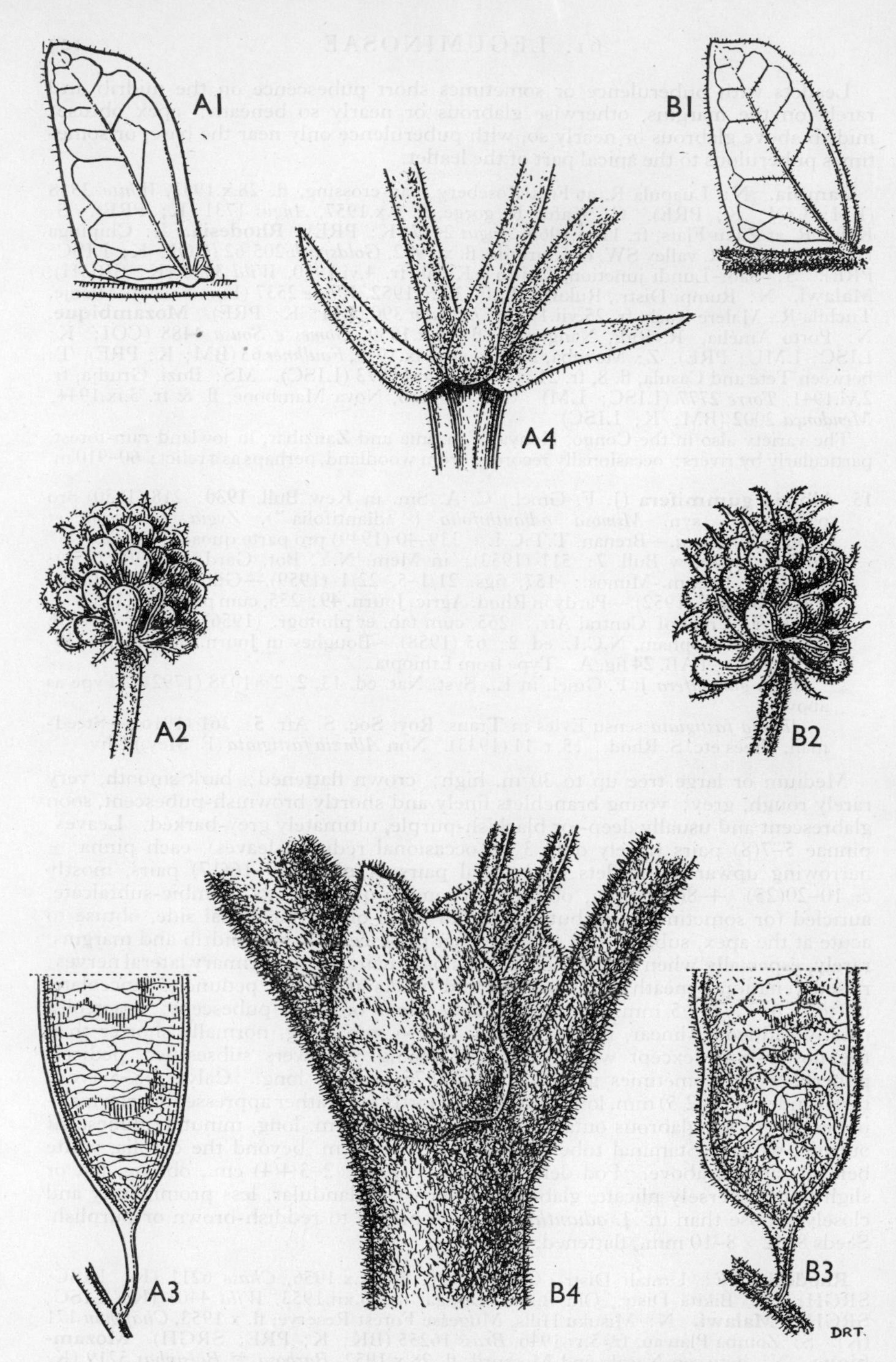

Tab. 24. A.—ALBIZIA GUMMIFERA VAR. GUMMIFERA. A1, leaflet, underside (×2); A2, young flower head (×4), all from *Lugard* 511; A3, basal part of pod (×1) *Jackson* 336; A4, stipules (×4) *Lugard* 511. B—ALBIZIA ADIANTHIFOLIA. B1, leaflet, underside (×2) *Burtt* 2894; B2, young flower head (×4) *Greenway* 3376; B3, basal part of pod (×1); B4, stipules (×4), all from *Burtt* 2894. From F.T.E.A.

All material from our area is referable to var. *gummifera* which has the leaflets auricled on the proximal side at the base.

The species appears to be more strictly confined to high altitudes with us than it is further north in E. Africa, where it may sometimes occur near sea-level.

16. **Albizia adianthifolia** (Schumach.) W. F. Wight in U.S. Dept. Agric. Bur. Pl. Industry, Bull. **137**: 12 (1909).—Gilbert & Boutique, F.C.B. **3**: 178 (1952).—Brenan in Kew Bull. **7**: 520 (1953); Mem. N.Y. Bot. Gard. **8**: 430 (1954); F.T.E.A. Legum.-Mimos.: 160, fig. 21/6–9, 22/2 (1959).—Torre, C.F.A. **2**: 295 (1956).—Codd in Bothalia, **7**: 79 (1958).—Fanshawe, Fifty Common Trees N. Rhod.: 16 cum tab. (1962).—F. White, F.F.N.R.: 88, fig. 16 E–F (1962).—Mitchell in Puku, **1**: 107 (1963).—Boughey in Journ. S. Afr. Bot. **30**: 158 (1964).—Gomes e Sousa, Dendrol. Moçamb. **1**: 239, t. 43 (1966). TAB. **24** fig. B. Type probably from Ghana,

Mimosa adianthifolia Schumach., Beskr. Guin. Pl.: 322 (1827). Type as above.

Zygia fastigiata E. Mey., Comment. Pl. Afr. 165 (1836). Syntypes from S. Africa.

Albizia fastigiata (E. Mey.) Oliv., F.T.A. **2**: 361 (1871).—Sim, For. Fl. Port. E. Afr.: 59, t. 58 (1909).—R.E.Fr. in Wiss. Ergebn. Schwed. Rhod.-Kongo Exped. **1**: 63 (1914). Types as for *Zygia fastigiata*.

Albizia fastigiata var. *chirindensis* Swynnerton ex Bak. f. in Journ. Linn. Soc., Bot. **40**: 65 (1911).—Eyles in Trans. Roy. Soc. S. Afr. **5**: 361 (1916). Type: Rhodesia, Chirinda Forest, *Swynnerton* 52 (BM, holotype; K; SRGH).

Albizia sassa sensu Bak. f., Legum. Trop. Afr. **3**: 868 (1930) pro parte excl. syn. *Sassa* et *Mimosa sassa*.

Albizia chirindensis (Swynnerton ex Bak. f.) Swynnerton ex Steedman, Trees etc. S. Rhod.: xv, 16, 180 (1933).—Swynnerton in Journ. of Bot. **74**: 317 (1939).

Tree (2·5)4–30 m. high; crown flattened; bark grey to yellowish-brown and rough (rarely smooth in our area); young branchlets densely rather coarsely and persistently rusty- to fulvous-pubescent; pubescence sometimes becoming grey as the branchlet ages. Leaves: pinnae 5–8 pairs (rarely only 3 on occasional reduced leaves), each pinna ± narrowing upwards; leaflets of 2 distal pairs of pinnae (8)9–17 pairs, mostly c. 7–17(24) × 4–9(15) mm., obliquely rhombic-quadrate or -oblong; proximal margin at base usually ± rounded into the pulvinus but not auriculate; apex of leaflet usually obtuse and mucronate, sometimes subacute, surface of leaflet thinly pubescent above, rather plentifully pubescent all over beneath, raised venation beneath close. Stipules and bracts at base of peduncles c. 5–12 × 3–6(8) mm., ovate. Peduncles clothed as the young branchlets; bracteoles variably persistent, 5–8 mm. long, exceeding the flower-buds, linear-spathulate to oblanceolate. Flowers subsessile; pedicels pubescent, 0·5–1(2) mm. long. Calyx 2·5–5 (rarely only 2) mm. long, pubescent outside. Corolla 6–11 mm. long, white or greenish-white, pubescent outside. Staminal tube exserted c. 1·3–2.5 cm. beyond the corolla, red to wholly greenish or pink. Pod dehiscent. 9–19 × 1·9–3·4(4·3) cm., usually pale brown, oblong, flat or slightly transversely plicate, ± densely and persistently pubescent, not glossy, prominently venose. Seeds 7–9·5 × 6·5–8·5 mm., flattened.

Zambia. N: Kasama Distr., Mungwi, fl. 5.x.1960, *Robinson* 3897 (K; SRGH). W: Ndola, fl. ix.1933, *Duff* 162/33 (K). S: S. of Namwala road c. 10 km. from Nambala Mission, fl. & fr. 17.ix.1947, *Brenan & Greenway* 7870 (K). **Rhodesia.** E: Inyanga Distr., Eastern Highlands Tea Estate, E. of Inyangani, fl. 17.x.1962, *Chase* 7857 (K; LISC; SRGH). S: Zimbabwe, fl. & fr. 4.x.1959, *Wild* 2991 (K). **Malawi.** N: Nkata Bay Distr., Mkuwadzi Hill, fl. 13.xi.1959, *Adlard* 313 (SRGH). C: Dowa, fl. & fr. 28.x.1941, *Greenway* 6369 (K; PRE). S: Cholo Mt., fl. & fr. 29.ix.1946, *Brass* 17859 (BM; K; PRE; SRGH). **Mozambique.** N: between Mueda and Nangade, fl. 19.x.1942, *Mendonça* 955 (BM; K; LISC). Z: between Quelimane and Gúruè, fl. 13.ix.1941, *Torre* 3437 (BM; K; LISC). MS: Cheringoma, Inhaminga, fr. 27.v.1948, *Mendonça* 4392 (BM; K; LISC). SS: Massinga, fl. ix.1937, *Gomes e Sousa* 2039 (COI; K). LM: Lourenço Marques, Costa do Sol, fl. & fr. 10.viii.1959, *Barbosa & Lemos* in *Barbosa* 8677 (COI; K; PRE; SRGH).

Widespread in tropical Africa from Gambia and Kenya southwards to Angola and extending to S. Africa (E. Cape Prov.). In rain-forest, woodland and wooded grassland; near sea-level–1680 m.

Further details of the synonymy will be found in Kew Bull. 7: 520–1 (1953). *A. adianthifolia* has a wide range of habitat, and ecotypes may be recognizable. It is not, however, a particularly variable plant. The colour of the staminal tube varies (see description) and also the surface of the bark. Although smooth-barked examqles have been recorded from Rhodesia and Mozambique, it is likely that rough bark is more usual in our area. It would be interesting to have information about the ecology and distribution of these two bark-variants.

16 × 15. **Albizia adianthifolia** (Schumach.) W. F. Wight × **gummifera** (J. F. Gmel.) C. A. Sm.—Brenan in Kew Bull. **7**: 528 (1952).

Differs from *A. adianthifolia* in having leaflets ± auriculate at the base and finely pubescent or puberulous beneath, narrower bracts and stipules, and shorter bracteoles. Differs from *A. gummifera* in having the leaflets finely pubescent or puberulous on the surface beneath and with sometimes less-marked auricles on the proximal side of the base.

Malawi. S: Mt. Mlanje, fl. x.1891, *Whyte* (BM). **Mozambique.** N: 10 km. N. of Nametil, fl. ix, x.1941, *Gomes e Sousa* 2271 (COI; K; PRE). MS: Chimoio, Pindangonga Forest (Gondola), fl. 18.x.1945, *Simão* 603 (LISC; LM).

Convincing hybrids between these two species seem very rare or very rarely collected. I have seen only one probable specimen from our area besides those cited above: *Andrada* 1380 (BM; K; LISC), from Mozambique, N: between Mueda and Namassa, fl. 21.ix.1948.

Very occasionally, specimens resembling *A. adianthifolia* except for having fine rather sparse indumentum on the leaflets have been collected. *Gomes e Sousa* 2060 (COI; K) from Mozambique, SS: Morrumbene, fl. x.1937 is an example. There seems no other evidence of hybridity, and the plants seem best considered as an unusual variant or form of *A. adianthifolia*.

17. **Albizia petersiana** (Bolle) Oliv., F.T.A. **2**: 362 (1871).—Sim, For. Fl. Port. E. Afr.: 60 (1909).—Bak. f., Legum. Trop. Afr. **3**: 867 (1930).—Brenan, T.T.C.L.: 340 (1949); F.T.E.A. Legum.-Mimos.: 162 (1959). TAB. **23** fig. E. Type: Mozambique, Boror and Sena, 16°–18° S. lat., *Peters* (B, holotype†; BM).

Zygia petersiana Bolle in Peters, Reise Mossamb. Bot. **1**: 2, t. 1 (1861). Type as above.

Albizia brachycalyx Oliv., F.T.A. **2**: 361 (1871).—Bak. f., Legum. Trop. Afr. **3**: 868 (1930).—Brenan, T.T.C.L.: 339 (1949).—Burtt Davy & Hoyle, rev. Topham, N.C.L., ed. 2: 65 (1958). Type from Tanzania.

Tree, sometimes shrubby, (2)3–21 m. high, deciduous; crown rounded or flat; bark smooth unless fire-scarred; young branchlets shortly crisped-pubescent to almost glabrous. Leaves: pinnae 2–5 pairs, each pinna broadening or not towards the apex; leaflets of 2 distal pairs of pinnae (2)3–12 pairs, 5–23(27) × 2·5–13(17) mm., oblong- to obovate-rhombic; base auricled or not on the proximal side; apex obtuse to subacute, sometimes acute; surface beneath subglabrous to rather densely pubescent, above glabrous to puberulous or shortly pubescent; stipules very quickly falling and usually absent, 1·75–3·5 × 0·6–1 mm., oblanceolate or triangular-acute. Peduncles subglabrous to pubescent; bracteoles falling while flowers are still in bud, 1 mm. long, oblanceolate, pubescent. Flowers on glabrous or almost glabrous pedicels 1–3 mm. long. Calyx 1–1·75 mm. long, usually glabrous except on the margins, rarely with some hairs on the tube. Corolla c. 7–10 mm. long, glabrous outside or sometimes with a few hairs near the apices of the lobes, white to pink. Staminal tube red, exserted c. 1–1·3 cm. beyond the corolla. Pod dehiscent, 4–17·5 × (1·4)1·7–3 cm., oblong, transversely plicate or sometimes flat, eglandular, venose, deep red-purple to brown when ripe, glabrous to ± puberulous over the surface. Seeds 9–12 mm. long or in diam.

Subsp. **petersiana**

Young branchlets shortly crisped-pubescent to almost glabrous. Leaves with petiole (1)1·5–4 cm. long and 2–5 pairs of pinnae; leaflets of 2 distal pairs of pinnae (4)5–12 pairs, usually glabrous or subglabrous beneath, sometimes appressed-pubescent, rarely with non-appressed pubescence. Peduncle (1)1·7–3(4) cm. long, usually glabrous or shortly crisped-pubescent. Pods glabrous or almost so.

Malawi. S: Ntondwe, fl. 1905, *Cameron* 139 (K). **Mozambique.** N: between Mocímboa da Praia and R. Messalo, fl. 13.v.1959, *Gomes e Sousa* 4461 (COI; K; LMJ; PRE; SRGH).

Also in Uganda, Kenya and Tanzania. In forest and woodland.

The number and spacing of the leaflets varies a good deal in subsp. *petersiana*. The indumentum also varies from subglabrous to quite densely pubescent, though the latter condition is uncommon.

Subsp. **evansii** (Burtt Davy) Brenan in Kew Bull. **21**: 482 (1968). Type from the Transvaal.
Albizia evansii Burtt Davy, F.P.F.T. **2**: xvii, 349 (1932).—Gomes e Sousa, Dendrol. Moçamb. **2**: 54, cum tab. (1951); Dendrol. Moçamb. Estudo Geral, **1**: 238, t. 42 (1966).—Codd, Trees & Shrubs Kruger Nat. Park: 54, fig. 47, 48 b (1951); in Bothalia, **7**: 79 (1958).—Boughey in Journ. S. Afr. Bot. **30**: 158 (1964). Type as above.

Young branchlets densely pubescent with pubescence longer than is usual in subsp. *petersiana*. Leaves with petiole 0·5–1·5 cm. long and 2–3(4) pairs of pinnae; leaflets of 2 distal pairs of pinnae (2)3–6 pairs, ± pubescent on both surfaces with non-appressed hairs. Peduncle 1–2(2·5) cm. long, ± densely pubescent. Pods shortly ± pubescent or puberulous on the surface.

Rhodesia. S: Nuanetsi Distr., Gona-re-zhou, fr. vii.1953, *Carter* 21/53 (K; SRGH). **Mozambique.** SS: Guijá, Aldeia da Barragem, near Xirúnso Mission, fl. 16.xi.1957, *Barbosa & Lemos* in *Barbosa* 8165 (COI; K; LISC). LM: between Umbeluzi and Moamba, fl. 16.x.1940, *Torre* 1786 (BM; K; LISC).
Also in the Transvaal and Natal. In woodland and bushland; 30–240 m.

Subsp. *evansii* is not clearly separable from subsp. *petersiana* by any single character, unless perhaps the indumentum of the pod. If, however, subsp. *evansii* is diagnosed by the combination of strong indumentum and few pairs of pinnae and leaflets, then few specimens will present any difficulty.
Subsp. *evansii* seems frequently to be several-stemmed from the base, but it can also be single-stemmed.

Cultivated Species

18. **Albizia chinensis** (Osbeck) Merr. in Amer. Journ. Bot. **3**: 575 (1916). Type from China.
Mimosa chinensis Osbeck, Dagbok Ostind. Resa: 233 (1757), not seen. Type as above.

Tree. Stipules large, obliquely ovate-cordate, soon falling. Leaves with 7–17 pairs of pinnae; leaflets c. 15–40 pairs, 6–13 × 2–3 mm., acute at the apex, ± pubescent, with midrib running along the distal margin from base to apex. Flowers capitate.

Rhodesia. E: Inyanga, st. 1960, *Chase* (K; SRGH).
A native of tropical Asia cultivated as a shade tree.

19. **Albizia odoratissima** (L.f.) Benth. in Hook., Lond. Journ. Bot. **3**: 88 (1844). Type from Ceylon.
Mimosa odoratissima L.f., Suppl. Pl.: 437 (1781). Type as above.

Tree. Stipules very small, soon falling. Leaves with 3–8 pairs of pinnae; leaflets 6–19(23) pairs, 1·5–3·7(4·4) × (3)4–12(17) mm., rounded to obtuse at the apex, pubescent to puberulous, with midrib very asymmetric but not marginal. Flowers capitate, with grey indumentum outside. Calyx 1–1·5 mm. long. Stamen-filaments short, c. 1–1·5 cm. long. Pod dehiscent, rather glossy, c. 9–28 × 2·5–4 cm.

Rhodesia. E: Inyanga, st. 1960, *Chase* (K; LISC; SRGH). **Malawi.** S: Limbe, fl. xii.1947, *Hayes* (PRE). **Mozambique.** N: Nampula, by R. Nicuta, fl. 14.xi.1948, *Andrada* 1461 (BM; COI; LISC; LMJ).
Native of tropical Asia.

20. **Albizia falcataria** (L.) Fosberg in Reinwardtia, **7**: 88 (1965). Type a plate by Rumphius of a plant growing in the Moluccas.
Adenanthera falcata L., Herb. Amboin.: 14 (1754) pro parte quoad Rumphii t. 111 sed excl. lectotypum, vide Fosberg (loc. cit.).
Adenanthera falcataria L., Sp. Pl. ed. 2: 550 (1762). Type as for *Albizia falcataria*.
Albizia falcata (L.) Backer, Voorl. Schoolfl. Java: 109 (1908) pro parte ut supra.

Tree. Leaves rather large, mostly with 5–10 pairs of pinnae; leaflets c. 8–15 × 2·5–6 mm., acute or subacute at the apex, puberulous, with midrib very asymmetric but not marginal. Flowers sessile, in paniculate spikes. Calyx and corolla densely

D.E.

Tab. 25. SAMANEA LEPTOPHYLLA. 1, flowering branch (×⅔) *Holmes* 1246; 2, leaflets (×4); 3, flower (×4); 4, calyx (×4); 5, corolla (×4); 6, stamen (×4); 7, anther, two views (×10); 8, ovary (×4); 9, 10, parts of pod (×⅔), all from *Angus* 499.

puberulous outside. Pod dehiscent, with one margin appearing as if narrowly winged.

Rhodesia. E: Umtali, fl. 14.xi.1952, *Chase* 4692 (BK; K; SRGH). **Mozambique.** LM: Lourenço Marques, fl. 18.xii.1946, *Gomes e Sousa* 3481 (COI; K; LISC).
Native of the Malay Islands.

14. SAMANEA (Benth.) Merr.

(By J. P. M. Brenan & R. K. Brummitt)

Samanea (Benth.) Merr. in Journ. Wash. Acad. Sci. **6**, 1: 46 (1916).—Brenan & Brummitt in Bol. Soc. Brot., Sér. 2, **39**: 192–201 (1965).
Pithecellobium (" Pithecolobium ") Mart. sect. *Samanea* Benth. in Hook., Lond. Journ. Bot. **3**: 197 (1844).

Differs from *Cathormion* as follows: unarmed; pods straight or slightly curved, transverse impressions between seeds visible externally or not, margins entire (sometimes with one or two constrictions at the base), mesocarp moderately to well developed.

See under *Cathormion* (p. 137) for notes on generic distinction. Three or four African species and at least three in C. and S. America are referable to *Samanea* in the present sense.

S. saman (Jacq.) Merr. (*Pithecellobium saman* (Jacq.) Benth.) is occasionally grown in our area (Zambia and Mozambique). It resembles an *Albizia*, particularly *A. versicolor* Welw. ex Oliv., but differs from that species by its yellowish not rusty indumentum, by the petiolar gland being smaller and placed at or near the insertion of the lowest pair of pinnae, not well below them, and from all species of *Albizia* by the thick straight indehiscent pod. See also p. 118.

Samanea leptophylla (Harms) Brenan & Brummitt in Bol. Soc. Brot., Sér. 2, **39**: 201 (1965). TAB. **25**. Type from the Congo (Kimuenza, S. of Kinshasa).
Albizia leptophylla Harms in Engl., Bot. Jahrb. **53**: 455 (1915). Type as above.
Pithecellobium dinklagei sensu Bak. f., Legum. Trop. Afr. **3**: 871 (1930) pro parte quoad syn. *Albizia leptophylla*.
Arthrosamanea leptophylla (Harms) Gilbert & Boutique in Bull. Jard. Bot. Brux. **22**: 182 (1952); in F.C.B. **3**: 191 (1952) pro parte excl. var. *guineensis*. Type as above.
Cathormion leptophyllum (Harms) Keay in Kew Bull. **8**: 489 (1954). Type as above.
Samanea sp. 1—F. White, F.F.N.R.: 94 (1962).

Tree c. 15–20 m. high, with smooth longitudinally striated dark-grey bark and spreading crown, apparently unarmed; young branchlets and leaf- and pinna-rhachides densely rusty-tomentose. Leaves: pinnae 12–16 pairs; leaflets 20–54 pairs, 4–7·5 × 1–1·3 mm., linear-oblong, ± obtuse at the apex, subglabrous except for ciliate margins or the basal leaflets more pubescent; basal pair of leaflets differentiated as in *Cathormion altissimum*. Inflorescences on peduncles 2–4·5 cm. long. Flowers whitish, subsessile or very shortly pedicellate. Calyx 1·5–3 mm. long, densely puberulous to short-pubescent outside. Corolla c. 4–5 mm. long, pubescent at least towards the ends of the lobes outside. Pod indehiscent or breaking transversely at irregular intervals, c. 10–18 × 1·6–2 cm., straight or slightly twisted, mostly straight-margined.

Zambia. W: Mwinilunga Distr., 6 km. N. of Kalene Hill, fl. 20.ix.1952, *Angus* 499 (BM; K; PRE).
Also in the Congo. In evergreen fringing forest.

We consider *Cathormion leptophyllum* var. *guineense* (Gilbert & Boutique) Gilbert & Boutique to be specifically distinct. It is, however, possible that *C. eriorhachis* (Harms) Dandy, which appears to differ in little but its larger leaflets, may not be specifically separable from *S. leptophylla*. If so, the range of this species would extend to Cameroon, the Central African Republic and the Sudan.

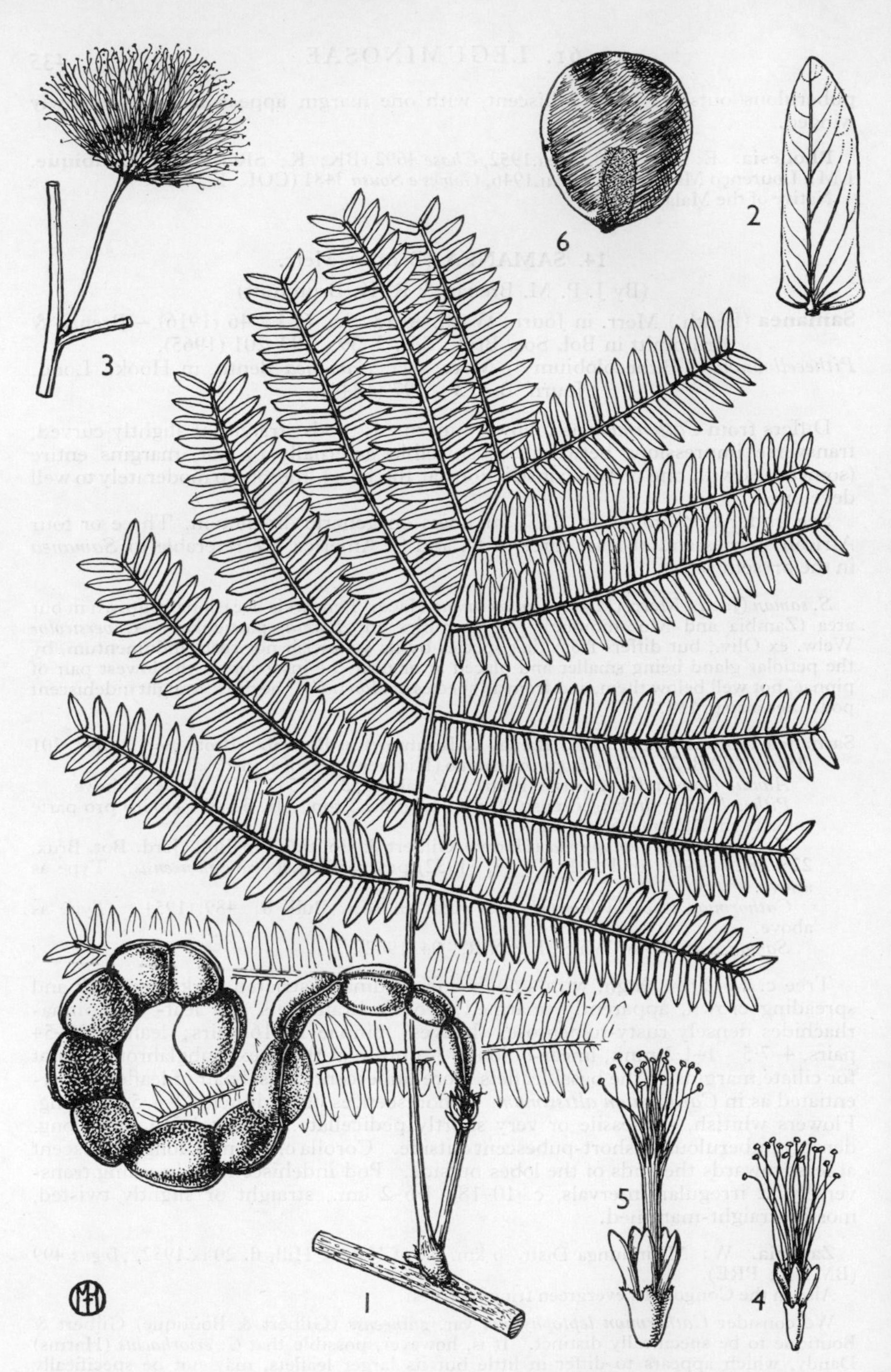

Tab. 26. CATHORMION ALTISSIMUM. 1, part of fruiting branch, showing pod (× ⅔); 2, leaflet (× 3), all from *Harris* 1126; 3, flower head (× 1); 4, flower from side of head (× 2); 5, flower from centre of head (× 2), all from *Linder* 933; 6, seed (× 3) *Unwin* 25. From F.T.E.A.

15. CATHORMION (Benth.) Hassk.

(By J. P. M. Brenan & R. K. Brummitt)

Cathormion (Benth.) Hassk., Retzia, **1**: 231 (1855).—Brenan & Brummitt in Bol. Soc. Brot., Sér. 2, **39**: 192–201 (1965).
Pithecellobium sect. *Cathormion* Benth. in Hook., Lond. Journ. Bot. **3**: 210 (1844).

Trees or shrubs, unarmed, at least the mature shoots. Leaves bipinnate, pinnae each with several to many pairs of leaflets; gland on upper side of petiole present or absent; glands often also present at insertion of the pinnae and on the upper part of the pinna-rhachis. Inflorescences of round heads which in the African species are pedunculate and mostly solitary or paired (sometimes in threes) in the axils. Flowers ☿, or said to be rarely ♂ and ☿; 1 to several central flowers in each head often modified and different in form from the others, and sometimes at least ☿. Calyx gamosepalous, shortly (4)5-dentate. Corolla gamopetalous, infundibuliform, (4)5-lobed. Stamens numerous (c. 16–22), fertile, their filaments united in their lower part into a slender tube not or scarcely projecting from the corolla (or very shortly so in the modified central flowers). Pod oblong, ± falcate or spirally curved, compressed, with the margins lobed, ± constricted between the seeds, at maturity breaking up into coriaceous or hard 1-seeded joints, or sometimes apparently indehiscent; mesocarp only slightly developed, septate. Seeds ± compressed, exarillate.

Delimitation of the genera in this group of the *Mimosoideae* is notoriously difficult (see Brenan & Brummitt in Bol. Soc. Brot., Sér. 2, **39**: 192–205 (1965)). The distinction between *Cathormion* and *Samanea* is somewhat tenuous, resting almost entirely on pod characters, and reconsideration of the group on a world-wide basis may result in their being regarded as congeneric. Some (such as Bak. f., Legum. Trop. Afr. **3** (1930)) consider that both these genera are best included in *Pithecellobium*. *Cathormion* in the present sense includes one species in Asia and Australia and one to three in Africa. Seven species from tropical America have been recently transferred to *Cathormion* by Burkart (in Darwiniana, **13**: 444–448 (1964)).

Cathormion altissimum (Hook. f.) Hutch. & Dandy, F.W.T.A. **1**: 364 (July 1928); in Kew Bull. **1928**: 401 (Dec. 1928).—Torre, C.F.A. **2**: 298 (1956).—Keay, F.W.T.A., ed. 2, **1**: 504 (1958).—Brenan, F.T.E.A. Legum.-Mimos.: 166, fig. 23 (1959).—F. White, F.F.N.R.: 91 (1962).—Brenan & Brummitt in Bol. Soc. Brot., Sér. 2, **39**: 203 (1965). TAB. **26**. Syntypes from Ghana and Nigeria.
Albizia altissima Hook. f. in Hook., Niger Fl.: 332 (1849). Syntypes as above.
Pithecellobium ("*Pithecolobium*") *altissimum* (Hook. f.) Oliv., F.T.A. **2**: 364 (1871).—R.E.Fr. in Wiss. Ergebn. Schwed. Rhod.-Kongo Exped. **1**: 63 (1914).—Bak. f., Legum. Trop. Afr. **3**: 870 (1930). Syntypes as above.
Arthrosamanea altissima (Hook. f.) Gilbert & Boutique in Bull. Jard. Bot. Brux. **22**: 182 (1952); in F.C.B. **3**: 193 (1952). Syntypes as above.

Shrub or tree 5–35 m. high, unarmed or (Eggeling & Dale, Indigenous Trees Uganda, ed. 2: 223 (1952)) often spinous on juvenile and sucker shoots; crown spreading; young branchlets puberulous. Leaves: rhachides ± densely and shortly crisped- or spreading-pubescent or -puberulous; pinnae 5–7 pairs; leaflets 11–22(25, *fide* Eggeling & Dale, loc. cit.) pairs, 7–16 × 2–6 mm. (to 22 × 10 on juvenile leaves), somewhat obliquely oblong, and widest near the base which is slightly auricled on both sides, narrowed to a usually obtuse apex, glabrous on both surfaces except for ciliolate margins at the base, with the lateral nerves rather close and fine; the basal pair of leaflets on each pinna characteristically represented by a single leaflet only, on the lower side, that on the upper side being replaced by a minute stipel. Inflorescences on peduncles 1·2–4·5 cm. long. Flowers white, sessile or subsessile. Calyx 2·5–3·5 mm. long, glabrous or subglabrous except on the teeth. Corolla 5–7 mm. long, glabrous outside. Pod c. 10–28 × 1·3–2 cm., ± regularly lobed along one or both margins which are puberulous, the pod otherwise subglabrous or sparingly puberulous. Seeds 6–9 × 6·5–7 mm., brown, ± flattened.

Zambia. N: Kawambwa, fl. 25.viii.1957, *Fanshawe* 3642 (K).

From Sierra Leone and the Sudan southwards to Angola, Zambia and Uganda. In fringing forest.

The species so far as at present known in our area is represented by var. *altissimum* only. The var. *busiraense* (Gilbert & Boutique) Gilbert & Boutique, differing mainly in the smaller flatter pod-segments and also in the deeper calyx-teeth, might also occur, however.

ADDITIONS AND CORRECTIONS

By A. W. Exell

VOLUME 1, PART 1

pp. 55–78. Butomaceae is to be added to the families due to the discovery of *Tenagocharis latifolia* (D.Don) Buchen. by H. Wild in Mozambique (Tete) and by E. A. Robinson in Zambia.

p. 85. Add under **Podocarpus milanjianus** Rendle:
Mozambique. T: Dómuè, fr. 9. iii. 1964, *Torre & Paiva* 11104 (LISC).

p. 90. Add under **Clematis simensis** Fresen.:
Rhodesia. C: Wedza Mt., fl. 14.v.1964, *Wild* 6567 (K).

p. 91. Add under **Clematis brachiata** Thunb.:
Mozambique. T: Mutarara, fl. & fr. immat. 9.iv.1943, *Torre* 5143 (LISC).

p. 95. Add under **Clematopsis uhehensis** (Engl.) Hutch. ex Staner & Léonard:
Malawi. N: Nyika. fl. & fr. 30.xii.1962, *Fanshawe* 7317 (K).

p. 96. Add under **Thalictrum rhynchocarpum** Dill. & Rich.:
Malawi. C: Nchisi Mt., fl. 19.ii.1959, *Robson* 1665 (K; SRGH).

p. 96. Add under **Thalictrum zernyi** Ulbr.:
Zambia. E: Nyika, fl. 24.xii.1962, *Fanshawe* 7236 (K).

p. 97. Add under **Ranunculus multifidus** Forsk.:
Botswana. SE: Moeng College, Palapye, fl. & fr. 29.xi.1957, *de Beer* 514 (SRGH).

p. 99. Add under **Ranunculus raeae** Exell:
Zambia. E: Nyika, fl. 24.xii.1962, *Fanshawe* 7240 (K).

p. 101. Add **Delphinium macrocentrum** Oliv. in Journ. Linn. Soc., Bot. **21**: 397 (1886). Holotype from Kenya.
Malawi. N: Mzimba, fl. 1948, *Kleinschmidt* (BM; SRGH). Also in Uganda and Kenya.

p. 166. Add under **Cissampelos hirta** Klotzsch:
Mozambique. N: Moçambique, fl. 9.x.1919, *Borle* 63 (COI).

p. 167. Add under **Cissampelos mucronata** A. Rich.:
Mozambique. SS: Chibuto, fr. 11.x.1957, *Barbosa & Lemos* 7997 (COI; LISC).

p. 175. Add under **Brasenia schreberi** J. F. Gmel.:
Rhodesia. C: Chilimanze, fr. 12.vii.1965, *Wild* 6577 (K).

p. 176. Add under **Nymphaea lotus** L.:
Zambia. N: mouth of Lufubu R., fl. 30.xii.1963, *Richards* 18719 (K). **Rhodesia.** S: Nuanetse, fl. 28.iv.1961, *Drummond & Rutherford-Smith* 7563 (K). **Mozambique.** SS: Chibuto, fl. & fr. 10.viii.1958, *Myre & Macedo* 3129 (LISC).

p. 176. Add under **Nymphaea petersiana** Klotzsch:
Rhodesia. S: Nuanetse, fl. 30.iv.1962, *Drummond* 7824 (K).

p. 177. Add under **Nymphaea caerulea** Savigny:
Mozambique. LM: Maputo, fl. 10.vii.1958, *Myre & Macedo* 2984 (LISC).

p. 184. Add under **Rorippa madagascariensis** DC.:
Zambia. C: Broken Hill, fl. & fr. 7.x.1963, *Fanshawe* 8031 (K).

p. 185. Add under **Rorippa nasturtium-aquaticum** (L) Hayek:
Mozambique. LM: Marracuene, fl. & fr. 2.x.1957, *Barbosa & Lemos* 7937 (LISC).

p. 186. Add under **Brassica juncea** (L.) Czern.:
Mozambique. T: Zumbo, fl. & fr. 11.vii.1889, *Sarmento* 103 (COI).

p. 188. Add under **Sisymbrium orientale** L.:
Rhodesia. E: Umtali, fr. 2.xi.1960, *Chase* 7399 (K).

p. 190. Add under **Lepidium bonariense** L.:
Zambia. C: Lusaka, fr. 7.xii.1964, *Fanshawe* 9041 (K).

p. 192. Add under **Lepidium divaricatum** Ait. subsp. **divaricatum**:
Mozambique. SS: Chibuto, fl. & fr. 8.ix.1952, *Pedro* 3968 (LMJ).

p. 193. Add under **Coronopus integrifolius** (DC.) Spreng.:
Zambia. B: Linwe Plain, 48 km. N. of Kalolo, fl. 14.xi. 1959, *Drummond & Cookson* 6458 (K).

p. 203. Add under **Cleome macrophylla** (Klotzsch) Briq. var. **macrophylla**:
Zambia. C: Katondwe, fl. 4.ii.1964, *Fanshawe* 8276 (K).

p. 205. Add under **Cleome macrophylla** var. **maculatiflora** (Merxm.) Wild:
Zambia. W: 50 km. N. of Mwinilunga, 11.xii.1963, *Robinson* 5889 (K).

p. 208. Add under **Cadaba termitaria** N. E. Br.:
Mozambique. LM: R. Matola, fl. 29.iv.1947, *Pedro & Pedrógão* 911 (LMJ)

p. 221. Add under **Maerua nervosa** (Hochst.) Oliv.:
Rhodesia. E: Odzani Valley, near Umtali, *Teague* 319 (BOL) (*fide* D. J. B. Killick).

p. 228. Add under **Maerua brunnescens** Wild:
Mozambique. T: Mutarara, fl. & fr. immat. 10.x.1944, *Torre* 6832 (LISC, paratype).

p. 231. Add under **Boscia corymbosa** Gilg:
Mozambique. SS: Massinga, fl. 27.vi.1950, *Chase* 2234 (LISC).

p. 234. Add under **Boscia matabelensis** Pest.:
Zambia. C: Kalenje, fl. & fr. 24.viii.1964, *Fanshawe* 8879 (K).

p. 238. Add under **Capparis sepiaria** L.:
Mozambique. SS: Inharrime, fl. 16.x.1957, *Barbosa & Lemos* 8068 (COI; LISC).

p. 239. Add under **Capparis lilacina** Gilg:
Mozambique. Z: Maganja da Costa, fl. 27.ix.1949, *Barbosa & Carvalho* 4218 (LMJ).

p. 245. Add to the Capparidaceae a probable record of **Bachmannia Woodii** (Oliv.) Gilg:
Mozambique. LM: Bela Vista, *Mogg* 27782 (J) (*fide* D. J. B. Killick).

p. 247. Add under **Rinorea convallarioides** (Bak. f.) Eyles:
Mozambique. LM: Namaacha, fr. 10.i.1947, *Barbosa* 121 (COI).

p. 250. Add under **Rinorea ilicifolia** (Welw. ex Oliv.) Kuntze:
Mozambique. N: Eráti, fl. & fr. 8.i.1964, *Torre & Paiva* 9888 (LISC).

p. 257. Add under **Hybanthus enneaspermus** (L.) F. Muell. var. **enneaspermus**:
Zambia. C: Katondwe, fl. & fr. 4.ii.1964, *Fanshawe* 8301 (K). And under var. **nyassensis** (Engl.) N. Robson:
Mozambique. T: Cazula, *Pimenta* (LISC).

p. 276. Add under **Oncoba spinosa** Forsk.:

Mozambique. T: road to Massanga, st. 5.vii.1949, *Andrada* 1698 (COI: LISC).

p. 278. Under **Scolopia zeyheri** (Nees) Harv.: *Torre* 7509 from Muchopes is from SS not ML, and add:

Mozambique. LM: Inhaca I., fl. & fr. 12.vii.1957, *Barbosa* 7686 (LMJ).

p. 286. Add under **Flacourtia indica** (Burm. f.) Merr.:

Mozambique. LM: Marracuene, fr. 25.ii.1950, *Macúcua* 21 (LMJ).

p. 291. Add under **Homalium abdessammadii** Aschers. & Schweinf. subsp. **abdessammadii**:

Mozambique. MS: Cheringoma, fl. 17.x .1943, *Torre* 5906 (COI).

p. 293. Add **Homalium mossambicense** Paiva in Bol. Soc. Brot., Sér. 2, **40**: 266, t. 2, 3 (1966). Type: Mozambique, between Palma and Mocímboa da Praia, *Gomes e Sousa* 4603 (COI, holotype).

Mozambique. N: Cabo Delgado, 15 km. from Palma towards Mocímboa da Praia, fl. & fr. 5.xi.1960, *Gomes e Sousa* 4603 (COI).

p. 298. Add under **Trimeria rotundifolia** (Hochst.) Gilg:

Mozambique. MS: Mandirgue, fl. 8.x.1946, *Simão* 1020 (COI; LM).

p. 298. Add to the Flacourtiaceae the genus **Grandidiera** Jaub. and the species **Grandidiera boivinii** Jaub.:

Mozambique. MS: R. Conduè, *Gomes e Sousa* 4722 (COI). See Paiva in Bol. Soc. Brot., Sér. 2, **40**: 264, t. 1 (1966).

p. 300. Add under **Pittosporum viridiflorum** Sims:

Mozambique. N: 20 km. from Vila Cabral, fr. 26.ix.1958, *Monteiro* 89 (LISC).

p. 304. Add under **Carpolobia conradsiana** Engl.:

Mozambique. MS: Beira, Macuti, fl. 23.ii.1946, *Simão* 1191 (LISC).

p. 312. Add under **Polygala goetzei** Gürke:

Mozambique. Z: Ile, fl. & fr. 10.x.1946, *Pedro* 2187 (LMJ).

p. 314. In the description of **Polgala adamsonii** Exell alter " Annual? " to " Shrublet " and add:

Mozambique. N: Ribáuè, fl. 28.i.1964, *Torre & Paiva* 10303 (LISC).

p. 315. Add under **Polygala senensis** Klotzsch:

Mozambique. T: Moatize, fl. and fr. 12.iii.1964, *Torre & Paiva* 11173 (LISC).

p. 316. Add under **Polygala erioptera** DC.:

Mozambique. T: Moatize, fl. & fr. 12.iii.1964, *Torre & Paiva* 11167 (LISC).

p. 317. Add under **Polygala sadebeckiana** Gürke:

Mozambique. N: Vila Cabral, fl. & fr. 1.xii.1934, *Torre* 450 (COI; LISC).

p. 318. Add under **Polygala virgata** var. **decora** (Sond.) Harv.:

Mozambique. T: Dómuè, fl. & fr. 9.iii.1964, *Torre & Paiva* 11101 (LISC).

p. 320. Add under **Polygala arenaria** Willd.:

Mozambique. N: Nampula, fl. & fr. 7.v.1937, *Torre* 1393 (COI; LISC).

p. 332. Add under **Polygala uncinata** E. Mey. ex Meisn.:

Mozambique. MS: Machanga, fl. & fr., 2.ix.1942, *Mendonça* 102 (LISC).

p. 333. Add under **Polygala producta** N. E. Br.:

Mozambique. SS: Muchopes, Manjacaze, fl. & fr. 5.xii.1944, *Mendonça* 3264 (LISC).

p. 333. Add under **Polygala xanthina** Chod.:

Mozambique. N: Ribáuè, fl. 23.iii.1964, *Torre & Paiva* 11320 (LISC).

VOLUME 1, PART 2

p. 339. Add under **Krauseola mosambicina** (Moss) Pax & K. Hoffm.:

Mozambique. SS: Chipenhe, fl. & fr. 13.x.1957, *Barbosa & Lemos* 8053 (COI; LMJ).

p. 341. Add under **Polycarpon prostratum** (Forsk.) Aschers. & Schweinf.:

Zambia. N: Fort Rosebery, fl. 13.xi.1964, *Mutimushi* 1063 (K). C: 100 km. E. of Lusaka, fl. 16.xi.1963, *Robinson* 5845 (K).

p. 354. Add under **Silene burchellii** var. **angustifolia** Sond.:

Mozambique. T: Dómuè, fl. & fr. 9.iii.1964, *Torre & Paiva* 11081 (LISC).

p. 360. Add under **Pollichia campestris** Ait.:

Zambia. S: Namwala, fl. & fr. 15.i.1965, *Van Rensburg* 2760 (K).

p. 362. Add under **Corrigiola litoralis** L.:

Mozambique. SS: Gaza, Limpopo at Dumela, fl. 30.iv.1961, *Drummond & Rutherford-Smith* 7633 (K).

p. 362. Add under **Corrigiola drymarioides** Bak. f.:

Mozambique. Z: Namuli, fl. & fr. 29.ix.1944, *Mendonça* 2272 (LISC).

p. 364. Add under **Portulaca oleracea** L.:

Mozambique. T: fr. viii.1931, *Pomba Guerra* 50 (COI). MS: Gorongosa, fl. & fr. 11.xi.1963, *Torre & Paiva* 9173 (LISC). LM: Lourenço Marques, fl. & fr. xii.1944, *Pimenta* 182 (LISC).

p. 364. Add under **Portulaca foliosa** Ker-Gawl.:

Rhodesia. N: Kariba, fr. 31.iii.1961, *Drummond & Rutherford-Smith* 7633 (K).

p. 370. Add under **Talinum crispatulatum** Dinter:

Zambia: C: Broken Hill, fr. 7.x.1963, *Fanshawe* 8037 (K).

p. 375. Add under **Bergia ammanioides** Heyne ex Roth:

Mozambique. SS: Gaza, Limpopo at Dumela, fl. 30.iv.1961, *Drummond & Rutherford-Smith* 7615 (K).

p. 376. Add under **Bergia polyantha** Sond.:

Mozambique. SS: Gaza, Limpopo at Dumela, fl. 30.iv.1961, *Drummond & Rutherford-Smith* 7624 (K).

p. 376. Add under **Bergia glutinosa** Dinter & Schulze-Menz:

Zambia. S: Machili, fr. 7.vi.1961, *Fanshawe* 6655 (K).

p. 385. Add under **Hypericum peplidifolium** A. Rich.:

Mozambique. N: Vila Cabral, fl. & fr. 25.ii.1964, *Torre & Paiva* 10797 (LISC).

p. 405. Add to the Theaceae the genus **Termstroemia** Mutis ex L. f. and the species **Termstroenia polypetala** Melchior

Malawi. N: above Nchenchena, fr. xiii. 1953, *Chapman* 125 (FHO).

p. 412. Add under **Monotes katangensis** (De Wild.) De Wild.

Mozambique. N: Marrupa, fl. 19.ii.1964, *Torre & Paiva* 10680 (LISC).

p. 426. Add under **Abelmoschus esculentus** (L.) Moench:

Zambia. N: Mpika Distr., fl. 7.v.1965, *Mitchell* 2872 (K). **Mozambique.** LM: Maputo, 10.v.1949, *Myre & Balsinhas* 710 (LISC).

p. 430. Add **Gossypium triphyllum** (Harv.) Hochr. in Bull. Herb. Boiss. Sér. 2, **2**: 1004 (1902). Type from SW. Africa.

Fugosia triphylla Harv. in Harv. & Sond., F. C. **2**: 588 (1862). Type as above.

Botswana. N: Aha Hills, fl. & fr. 12.iii.1965, *Wild & Drummond* 6952 (K; SRGH).

Also in Angola and SW. Africa.

p. 434. Add under **Azanza garckeana** (F. Hoffm.) Exell & Hillcoat:

Mozambique. Z: Morrumbala, fr. 28.iv.1943, *Torre* 5227 (LISC).

p. 438. Add under **Hibiscus surattensis** L.:

Zambia. C: Iolanda, Kafue, fl. & fr. 4.vi.1965, *Robinson* 683 (K).

p. 443. Add under **Hibiscus mechowii** Garcke:

Caprivi Strip. 19 km. N. of Shakawe, fl. 16.iii.1965, *Wild & Drummond* 7085 (K; SRGH).

p. 445. Add under **Hibiscus sidiformis** Baill.:

Mozambique. LM: Magude, fl. & fr. 7.i.1948, *Torre* 7065 (LISC).

p. 446. Add under **Hibiscus lobatus** (Murr.) Kuntze:

Zambia. C: Iolanda, fl. & fr. 14.iii.1965, *Robinson* 6422 (K). **Mozambique.** N: Mossuril, fr. 1884–85, *Carvalho* (COI).

p. 451. Add under **Hibiscus shirensis** Sprague & Hutch.:

Mozambique. LM: Namaacha, fl. vii.1918, *Moreira* 28 (LMM).

p. 453. Add under **Hibiscus micranthus** L. f.:

Mozambique. MS: Chemba, fl. & fr. 29.ix.1949, *Pedro & Pedrógão* 8419 (LMJ).

p. 454. Add under **Hibiscus migeodii** Exell:

Zambia. N: Mpika Distr., fl. 4.v.1965, *Mitchell* 2831 (K).

p. 462. Add under **Hibiscus caesius** Garcke:

Caprivi Strip. 19 km. N. of Shakawe, fl. 16.iii.1965, *Wild & Drummond* 7083 (K; SRGH).

p. 463. Add under **Hibiscus articulatus** Hochst. ex A. Rich.:

Mozambique. MS: Chimoio, Vila Machado, fl. & fr. 19.iv.1948, *Mendonça* 4042 (LISC).

p. 465. Add under **Hibiscus physaloides** Guill. & Perr.

Mozambique. T: Boroma, fr. 25.vii.1950, *Chase* 2797 (COI; LISC).

p. 466. Add under **Hibiscus schinzii** Gürke:

Mozambique. SS: Muchopes, fr. 3.xii,1942, *Mendonça* 1590 (LISC).

p. 467. Add under **Hibiscus engleri** K. Schum.:

Mozambique. N: Nampula, fl. 2.iii.1936, *Torre* 844 pro parte (LISC). Z: Ile, fl. & fr. 2.iv.1943, *Torre* 5039 (LISC).

p. 472. Add under **Hibiscus vitifolius** subsp. **vulgaris** Brenan & Exell:

Mozambique. SS: Guijá, fl. & fr. 16.xi.1957, *Barbosa & Lemos* 8153 (COI; LISC; LMJ).

p. 477. Add under **Sida alba** L.:

Mozambique. SS: Guijá, fl. & fr. 3.v.1957, *Carvalho* 150 (LISC).

p. 478. Add under **Sida acuta** Burm. f.

Mozambique. N: Nampula, fl. & fr. 30.iii.1937, *Torre* 1247 (COI; LISC).

p. 480. Add under **Sida rhombifolia** L.;

Mozambique. N: Maniamba, 29.iv.1948, *Pedro & Pedrógão* 4104 (LMJ).

p. 481. Add under **Sida chrysantha** Ulbr.:

Mozambique. LM: Namaacha, fr. 19.xii.1952, *Myre & Carvalho* 1397 (LISC).

p. 482. Add under **Sida hoepfneri** Gürke:

Mozambique. LM: Namaacha, fl. & fr. 18.ii.1945, *Esteves de Sousa* 41 (LISC).

p. 488. Add under **Abutilon hirtum** (Lam.) Sweet:

Mozambique. T: Tete, fl. vi.1930, *Pomba Guerra* 25 (COI).

p. 488. Add under **Abutilon angulatum** (Guill. & Perr.) Mast. var. **angulatum**:

Zambia. N: Mpika Distr., fr. 7.v.1965, *Mitchell* 2868 (K).

p. 489. Add under **Abutilon angulatum** var. **macrophyllum** (Bak. f.) Hochr.:

Mozambique. MS: Gorongosa, fl. 1884–85, *Carvalho* (COI). SS: Guijá, fl. 9.vi.1947, *Pedrógão* 265 (LMJ).

p. 491. Add under **Abutilon fruticosum** Guill. & Perr.:

Botswana. SE: 18 km. E. of Botletle channel, fr. 23.iii.1965, *Wild & Drummond* 7232 (K; SRGH).

p. 498. Add under **Abutilon engleranum** Ulbr.:

Mozambique. N: Vila Cabral, fr. i.vi.1933, *Gomes e Sousa* 1463 (COI). MS: Chimoio, near Bandula, fl. & fr. 21.v.1949, *Pedro & Pedrógão* 5780 (LMJ).

p. 498. Add under **Abutilon ramosum** (Cav.) Guill. & Perr.:

Mozambique. LM: Namaacha, fl. & fr. 23.v.1957, *Carvalho* 240 (LISC).

p. 499. Add under **Wissadula rostrata** (Schumach.) Hook. f.:

Zambia. N: Mpika Distr., fr. 1.v.1965, *Mitchell* 2762 (K).

Mozambique. T: between Matundo and Massamba, fr. 5.vii.1949, *Barbosa & Carvalho* 3437 (LISC; LMJ).

p. 510. Add under **Pavonia columella** Cav.:

Mozambique. LM: Namaacha, fl. & fr. 25.iv.1947, *Pedro & Pedrógão* 752 (LMJ).

p. 511. Add under **Pavonia urens** Cav.:

Mozambique. T: Angónia, near Metengobalame, fl. & fr. 17.vii.1949, *Barbosa & Carvalho* 3657 (LMJ).

p. 512. Add under **Adansonia digitata** L.:

Mozambique. T: Chioco, fr. 26.ix.1942, *Mendonça* 434 (LISC).

p. 513. Add under **Bombax rhodognaphalon** K. Schum. var. **rhodognaphalon**:

Mozambique. LM: Maputo, fl. 14.viii.1948, *Myre & Carvalho* 112 (LISC).

p. 515. Change **Bombax oleagineum** (Decne.) A. Robyns to **Bombax glabrum** (Pasq.) A. Robyns in Bull. Jard. Bot. Brux. **30**: 474 (1960).

p. 523. Add under **Dombeya nyasica** Exell:

Mozambique. LM: Marracuene, fl. & fr. 27.iii.1957, *Barbosa & Lemos* 7514 (LISC); LMJ).

p. 527. Add under **Dombeya kirkii** Mast.:

Mozambique. LM: Namaacha, fl. & fr. 30.iv.1947, *Pedro & Pedrógão* 1029 (LMJ).

p. 531. M. L. Gonçalves (in Mem. Junta Invest. Ultram. Sér. 2, **41**: 136 (1963)) refers a Mozambique specimen (SS: Guijá, fl. & fr. vii.1915, *Gazaland Expedition* 337 (LMM)) to **Melhania griquensis** Bolus. If correct this is a very unusual distribution as the species is known in our area only from Botswana.

p. 532. Add under **Melhania forbesii** Planch. ex Mast.:

Mozambique. N: Porto Amélia, fl. & fr. 14.iii.1960, *Gomes e Sousa* 4533 (COI).

p. 535. Add under **Melochia corchorifolia** L.:

Botswana. N: 30 km. from Gomare on the river road to Nakaneng, fr. 17.iii.1965, *Wild & Drummond* 7120 (K; SRGH). **Zambia.** N: Mporokoso, fl. 17.iv.1961, *Phipps & Vesey-FitzGerald* 3252 (K). **Rhodesia.** N: Kariba, fl. 31.iii.1961, *Drummond & Rutherford-Smith* 7504 (K).

p. 547. Add under **Hermannia kirkii** Mast.:

Mozambique. Z: R. Zambeze, fl. & fr. 1884–85, *Carvalho* (COI).

p. 549. Add **Hermannia helianthemum** K. Schum. in Engl. Bot. Jahrb. **10**: 44 (1888). Type from SW. Africa.

Botswana. SE: 60 km. NW. of Serowe, fl. 24.iii.1965, *Wild & Drummond* 7293 (K).

p. 554. Add under **Sterculia africana** (Lour.) Fiori:

Mozambique. Z: Ile, 31.v.1949, *Barbosa & Carvalho* 2934 (LM; LMJ).

p. 555. Add under **Sterculia appendiculata** K. Schum.:

Mozambique. Z: Morrumbala, near Megaza, 14.vi.1949, *Barbosa & Carvalho* 3073 (LM; LMJ).

VOLUME 2, PART 1

p. 75. Add under **Triumfetta annua** L. forma **annua**:

Malawi. S: Fort Johnson, 950 m., fr. 2.v.1960, *Leach* 9896 (K).

p. 89. Add under **Corchorus saxatilis** Wild:

Zambia. N: 6 km. N. of Lusaka, fr. 20.iii.1965, *Robinson* 6443 (K).

p. 93. Transfer **Hugonia busseana** Engl. to the synonymy of **H. orientalis** Engl. (p. 95) and transfer the specimens cited to that species. As a result of the study of recent collections it has been found impossible to maintain *H. busseana*.

p. 97. Add under **Linum thunbergii** Eckl. & Zeyh.:

Zambia. W: Kitwe, fl. 6.xi.1962, *Mutimushi* 193 (K). E: Nyika, fl. 30.xii.1962, *Fanshawe* 7310 (K).

p. 100. Change **Ochthocosmus** Benth. to **Phyllocosmus** Klotzsch and the species **O. lemaireanus** De Wild. & Dur. to **P. lemaireanus** (De Wild. & Dur.) T. & H. Dur. The genus *Phyllocosmus* is now considered to be distinct from the American genus *Ochthocosmus*.

p. 106. Add under **Erythroxylum emarginatum** Thonn.:

Zambia. B: 88 km. from Monga on the Mankoya road, fl. 19.xi.1959 *Drummond & Cookson* 6614 (K; SRGH).

p. 118. Add under **Triaspis suffulta** Launert:

Mozambique. MS: Cherine, edge of mangrove swamp, fl. 2.ix.1961, *Leach* 11248 (K).

p. 135. Add under **Geranium vagans** Bak. subsp. **vagans**:

Zambia. E: Nyika Plateau, fl. 2.i.1964, *Benson* 418 (K).

p. 140. Add under **Monsonia angustifolia** E. Mey.:

Botswana. SE: 60 km. NW. of Serowe, fr. 24.iii.1965, *Wild & Drummond* 7288 (K; SRGH). **Zambia.** C: N. bank of Kafue R., near Kafue town, fr. 14.iii.1965, *Robinson* 6424 (K). S: Mazabuka, fr. 6.ii.1963, *Van Rensburg* 1335 (K).

p. 151. Add under **Oxalis trichophylla** Bak.:

Zambia. W: Mwinilunga Distr., 14.xii.1963, *Robinson* 5976 (K). **Malawi.** N: Nyika Plateau, fl. 30.xii.1962, *Fanshawe* 7345 (K).

p. 158. Add under **Biophytum abyssinicum** Steud. ex A. Rich.:

Rhodesia. S: Chibi, fl. 3.v.1962, *Drummond* 7894 (K).

p. 158. Change **Biophytum sensitivum** (L.) DC. to **B. helenae** Buscal. & Muschl. See Veldkamp in Blumea, **16,** 1: 137 (1968).

p. 166. Add under **Impatiens prainiana** Gilg:

Zambia. E: Nyika, fl. 31.xii.1962, *Fanshawe* 7386 (K).

p. 173. Add under **Impatiens limnophila** Launert:

Zambia. C: Chiwefwe, fl. 19.ix.1964, *Mutimushi* 1053 (K).

p. 191. Under **Fagaropsis angolensis** var. **mollis** (Suesseng.) Mendonça: note that *Mc.Gregor* 48/38 (FHO) is from Wedza in Rhodesia, not from Dedza in Malawi.

p. 200. Add under **Vepris undulata** (Thunb.) Verdoorn & C. A. Sm.:

Mozambique. N: Querimba I., *Peters* (K).

p. 206. Add under **Teclea grandifolia** Engl.:

Zambia. N: Kashiba, fl. 17.ix.1963, *Fanshawe* 7988 (K).

p. 209. Add under **Teclea fischeri** (Engl.) Engl.:

Rhodesia. Lonely Mine, fl. xii.1955, *Goldsmith* 84/56. (K).

p. 216. Add under **Kirkia acuminata** Oliv.:

Botswana. SE: Mahalapye, fr. 2.i.1962, *Yalala* 137 (K).

p. 223. Add under **Balanites aegyptiaca** (L.) Del.:

Rhodesia. S: Nuanetsi, st. 25.iv.1962, *Drummond* 7745 (K; SRGH).

p. 250. Add under **Ochna confusa** Burtt Davy & Greenway:

Zambia. W: Mwinilunga, 13.xii.1963, *Robinson* 5976 (K). C: Chakwenga Headwaters, E. of Lusaka, fr. 28.x.1963, *Robinson* 5798 (K).

p. 268. Add under **Commiphora pyracanthoides** subsp. **glandulosa** (Schinz) Wild:

Zambia. C: Katondwe, fr. 23.ii.1965, *Fanshawe* 9267 (K).

p. 273. Add under **Commiphora mollis** (Oliv.) Engl.:

Zambia. N: Abercorn, fr. 28.xii.1963, *Richards* 18695. (K).

p. 277. Add under **Commiphora africana** var. **rubriflora** (Engl.) Wild:

Rhodesia. C: Gwelo, fr. 21.i.1963, *Loveridge* 598 (K).

p. 282. Add under **Commiphora caerulea** B. D. Burtt:

Rhodesia. W: Wankie, fr. 28.ii.1963, *Wild* 6045 (K).

p. 282. Add under **Dacryodes edulis** (G. Don) H. J. Lam:

Zambia. W: Mwinilunga, fl. 6.ix.1955, *Holmes* H. 117 s (K).

p. 324. Add under **Dichapetalum cymosum** (Hook.) Engl.:

Rhodesia. S: Nuanetsi Distr., recorded by Cleghorn & Hill in Rhod. Agric. Journ. **62,** 5: 99–100 (1965).

p. 336. Add under **Strombosia scheffleri** Engl.:

Mozambique. MS; Chimoio, veg. iv.1936, *Gilliland* 1818 (BM).

VOLUME 2, PART 2

p. 436. Add under **Helinus integrifolius** (Lam.) Kuntze:
Zambia. B: 64 km. S. of Balovale, fl. i.iv.1952, *Holmes* 1069 (K).

p. 442. Add under **Ampelocissus africana** (Lour.) Merr.:
Zambia. C: Kafue, fr. 2.iii.1963, *Van Rensburg* 1569 (K).

p. 466. Add under **Cyphostemma crotalarioides** (Planch.) Descoings:
Zambia. E: Fort Jameson, f. 15.xii.1964, *Robinson* 6284 (K).

p. 467. Add under **Cyphostemma chloroleucum** (Welw. ex Bak.) Descoings:
Zambia. N: Mafinga Mts., Isoka Distr., fl. & fr. 18.xii.1964, *Robinson* 6289 (K).

p. 474. Add under **Cyphostemma congestum** (Bak.) Descoings:
Zambia. C: Broken Hill, fr. 13.iii.1962, *Morze* 254 (K). **Mozambique.** LM; Umbeluzi, fl. 12.x.1940, *Torre* 1761 (K).

p. 476. Add under **Cyphostemma setosum** (Roxb.) Alston:
Zambia. W: Kalene Hill, fl. 16.xii.1963, *Robinson* 6119 (K).

p. 477. Add under **Cyphostemma gigantophyllum** (Gilg & Brandt) Descoings:
Zambia. N: Mbeereshi, fl. 12.i.1962, *Richards* 12339 (K). **Rhodesia.** E: Umtali Distr., fr. 18.iv.1962, *Chase* 7778 (K; SRGH).

p. 480. Add under **Cyphostemma kirkianum** (Planch.) Descoings:
Rhodesia. W: Wankie Distr., fl. 28.ii.1963, *Wild* 6063 (K; SRGH).

p. 481. Add under **Cyphostemma buchananii** (Planch.) Descoings:
Zambia. C: Chakwenga Headwaters, fl. 7.i.1964, *Robinson* 6151 (K). S: Kafue Gorge, S. side, Mazabuka Distr., fl. 24.xi.1963, *Robinson* 5855 (K).

p. 485. Add under **Cyphostemma cirrhosum** subsp. **rhodesicum** Wild & Drummond:
Zambia. N: Abercorn Distr., fl. & fr. 17.ii.1962, *Richards* 16132 (K).

p. 492. Add under **Cayratia ibuensis** (Hook. f.) Suesseng.:
Zambia. N: Abercorn Distr., Kawimbe Rocks, fl. 17.i.1964, *Richards* 18812 (K).

p. 502. Add under **Allophylus rubifolius** (Hochst. ex A. Rich.) Engl.:
Zambia. C: Kafue, fl. 24.i.1964, *Van Rensburg* 2825 (K).

p. 506. Add under **Allophylus africanus** Beauv. **Group B**:
Zambia. W: Mindolo, Kitwe, fr. 6.iii.1964, *Mutimushi* 653 (K).

p. 512. Add under **Cardiospermum grandiflorum** Sw.:
Zambia. N: Lake Kashiba, fr. 17.ix.1963, *Fanshawe* 7989 (K)

p. 537. Add under **Filicium decipiens** (Wight & Arn.) Thw.:
Mozambique. N: Vila Cabral, fl. xii.1932, *Gomes e Sousa* 1047 (COI).

INDEX TO BOTANICAL NAMES